Microspectroscopic Imaging of Polymers

Microspectroscopic Imaging of Polymers

Jack L. Koenig

Case Western Reserve University

ACS Professional Reference Book

American Chemical Society
Washington, DC

Chem

Library of Congress Cataloging-in-Publication Data

Koenig, Jack L.
Microspectroscopic Imaging of Polymers / Jack L. Koenig
      p.   cm.
Includes bibliographical references and index.

ISBN 0–8412–3493–0

1. Polymers–Structure   2. Microspectrophotometry. I. Title

QD381.9.S87K64   1997
547′.70442—dc20   97–41362
                                            CIP

The paper used in this publication meets the minimum requirements of American National Standard for Information Sciences—Permanence of Paper for Printed Library Materials, ANSI Z39.48-1984.

PRINTED IN SOUTH KOREA

Advisory Board

To My Students

About the Author

Jack L. Koenig is the Donnell Institute Professor of the Department of Macromolecular Science and Chemistry at Case Western Reserve University in Cleveland, Ohio. Dr. Koenig received his B.A. in chemistry and mathematics from Yankton College, Yankton, South Dakota, in 1955. His PhD. in theoretical spectroscopy was granted in 1959 by the University of Nebraska.

After receiving his PhD., Dr. Koenig was first employed as a research chemist in the Plastics Department at the E. I. du Pont Nemours and Company in Wilmington, Delaware, from 1959 to 1963. In 1963, Dr. Koenig became an assistant professor of polymer science at Case Institute of Technology, and three years later he became an associate professor. In 1979, Dr. Koenig was promoted to professor in the Department of Macromolecular Science at Case Western Reserve University. He was awarded the Donnell Institute Endowed Chair in 1990.

Dr. Koenig has been active in microspectroscopy from its inception. He used one of the first commercial infrared microscopes while at Dupont in the 1960s and later used one of the first Fourier transform infrared spectroscopy (FTIR) microscopes in the 1980s. In Raman microscopy, he was the first to study glass fibers and their surface treatments. In NMR imaging, he initiated programs in analysis of the heterogeneous structures of elastomers, composites, and polymer liquid crystals, as well as diffusion and desorption measurements.

Dr. Koenig has received many awards and honors during his career. In 1984, he received the prestigious Pittsburgh Society Spectroscopy Award. In 1986, he won the Alexander von Humbolt Award for Senior U.S. scientists and spent the year in Germany as a visiting professor at the Institute for Macromolecular Chemistry at the University of Freiburg. In

1991, he received the Society of Plastics Engineers Research Award. In 1992, he received the Gold Medal of the Eastern Analytical Society. In 1993, he won the Bomem-Michelson Award. In 1994, he was awarded the George Stafford Whitby Award by the ACS Rubber Division, Inc. In 1995, he was designated an Honorary Member of the Society of Applied Spectroscopy. In 1996, he received the Society of Plastics Engineers (SPE) Education Award, and in 1997, he received the ACS Phillips Award in Applied Polymer Science.

Preface

"It's a small, small world."

—Walt Disney

The merging of spectroscopy with microscopy has generated an entirely new discipline which can be termed microspectroscopy. The utilization of the computer for control and analysis has extended the field of microspectroscopy to the discipline of spectroscopic imaging or mapping. This new field is a technical and intellectual stimulus to many of us and has the potential of becoming a dominant microanalysis technique.

My aim in writing this book has been to introduce the basics of the essential components—infrared, and NMR spectroscopy; microscopy; and imaging. It is an endeavor to bring these different disciplines together in a coherent and unified fashion to be useful for the beginner and as a reference for the experienced. Finally, I illustrate the integration of these analytical concepts into materials characterization, and demonstrate the impact of this discipline on our understanding of the performance of materials, particularly polymers.

Classically, spectroscopic techniques generate an "average" structure over the dimensions of the sample, and no information about the distribution of the structure is obtained. Microspectroscopy allows the measurement of the spatial distribution of chemical structures in the characterization of materials. As such, new concepts in the spatial distribution of structure are being developed which encompass virtually all materials.

The subject is truly interdisciplinary, and although this book presents a view of one person written at a particular moment in the rapid evolution of the discipline of microspectroscopic imaging, it is clear that advances in instrumentation, ie., sources, detectors, high-speed computation control and positioning devices, will in a very short time give the discipline of microspectroscopic imaging a whole new look. The important new advances are in the areas of sources, detectors, high-speed computation, and control and positioning devices. Real-time spectral images will be acquired, processed, analyzed, and presented to the user—who may have little knowledge of the basis of the imaging system.

My hope in writing this book is that I can capture the flavor and excitement of the field and stimulate others to become involved by providing the basic knowledge required to benefit from the use of spectral imaging techniques. Of course, individually, the disciplines of optical microscopy and spectroscopy, whether infrared, Raman, or NMR, have a rich history. The special knowledge required to practice these individual disciplines has been documented in many places, but relatively little has been written about how to successfully use these spectroscopic techniques in combination with microscope and mapping. The ideas outlined in this book are the sum of the experiences of many previous investigators who are eager to share their knowledge and enthusiasm for the field.

A book like this is not the undertaking of a single individual, but the cumulative efforts of a number of people including students, departmental colleagues, and distant colleagues as well as individuals who I have not met, but whose work I have read. To all of them I express my appreciation. But several individuals deserve special recognition. Constantinos Arvanitoupolos (the Greek) drew many of the marvelous figures that enhance this book, and his efforts are appreciated. Dallas Parker and Bentley Wall proofread and made a number of valuable suggestions. Long discussions with John Reffner, one of the premier microspectroscopists in the field are acknowledged, as well as his valuable criticism of the manuscript at various stages in its drafting. Barbara Leach and June Ilhan have carried the burden of producing a manuscript in readable form. Betty Hodges has had the responsibility of relating the figures to the publisher and has done a commendable job. Finally, my wife, Jeanus, deserves to be acknowledged for her patience. She was surprised that a book came from the hours that I spent "playing" at my computer. Or, to put it in basic terms—as was eloquently said by a child—"writing is only rearranging the alphabet."

To the reader, I hope that my efforts at reducing a complicated subject to simpler terms are helpful. I once heard that a "monograph expounds but a textbook explains." This textbook is a collection of explanations, and if they are useful in developing your understanding, I have received my reward.

I wish to express my appreciation to my editor, Barbara Pralle of the American Chemical Society for her persistence in pushing my efforts and those of the ACS. My copy editors, Zeki Erim, Jr., and Jay C. Cherniak have done a commendable job of translating my complex writing style into readable form. I also benefited from the constructive comments of two reviewers of the Raman and NMR chapters. Their comments generated extensive rewriting and substantial improvements in the content of those chapters.

Finally, let me express my admiration to the many students that I have taught. One learns most from one's students, and I have been fortunate to be surrounded by outstanding students who challenge me daily. Their research work has contributed substantially to this book, whether in infrared, Raman, or NMR spectroscopy, and I hope they can recognize their role in this book in terms of research output and also in scrutinization of the fundamentals.

Jack L Koenig
Department of Macromolecular Science
Case Western University

Contents

1

Microspectroscopy and Imaging of Polymers

The search for perfection begins with detecting imperfection.

—Advertisement by A. Finkl & Sons Company
(first to receive ISO 9002 certification),
Plastics World, May 1996, p 20

The microscope is the single most important instrument in the whole of scientific research. The reason for this is clear: microscopes enable us to observe, analyze, and understand the living and material worlds at increasing levels of resolution and detail.

—Chris Hammon and Peter Evenett,
Royal Microscopical Society, Oxford,
USA Microscopy and Analysis, Nov. 1994, p 20

Microspectroscopy and Imaging of Polymers

Advances in materials science play a key role in virtually all aspects of society. Materials science deals with the relationships among the processing, structure, and properties of materials. The goal of materials engineering is to control the manufacturing and processing to yield materials with exceptional properties and enhanced performance. The ability of process engineering to alter the properties of a material is rooted in the understanding that the properties of most materials are dictated by the microstructure and morphology of the material, particularly the nature and distribution of defects.

The Role of Defects in Materials Performance

Engineer's rule: Materials defects can lead to premature failure.

No ideal or perfect material free from defects has ever been formed in nature or in a laboratory. There is an increasing awareness of the deleterious effects of inhomogeneities on the useful physical and mechanical properties of engineering materials (1). For example, the presence of only 0.1% of impurity in the crystal lattice of germanium or silicon has no practical effect on the structure of the crystal but changes its electrical resistance substantially (i.e., 1000-fold).

The continuing development of new extraordinary materials requires more and more stringent control of the defect structures. Mechanical, thermal, hydrolytic, and photolytic processes are the basis for failure, but internal stresses are the driving forces that are accentuated by the presence of defects. The knowledge of the chemical composition of the inhomogeneities, how many exist, their morphology, and their spatial distribution plays an important role in efforts to improve materials.

In addition, it is now recognized that the preservation of quality is an essential requirement for commercial success. Improvement in the quality of the manufactured product will make that product more competitive in the marketplace.

The rate of technological progress is limited by the availability of advanced materials. Often, advanced materials are engineered to exhibit desirable properties by synthesizing them from multiple components. The use of multicomponent materials such as composites, blends, and alloys has increased in recent years as the high-performance properties have improved and the costs have decreased. Consequently, efficient and reliable quality control of these products appears to be one of the most important problems for these industries. Materials inspection methodologies are required for optimizing new manufacturing technologies, for quantitative quality-control analysis to produce reliable and reproducible products, and for damage assessment to determine endurance parameters of the materials. The processing of materials is evolving from an empirical to a more predictive science. Nevertheless, a fully predictive model of the relationships between the microstructure of a material and the technique used to process the material remains an elusive goal.

To study imperfections and inhomogeneities, the analysts need microanalytical techniques that can yield information not only about the number, size, and shape of the inhomogeneities but also their chemical composition and structure (2). *Defectoscopy*, which is the determination of the structure, composition, and spatial distribution of defects or inhomogeneities, requires the use of spatially resolved, microspectroscopic techniques. The construction of the actual spatial or chemical structure of a multicomponent specimen identifies the faults in the physical and chemical processing steps leading to a defective product.

Processing to Control Materials Properties

Materials generally exhibit an internal microenvironmental heterogeneity. All of the important phenomena that determine materials structure and properties during processing are controlled by heat, mass, and momentum transport. For each of these effects, the nature of the transport is determined by the relative importance of convective and diffusive transport. The generation and propagation of defects during the fluid-to-solid phase change are generally understood on only an empirical basis. The size, morphology, composition, and concentration of the inhomogeneities and their distribution arise from their response to microphenomena such as microflow, microdeformation, and microdiffusion arising from internal chemical, thermal, and stress field gradients (*3*). For a material to transform to a more ordered phase, it is necessary to first form an aggregate or cluster of molecules above a critical size to initiate the process. Such an aggregate may form on a foreign surface, such as a container wall or an active nucleating site (heterogeneous nucleation), or may form spontaneously from random internal fluctuations (homogeneous nucleation). Homogeneous nucleation can occur only if the melt is cooled well below its normal freezing temperature without solidifying. Heterogeneous nucleation will almost always occur first if there are any impurities that can act as nucleation sites. We are interested in the end result of these physical processes that modify the internal morphology through local action. Such processes will, after innumerable repetitions, typically produce particular morphological forms with specific surface shapes.

Materials scientists postulate processes such as phase transitions involving symmetry breaking that are, at least in part, responsible for most of the properties of matter (*4*). A particularly simple and notable example is the simple process of the freezing of water. As water molecules settle into position to create an ice crystal, they lose their freedom to rotate about an axis or to move in a straight line (i.e., to "translate") in any direction. In other words, the phase transition from liquid to solid removes or "breaks" a water molecule's rotational and translational symmetries. Water molecules locked in ice thus show a lower degree of symmetry than do water molecules in the liquid. Even so, phase transitions rarely occur uniformly. When water freezes, the orientation of a crystal formed in one region of the liquid often differs from those of crystals formed elsewhere. This produces a mismatch, or defect, wherever the crystals eventually meet. In ice, such defects show up visibly as milky filaments or frosty sheets suspended in the clear solid.

This type of phase transition is typical of materials, and control of these processes is required to produce engineering structures of practical utility. Because different microstructures can be developed in a given material by changing the processing conditions, it is critical to understand the fundamental phenomena that dictate microstructure that yields a desired set of properties to the material. Characteristic dimensions of a given phase and a given morphology determine the mechanical properties of a material. The processing variables are the growth rate at which the solidification takes place and the temperature profile in the region near the solid–melt interface. Changes in growth rate and thermal gradient alter the relative importance of thermal or mass transport and interfacial energy effects, and the magnitude of this partitioning effect is difficult to evaluate without detailed morphological information.

Materials Inspection Techniques

In some industrial fields such as electronics, photonics, and medicine, microfabrication techniques are being used to prepare devices with the dimensions of nanometers. To examine these devices and processes, microspatial spectroscopic techniques are required.

Traditionally, engineering devices are inspected visually. The surface is illuminated at different angles, and the trained eye can detect glints or shadows that indicate cracks, scratches, corrosion, or other underlying structural problems. Flaws such as pits, particles, or scratches can merely be a nuisance or the source of "killer" defects—defects that can catastrophically lead to degradation of the performance of the material in a device. Identifying many types of faults is important because the reduction of these faults leads to a better performing device and an increase in the manufacturer's profitability. In general, chemical defects can be generated from a variety of sources such as handling, cleaning, equipment, and solvents. It is usually assumed that the majority of defects in modern fabrication plants are caused by equipment; particulate contamination is the major source of problems.

Manufacturers and users require tests to be performed to validate specifications and to grade performance of a product. The presence of defects leads to poor performance. Microspectroscopic images can count the defects and measure the size and distribution for the manufacturers. Microspectroscopic imaging of defects can also be useful in development of new manufacturing processes leading to improved products and greater competitiveness.

Traditionally, in an effort to understand the behavior of materials, analysts have reported average concentrations of components in samples measured by macroscopic techniques. However, concentration is a static

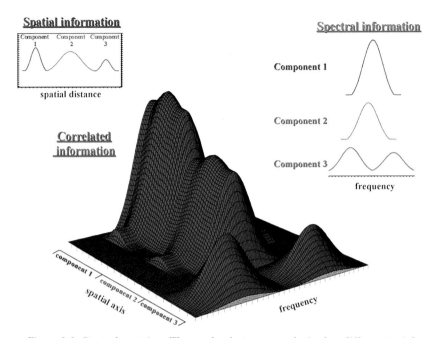

Figure 1.1. Spectral mapping. The correlated spectra are obtained at different spatial positions. Each component can have its specific frequencies as indicated for components 1, 2, and 3. The concentrations of each component can be determined at each spatial position.

concept and is not useful in establishing the nature of the microprocesses occurring in the specimen. Gradients are dynamic and reflect the nature of the physical processes involved. In spatially resolved microanalysis, measurements are obtained for many small neighboring locations, and spatial and gradient relationships can be determined. These microspectral methods are ultimately "chemical microscopic" techniques because they couple directly the spectroscopic information (chemical entities) with space information (Figure 1.1). The maps or images display the spatial distribution of the local chemical environments arising from concentration and structural gradients in the sample. In many cases very good spectra can be obtained from volumes of a few cubic micrometers, typically 10^{-17} g of matter (5).

Developments in Microscopic Imaging

Microscopes have been an invaluable tool for the characterization of materials. A monochromatic or gray-scale image obtained, for example, from an optical microscope contains only luminance information (i.e., how light intensity varies with position in the image and object planes). This type of imaging can judge the size, shape, location, and movement of objects. A polychromatic or spectroscopic image contains luminance and chrominance information, allowing objects to be characterized by the wavelengths of light that they reflect, emit, or absorb. The multivariate images contain substantially more compositional and structural information than monochromatic images. Such spectroscopic images are invaluable for differentiating between defects arising from chemical sources.

Only recently have these microscopic spectroscopic methods been applied routinely to materials characterizaton because of limitations in spectral sensitivity, display, and computational techniques. However, a number of key technologies have combined to make spectroscopic imaging routine. First, the development of microelectronic systems has allowed the construction of smaller components that are faster and cheaper. Next, new developments in sources, detectors, and optics have made it possible to make spatially resolved, spectroscopic measurements of high sensitivity in a reasonable experimental measurement time. Finally, computer control and manipulation of beams and sample stages have proven invaluable to obtaining reproducible spatial measurements. The availability of fast computers with large memories and storage capacities has allowed the development of new image processing and display software, which has made the human–machine interface more friendly.

Computers are the key enabling component in imaging today. Major recent developments in imaging have been due to the availability of powerful and cost-effective electronic computing systems. By adding electronic control and micropositioning with fast analysis capabilities to a microscopy system, one can see and use more of the image information. Image processing with fast workstations can occur almost in real time with dedicated computer hardware on the imaging instrumentation.

Sources of Structural Inhomogeneities

For some materials systems, such as polymers, a hierarchy of microscopic structures contributes to the bulk macroscopic sample (6). Although it is easy to perceive that the samples are uniform, as in Figure 1.2a, in fact they are probably more like the computer-generated spatial distribution of a random sample shown in Figure 1.2b or a nonrandom patchy sample, as in Figure 1.2c (7).

In fact, in most polymer samples, a range of microstructures is built up by association and aggregation of the polymer chains. These microstructures for polymers are shown in Figure 1.3 (8). In addition to these chain structures, higher order structures for crystalline polymers can be classified into several groups (Figure 1.4) (9):

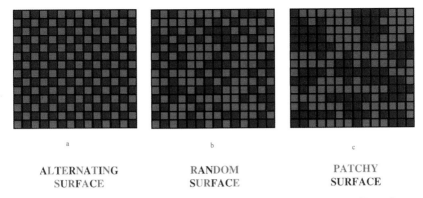

a   b   c

ALTERNATING SURFACE   **RANDOM SURFACE**   **PATCHY SURFACE**

Figure 1.2. Uniform and nonuniform distributions from (a) alternating, (b) random, and (c) patchy surfaces.

- the internal structure of a spherulite such as the ring pattern, the thickness of the fibril, and the spatial distribution of orientation;

- the overall structure of the spherulites;

- the structure formed by phase separation;

- the shape of the spherulite; and

- the orientation order in polymer systems.

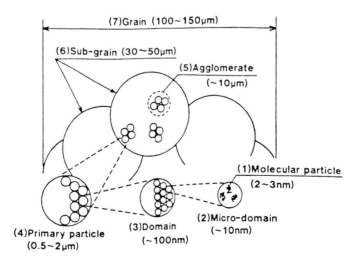

Figure 1.3. Components of poly(vinyl chloride) microstructures. (Reproduced from ref 8. Copyright 1991 American Chemical Society.)

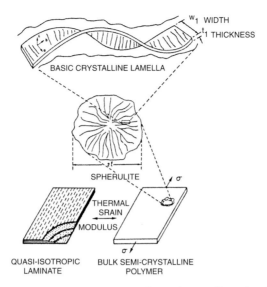

Figure 1.4. Spherulite structure of semicrystalline polymers. (Reproduced with permission from ref 9. Copyright 1975 Society of Plastics Engineers.)

An understanding of the interrelationships among these internal and morphological features is essential to understanding the ultimate materials performance.

Information Content of Spectroscopic Images

Two basic types of information exist in spectroscopic images: spatial resolution of internal objects and dynamic range. Spatial resolution determines the detectable feature size, and dynamic range determines the amplitude over which those features are detected. *Digital imaging microscopy* usually involves four basic steps:

- image acquisition;
- image restoration, to reverse the optical distortion induced by the object;
- image feature extraction, to identify and isolate important characteristic features of an image; and
- visualization and interactive analysis of the image.

These steps may differ, depending on the nature of the sample and the basis of the investigation. Each of these components will be the subject

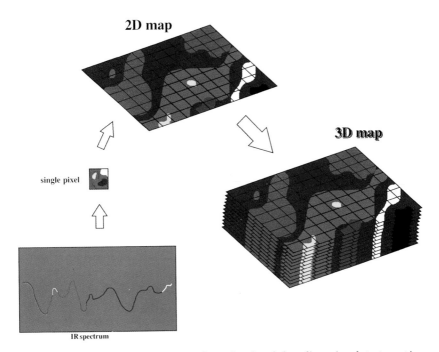

2D map

3D map

single pixel

IR spectrum

Figure 1.5. Process of generating two-dimensional and three-dimensional spectroscopic images.

of discussion in ensuing chapters. The versatility of digital imaging microscopy makes it a powerful tool (*10*).

Digitized spectral images allow the generation of functional maps of the distribution of the contributing object (Figure 1.5). The spectrum of an individual pixel contains contributions from the spectrum from each component. The pixels are scanned to form a two-dimensional map, and multiple maps can be combined to form a three-dimensional map.

Digitized images are a convenient way to communicate results. They are easy to read, understand, and analyze and can provide a permanent record. Images have the characteristics of resolution, contrast, brightness, and physical dimensions. All of these properties yield information about the sample.

The size and spatial distribution of objects in samples are unpredictable. It is reasonable to regard a sample as a spatial arrangement of two- or three-dimensional morphological objects superimposed on a background. The spatial structure of real samples induces a measurable spatial structure in the images derived from them.

For spectroscopic imaging of materials, distinguishable inhomogeneities or objects tend to be spectrally homogeneous and separable from other components and from the background. Adjacent points in the samples are consequently more similar spectrally than might be expected on average, and the mean distance at which this "spatial autocorrelation" disappears is related to the size, spacing, and shape of the inhomogeneities or objects in the sample (see Figures 1.2b and 1.2c).

The observed spatial structure will also depend on the spatial resolution of the imaging probe as compared to the objects that are being imaged. If the spatial resolution is small with respect to the objects, then it will be possible to identify the size and shape of each object as well as to examine the spectroscopic values. If the spatial resolution is larger than the objects in the sample, each digital value represents a brightness for a combination of objects and background. These simple considerations show that the spatial structure and statistics of digital images will be greatly influenced by (1) the size, shape, and spectral character of inhomogeneities in the sample and (2) the relationship between the size of the inhomogeneities and the size of the resolution elements in the digital image of the sample.

Basis of Microspectroscopic Images

Photons from the probe beam can interact with the local features of a specimen in several ways. Typical interactions include absorption, scattering, diffraction, and photofluorescence. Each of these interactions occurs with a probability determined by the chemical and physical characteristics of the specimen under examination. In microspectroscopic imaging systems, the strength of the interaction between the photons in the probe beam and the specimen section illuminated is recorded and assigned a pixel location. In imaging the sample, one can scan the probe beam over a fixed sample or fix the probe beam and mechanically move the sample position relative to the beam in a regular two-dimensional manner. Scanning the sample grid yields a map of the sample's spectroscopic responses as a function of spatial position. Two modes of operation are possible. In the *point mode*, an area of interest on the sample is selected. Then the probe beam is positioned to this point and a spectrum is obtained. In the *global mode*, the spectrometer is set to a particular frequency characteristic of a species of interest, and the beam is scanned over the sample. The light from the selected portion of the sample is directed through the microspectrometer to its monochannel detector, or to a microspectrograph when the detector is a two-dimensional array system. Finally, the image (the two-dimensional spectrum) is visualized on a TV monitor or a CRT screen.

Every interesting new material is available initially in small quantities that require microspectroscopic examination. There is an increasing tendency for these new materials to be structurally complex, and a knowledge of their structure is an important first step in the approach to development. Microspectroscopic measurements can carry out this initial task effectively and rapidly.

For spatially resolved microscopic techniques, we are ultimately interested in the measurement of dimensions and the accuracy of these measurements. Measurement accuracy depends on the microscopic spatial resolution, the amount of image noise, and the edge detection method. We will discuss these matters in Chapter 3.

Super-Resolution Near-Field Imaging Spectroscopy

"One can't believe impossible things," Alice said.

"I daresay," said the Queen, "you haven't had much practice. . . . Why sometimes I've believed as many as six impossible things before breakfast."

—Lewis Carroll, *The Complete Works of Lewis Carroll;*
Bracken Books: London, 1994

The invention of the optical microscope occurred almost 500 years ago. Numerous advances have been made over the years with improved lenses, new contrast mechanisms, and the use of lasers, but the magnification power of the optical microscope has remained the same. Diffraction limits the resolution, and the maximum usable magnification is approximately one-half the wavelength of light. However, a "near-field" microscope was postulated by E. H. Synge in 1928 (*11*) to overcome this diffraction barrier to optical resolution. The super-resolution microscope would illuminate the specimen with a tiny aperture, scanning the spot of light across the sample. If the aperture was smaller than the wavelength of the illumination and was held within a few nanometers of the sample (in the near field), the resolution limit of this type of microscope would be primarily defined by the diameter of the aperture.

It was not until 1984 that this concept was demonstrated by Dieter Pohl of IBM (*12*), using a metal-coated glass fiber on a type of scanning tunneling microscope. Currently, it is possible to obtain resolution of 1/20 with a near-field scanning optical microscope (NSOM) rather than the Rayleigh diffraction limit of 1/2 of a normal optical microscope. Additionally, it is possible to do super-resolved optical spectroscopy using a near-field instrument (*13*), and it has been possible to record the optical spectra of individual fluorescent molecules (*14*). High-resolution near-field spectroscopic imaging was also reported and applied to the examination of semiconductor structure (*15*). In 1994, TopoMetrix introduced the Aurora, the first commercial NSOM that is capable of super-resolved transmission, reflection, polarization, refractive index, and fluorescence imaging (*16*).

Molecular Resolution Techniques for Materials Mapping

After all that we now know from the science of our time, there is a limit to our vision which we cannot exceed. This limit consists of the nature of light itself. But we still have the consolation that there are lots of things between heaven and earth which we cannot imagine at this moment. Perhaps one day, the human genius will find a way to transcend these limits which we cannot exceed. In my opinion, however, the instruments which one day will serve to observe the ultimate details of nature will only share the term microscope with the instruments of our time.

—E. Abbe, *Arch. Microskop. Anal.* **1876**, *IX*, 413

A large number of microscopic techniques are currently in use, and information at molecular resolution in the range from interatomic spacing to a tenth of a millimeter can be obtained by direct imaging with transmission or scanning electron microscopy (TEM or SEM), or using scanning probe microscopic techniques such as scanning tunneling microscopy (STM) or atomic force microscopy (AFM) (*17*). However, these molecular resolution techniques are usually used to measure properties of surfaces and are applicable to only a limited number of materials systems.

With electron microscopes, resolution is limited by the properties of the electromagnetic lens to about 50 Å. They image and measure the surface of the sample. The contrast mechanisms mimic the play of light and shadow on solid surfaces, simplifying the interpretation process (*18*). TEM requires sophisticated sample preparation involving ultramicrotoming or fracturing, in addition to chemical fixation such as staining, etching, or replication. These complex sampling techniques can severely limit the structural details present in the image as well as introducing new structures through deformation in the sample itself (*19*). One of the limitations of TEM is sample size because large samples cannot be examined. Organic compounds, like most polymers, cannot withstand the irradiation in TEM or SEM, so films must be prepared on special electron microscope (metal) grids shadowed with carbon, which involves significant sampling effort and has the potential for inducing structural changes in the sample itself during the preparation. The principal limitations of SEM are the cost and complexity of the instruments because a vacuum system is required.

The main feature of scanning probe microscopes (SPM) is that the measurements are performed with a sharp probe operating in the near field (i.e., scanning over the surface while maintaining a very close spacing to the surface). With STM, a sharp conducting tip is brought sufficiently close to a conducting surface so that electron tunneling can occur between the tip and the surface. The magnitude of the tunneling current decreases exponentially as the tip–surface separation is increased. Measurements of separations of a few angstroms are possible, yielding a lateral resolution of 2 Å and a vertical resolution of 0.01 Å for the right sample on the best instruments. STM images can only be obtained for organic samples sufficiently thin to allow electrons to tunnel through the material. Specimens thicker than 1 mm must be coated with a metal to a thickness of at least 1 nm to obtain an STM image. Polymeric materials are generally insulating materials, and a conducting layer (gold) on the polymer is required for STM (*17*). Recently, it has been observed that an ultrathin water film coating on an insulator's surface provides sufficient electrical conductivity to image the surface by STM (*20*).

The AFM is based on the repulsive van der Waals forces experienced by the probe when it comes in contact with the specimen surface. In the contact mode, the tip is attached to a cantilever, which is scanned across the surface. An interaction force is measured between the tip and the surface (*21*). The force is in the range of interatomic forces (about 10^{-9}), thus the name "atomic force microscopy." The deflections of the cantilever are monitored using a laser beam. Because no current flows through the specimen, the thickness and the conductivity of the specimen are not limiting factors as for STM. The AFM systems can measure the true surface roughness on samples that are as smooth as 0.3 Å root mean square (rms) roughness. Imaging with AFM by the contact mode requires that the molecules of interest be rigidly mounted and immobilized with well-defined orientation to avoid damage by physical contact. The main limiting factor in AFM is the quality of the scanning tip. Image artifacts are introduced due to surface deformation and irregularities of the probe. AFM can also be performed in the noncontact mode in which the tip is made to scan the sample at a constant distance from the surface. In this mode, the cantilever is made to vibrate at its resonant frequency, and the interaction damps the amplitude of the vibration. The noncontact mode of AFM can be used on soft samples; however, the spatial resolution is lower than for the contact mode.

There have been other variations on the SPM, including magnetic force microscopy (MFM) and chemical force microscopy (CFM). In MFM, a ferromagnetic probe is scanned over a sample to detect the forces exerted on the tip by the sample's stray magnetic fields, producing a three-dimensional magnetic force image. CFM involves using a chemically reacting species that is scanned over the surface and generates an image (*22*). Probe tips are functionalized with hydrophobic or hydrophilic molecules to show interactions between simple functional groups, which correlate directly with friction images of sample surfaces patterned with these groups (*22*).

These molecular resolution techniques have found a number of applications in the study of polymeric materials, but they have severe experimental limitations. In contrast, the microspectroscopic optical techniques are noninvasive and nondestructive, have fast response times and high sensitivity, and can be used to perform measurements under ambient conditions (*23*).

Where Are We Going?

Real-Time Imaging

High-speed imaging is important to a number of scientific applications. When structural changes in images occur at rates in excess of 100 Hz, extremely high-speed readout rates are required to capture the information. However, traditionally high-speed imaging systems have produced images inferior to those obtained with precision low-scan imaging systems in terms of readout noise, low contrast capability, and low light sensitivity. In conventional detectors, the need for low readout noise is in direct conflict with a high-speed operation, and readout rates are of the order of seconds per frame.

Imaging processing also becomes a problem in real-time data-input speeds. A digitized monochrome image consumes 1 byte per pixel, so a 1024 × 1024 image occupies 1 megabyte of frame store memory. To send a monochrome image in real time (30 frames/s) to system memory, the bus must be capable of at least 9 megabyte/s throughput. For color images, each pixel must be assigned a color value requiring 3 bytes per pixel for 24-bit color.

Most image applications require a frame grabber board to digitize a camera's analog signals and pass the digital data to a computer or other processor for analysis. This requirement was met by the creation of a fast bus, a dedicated (input/output) I/O interface that can pass up to 40 Mb/s. This should be compared to the personal computer's industry standard architecture (ISA) bus, which has a maximum speed of 0.5 Mb/s.

Intel has developed a new interface, dubbed PCI, as well as the Pentium chip. The PCI has a maximum theoretical speed limit of 132 Mb/s. Thus the PCI can transmit the information into the personal computer's central processing unit, which in the case of a 75-MHz Pentium chip offers industrial-strength processing speed. In addition, on-board image processors can be added and are programmable. The on-board processors are up to 1000 times faster than host-based systems. On-board processing can be used to control devices and to synchronize events during the experiment.

Real-time spectroscopic imaging is on the horizon as key technologies continue to develop.

Three-Dimensional Images

Typically, two-dimensional images or maps are obtained that limit the information content of the image. Three-dimensional images can be obtained by using special optical techniques (i.e., confocal or slice selection and reconstruction techniques derived from multidimensional measurements). The reconstruction techniques for three-dimensional images are computationally intensive, requiring special software and display devices.

Foreseeing a Future for Microspectroscopic Imaging in Materials Characterization

In an age of increasing competition in the materials industry, engineers must improve the quality and decrease the production costs of materials. The key to quality improvement and cost reduction is a better understanding of processing–microstructure–property relations. Spectroscopic imaging is one of the microstructural characterization tools that can provide information about the microstructural link between processing and properties. Spectroscopic imaging can provide quantitative chemical analysis. Some samples can be analyzed remotely. The measurements are rapid and in some cases can be made in real time. Data acquisition and analysis are not difficult and can be carried out using standard software like NIH image (National Institutes of Health).

Spectroscopic imaging can have a significant role in the future characterization of materials. The materials industry is heading in the direction of micromachining, nanotechnology, and gigabyte chips. These two-dimensional microcosms contain billions of devices and miles of interconnect lines. The defect densities of future technologies will have to be very low, and we will be concerned more about the number of perfectly processed systems per defect than the density of defects themselves. Defects will not be accepted as part of the technology in the future. This situation creates a serious need for understanding the source of defects and their formation mechanisms in materials. First, we need to understand the chemistry, the sources of impurities, and the transport mechanisms. Second, we will need tools for real-time analysis in the processing and fabrication. More sophisticated analytical work will be required to gather data for defect statistics in a reasonable time. One of these analytical tools will surely be microspectroscopic imaging.

References

1. Pantelides, S. T. *Phys. Today* **1992**, 67.
2. Sarikaya, M.; Aksay, I. A. *Mater. Res. Soc. Symp. Proc.* **1992**, *255*, 293.
3. Vilgis, T. A.; Heinrich, G. *Kautsch. Gummi Kunstst.* **1992**, *45*,12.
4. *Phase Transformations;* Aaronson, H. I.; Wayman, C. M., Eds.; American Society of Metals: Metals Park, OH, 1982.
5. Hirschfeld, T. In *Microbeam Analysis 1982;* Heinrich, K. F. J., Ed.; San Francisco Press: San Francisco, CA, 1982, p.42.
6. Aksay, I. A.; Baer, E.; Sarikaya, M.; Tirrell, D. A. *Mater. Res. Soc. Symp. Proc.* **1991**, p., *255*.
7. Termonia, Y. *Macromolecules* **1991**, *24*, 1128.
8. Fukumori, K.; Sato, N.; Kuravchi, T. *Rubber Chem. Technol.* **1991**, *64*, 522–533.
9. Kardos, J. L.; Raisoni, J. *Polym. Eng. Sci.* **1975**, *15*, 3.
10. Tredo, P. J.; Levin, I. W.; Lewis, E. N. *Appl. Spectrosc.* **1992**, *46*, 553.
11. Synge, E. H. *Philos. Mag.* **1928**, *6*, 356.
12. Pohl, D. W.; Denk, W.; Lanz, M. *Appl. Phys. Lett.* **1984**, *44*, 651.
13. Bertzig, E.; Trautman, J. *Science (Washington, D.C.)* **1992**, *257*, 189.
14. Bertzig, E.; Chichester, R. J. *Science (Washington, D.C.)* **1993**, *262*, 1422.
15. Harris, T. D.; Grober, R. D. et al., *Appl. Spectrosc.* **1994**, *48*, 14A, 1.
16. Higgins, D.; Barbara, P. *J. Phys. Chem.* **1995**, *99*, 3.
17. Leggett, G. H.; Davies, M. C.; Jackson, D. E.; Roberts, C. J.; Tendler, S. J. B. *Trends Polym. Sci. (Cambridge, U.K.)* **1993**, *1*, 115.
18. Buseck, P. R.; McCowling, J.; Eyring, L. *High-Resolution Transmission Electron Microscopy and Associated Techniques;* Oxford University: Oxford, England, 1988.
19. Thomas, E. L. *Encycl. Polym. Sci. Eng.* **1986**, *5*, 644.
20. Guckenberger, R.; Heim, M.; Ceyc, G.; Knapp, H. F.; Wiegrade, W.; Hillebrand, A. *Science (Washington, D.C.)* **1994**, *266*, 1538.
21. Sarid, D. *Scanning Force Microscopy with Applications to Electric, Magnetic, and Atomic Forces;* Oxford University: Oxford, England, 1991.
22. Frisbie, C. D.; Rozsnyai, L. F.; Noy, A.; Wrighton, M. S.; Lieber, C. M. *Science (Washington, D.C.)* **1994**, *265*, 2071.
23. Sheppard, C. J.; Wilson, T. *Theory and Practice of Optical Scanning Microscopy;* Academic: New York, 1985.

2
Basics of Imaging

The general imaging properties of a spectroscopic microscope can be derived by following the paths of the light rays geometrically when the image produced by the objective lens is conjugate with the specimen (*1*). Each image point is then geometrically related to a corresponding point in the specimen. There are two approaches to understanding the imaging process: one involves the use of classical optics and ray tracing and the other the use of Fourier transform mathematics to arrive at the same conclusions (*2*). Each approach has its advantages and is invoked as appropriate. We will examine both approaches and their interrelationships.

Elementary Optical Considerations for Imaging

Light can be viewed as a train of waves, or alternatively, the propagating waves can be represented by rays that are oriented normal to the wave fronts (*3*). Oscillatory phenomena like light are described by an amplitude and phase. In the time domain, the oscillation is given by $A \exp(i\omega t)$, where A is the amplitude and ω the phase. The real and imaginary parts of $A \exp(i\omega t)$ oscillate in quadrature as cosine and sine variations, respectively. The angular frequency and time have inverse dimensions and are known as *conjugate variables*. In the spatial domain, the equivalent expression is $A \exp(i\mathbf{k} \cdot \mathbf{r})$, where the wave vector $\mathbf{k}$ and spatial coordinate $\mathbf{r}$ are conjugate. In light both components are present, and a plane wave propagating in the direction $\mathbf{k}$ can be described as $A \exp(i\mathbf{k} \cdot \mathbf{r} - i\omega t)$. Note that $\omega = 2\pi/T$ and $\mathbf{k} = 2\pi/\lambda$, where T and λ are the time period and wavelength, respectively.

To design an optical system, two optical invariants must be considered—optical conductance and radiance. The optical conductance is the maximum optical energy transmitted through an optical system and is the primary consideration of the optical system designer when a ray trace

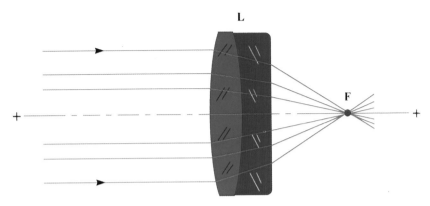

Figure 2.1. Ray diagram through a perfect lens. L is the lens and F is the focus. (Reproduced with permission from ref 4. Copyright 1986 Plenum.)

analysis is performed. Additional considerations are the field-of-view, object distance, and other system requirements. Optical ray trace analysis ensures that the objects located within the instrument field-of-view are imaged with the same degree of fidelity. This means that a one-to-one geometric correspondence exists between the points in the object space and the image space, that is, the *conjugate points.*

The information about the object is carried by the photons traveling along the ray trajectories. In terms of the information transfer, a complete ray trace analysis ensures that the photons carry the information from the object to the image point.

A perfect *lens* converts a paraxial, parallel beam of light into a concentrated, fine spot of light at its focus *F* (Figure 2.1). Conversely, a point source of light located at *F* emerges as a paraxial, parallel beam of light from the lens (*4*).

Consider a plane wave front incident on the lens (Figure 2.2). The incident wave of wavelength λ travels along the lens axis, and the effect in the lens medium is to slow the wave more at its center (thicker) than at the outer edges (thinner). The result is to cause points of equal phase to form at point *F* with identical phase despite having traveled different distances.

Real lenses have *aberrations* (imperfections) that prevent an optical lens from producing an exact geometrical (and chromatic) correspondence between an object and its image. Possible optical aberrations include

- spherical aberration, in which the spherical symmetry of the beam is destroyed;

- chromatic aberration, in which the different frequencies result in different focused spots;

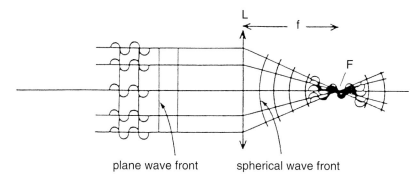

plane wave front   spherical wave front

Figure 2.2. Wave diagram through a perfect lens. L is the lens (vertical double-headed arrow represents the lens), F is the focus, and f is the focal length. (Reproduced with permission from ref 4. Copyright 1986 Plenum.)

- coma aberration, in which the off-axis beams do not form a single focused spot but rather comet-shaped patterns; and

- field curvature aberration, in which the electric field of the light is distorted by curvature.

In all of these aberrations the light from a point is not, in fact, focused to an infinitely small point in the conjugate plane. Rather, upon passing through a lens, the waves converge together and interfere with each other around the focus to produce a three-dimensional diffraction pattern. The diffraction pattern is periodic along the axis of observation (longitudinal) as well as in planes perpendicular to that axis (transverse). When the three-dimensional diffraction pattern is sectioned in the focal plane, it is observed as a two-dimensional diffraction pattern termed the *Airy disk (4)* (Figure 2.3). In the image with optical aberrations, each point of the specimen is thus represented by an Airy disk diffraction pattern, rather than by an infinitely small conjugate point.

The spatial resolving power of a real objective lens can be determined by examining the size of the Airy disk formed by that lens. Let D equal the peak-to-peak distance of the intensity distribution curves (or the center-to-center distance of the Airy disk images) and let r equal the disk radius. The sum of the intensities of the pair of Airy disks shows two peaks when $D > r$, and barely two peaks when $D = r$ (the condition known as the *Rayleigh criteria*). When $D < r$, the sum of the peaks merges into a single peak, and the images of the two points are said not to be resolved. Therefore, for well-corrected optical objective lenses with a uniform circular aperture, two adjacent points are just resolved when the centers of their Airy disks are separated by r given by

$$r = 1.22\lambda_0/(2\mathrm{NA}_{obj}) \tag{2.1}$$

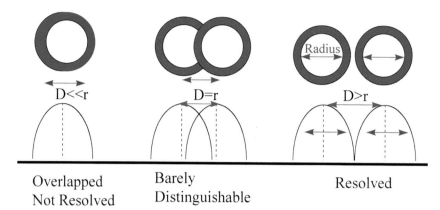

| Overlapped Not Resolved | Barely Distinguishable | Resolved |

Figure 2.3. The Airy disk and intensity distribution in the diffraction pattern for different values of D. On the right is the case where the diffraction pattern is greater than the radius of the Airy disk and the peaks are spatially resolved. In the middle, the diffraction patterns overlap to the extent that D = r and the lines are barely distinguishable. When D ≪ r, the bands are overlapped and not resolved.

where λ_o is the wavelength of light in air and NA_{obj} is the *numerical aperture* of the objective lens

$$NA_{obj} = n \sin \theta \qquad (2.2)$$

Here, n is the refractive index of the medium between the specimen and the objective lens, and θ is the half-cone angle of light captured by the objective lens.

For an optical microscope λ is the wavelength of visible light (about 0.5 μm). This means that an optical microscope with a 1.4 NA_{obj} objective lens can (at best) resolve spatial details of about 0.22 μm.

Elementary Fourier Transform Considerations

To appreciate the Fourier approach to imaging, let us start with the input function and follow the imaging process through the imaging system. The output will then be the sum of the responses of each individual component (*3*).

We will consider a linear model of a digital imaging system (*3*). The optical system consists of an imaging subsystem that produces an image that is a blurred sampled version of the sample. All spatial coordinates (*x,y*) are referenced to a common orthogonal coordinate system that is normalized so that the sampling interval in both directions is unity.

The *delta function*, $\delta(x,y)$, is an idealized two-dimensional input function since it has infinitesimal width in both dimensions and has an integrated volume of unity, that is,

$$\int_{-\infty}^{\infty} \int \delta(x,y) = 1 \tag{2.3}$$

Our *input function* $g_1(x_1,y_1)$ will be an array of two-dimensional delta functions given by

$$g_1(x_1,y_1) = \int_{-\infty}^{\infty} \int g_1(x',y') \delta(x_1 - x', y_1 - y') \, dx' \, dy' \tag{2.4}$$

The delta function here is exhibiting its *sifting property*. That is, the delta function at $(x_1 = x', y_1 = y')$ has sifted out the particular value of g_1 at that point. The *output function* $g_2(x_2,y_2)$ is given by the system operator G operating on the input function:

$$g_2(x_2,y_2) = G\{g_1(x_1,y_1)\} \tag{2.5}$$

Substituting, we have

$$g_2(x_2,y_2) = G\left[\int_{-\infty}^{\infty} \int g_1(x',y') \delta(x_1 - x', y_1 - y') \, dx' \, dy'\right] \tag{2.6}$$

We can consider $g_1(x',y')$ as a weighting function for each delta function so the output can be written

$$g_2(x_2,y_2) = \int_{-\infty}^{\infty} \int g_1(x',y') G[\delta(x_1 - x', y_1 - y') \, dx' \, dy'] \tag{2.7}$$

We will describe the term in the integral as the *impulse response* or *point-spread function* (PSF) and it is given by

$$\text{PSF} = h(x_2,y_2; x',y') = G[\delta(x_1 - x', y_1 - y') \, dx' \, dy'] \tag{2.8}$$

which is the output function at x_2,y_2 due to an impulse or two-dimensional delta function at $x_1 = x', y_1 = y'$. The effect of the PSF can be simply demonstrated by the example shown in Figure 2.4, where a planar array of sources $I_1(x_1,y_1)$ is separated from a detector in which the output image intensity is $I_2(x_2,y_2)$.

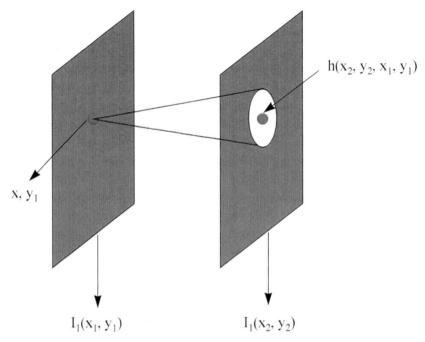

Figure 2.4. Elementary imaging system. $I_1(x_1, y_1)$ is the intensity at (x, y), and $h(x_2, y_2, x_1, y_1)$ is the transfer function. (Reproduced with permission from ref 5. Copyright 1986 Prentice Hall.)

The output is blurred due to the spreading of the radiation over the distance between the planes. Substitution of the PSF in the integral, and changing the coordinate notation ($x_1 - x'$ becomes x_1, etc., for simplicity) yields

$$g_2(x_2,y_2) = \int_{-\infty}^{\infty} \int g_1(x_1,y_1)\, h(x_2,y_2; x_1,y_1)\, dx'dy' \qquad (2.9)$$

Once we know $h(x_2,y_2; x_1,y_1)$ for all input coordinates, we can find the output due to any input function $g_1(x_1,y_1)$. If the impulse function is the same for all input points, the PSF becomes dependent solely on the difference between the output coordinates and the position of the impulse.

The spectroscopic PSFs should be linear. That is, they should have the properties of *scaling* and *shift invariance*. Scaling invariance simply implies that if we record a spectral measurement for a first voxel and then another due to a second voxel, the image intensity due to both voxels acting simultaneously is the sum of the individual image intensities. This scaling can be expressed formally as

$$G\{aI_1(x,y) + bI_2(x',y')\} = aG\{I_1(x,y)\} + bG\{I_2(x',y')\} \quad (2.10)$$

The PSFs should be *spatially invariant* (or isoplanatic). *Isoplanatic* means that the output image does not depend on the location of the input object. To be spatially invariant, the impulse response must be the same for all input points; then the impulse response merely shifts its spatial position for different spatial input points but does not change their functional behavior.

Actually, no spectroscopic microscopy system is strictly linear or shift-invariant. All optical components of the system contribute to G yielding a range of responses spatially. However, the linear, shift-invariant model is an excellent approximation of the system and is different for each spectroscopic technique. The PSFs can be different for each set of measurements since they depend on instrument performance. The PSFs can also have different components that reflect the different sampling techniques (i.e., transmission vs. attenuated total reflectance (ATR). The imaging system and the sampling function are cascaded (convoluted):

$$G(x,y) = G_o(x,y) * G_s(x,y) \quad (2.11)$$

where $G_o(x,y)$ is the optical PSF, $*$ is the convolution function, and $G_s(x,y)$ is the sampling PSF.

An Example of the Impulse Function: Magnification

Consider the typical system of magnification using a pinhole aperture as shown in Figure 2.5 (5). In this case, the impulse function is space-variant.

With the pinhole on axis, the impulse response by geometry becomes

$$h(x_2,y_2; x_1,y_1) = h(x_2 - Mx_1, y_2 - My_1) \quad (2.12)$$

where $M = (-b/a)$. Thus, the output intensity is given by

$$I_2(x_2,y_2) = \int_{-\infty}^{\infty} \int I_1(x_1,y_1) h(x_2 - Mx_1, y_2 - My_1) dx_1 dy_1 \quad (2.13)$$

Let us substitute

$$x'' = Mx_1 \quad \text{and} \quad y'' = My_1 \quad (2.14)$$

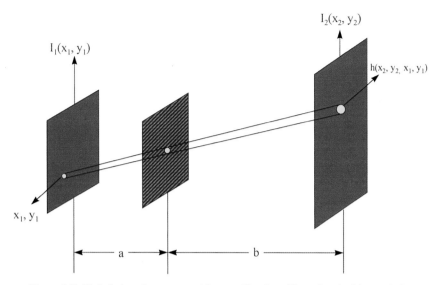

Figure 2.5. Pinhole imaging system with magnification. (Reproduced with permission from ref 5. Copyright 1986 Prentice Hall.)

giving

$$I_2(x_2,y_2) = (1/M^2) \int_{-\infty}^{\infty} \int I_1(x''/M),(y''/M)\, h(x_2$$

$$- x'', y_2 - y'')\, dx''dy'' \quad (2.15)$$

Thus the infinitesimal pinhole camera experiences magnification but no blurring. The $1/M^2$ factor is the loss of intensity due to the magnification of the image. In this case, the impulse response $h(x_2 - x'', y_2 - y'')$ is a magnified version of the aperture function where the magnification factor is $(a + b)/a$.

The Optical Transfer Function

Suppose we have an initial input function $g_i(x,y)$ with Fourier transform $G_i(f_x,f_y)$, where f_x and f_y are the spatial frequencies (the frequency in space in which a recurring feature appears) and we multiply G_i by another function $H(f_x,f_y)$. The function H defines a linear spatial filter and is called the *modulation transfer function* (MTF) of the filter. The product is the Fourier transform of the output of the filter.

$$G_0(f_x,f_y) = G_i(f_x,f_y)H(f_x,f_y) \quad (2.16)$$

The output function $g_0(x,y)$ is the inverse Fourier transform of G_0. By the convolution theorem, we have

$$g_0(x,y) = F^{-1}\{G_0(f_x, f_y)\} = F^{-1}\{G_i(f_x, f_y) H(f_x, f_y)\} \qquad (2.17)$$

$$g_0(x,y) = g_i(x,y) * h(x_2, y_2; x_1, y_1) \qquad (2.18)$$

where

$$h(x_2, y_2; x_1, y_1) = F^{-1}\{H(f_x, f_y)\} \qquad (2.19)$$

and the function $h(x_2, y_2; x_1, y_1)$ is the impulse function or PSF.

The MTF is simply the Fourier transform of the PSF:

$$H(f_x, f_y) = F\{g_1(x,y)\} \qquad (2.20)$$

$$= \int_{-\infty}^{\infty} \int_{-\infty}^{\infty} h(x_2, y_2; x_1, y_1) \exp[-i2\pi(f_x + f_y)] \, dx \, dy$$

where F is the Fourier transform operator. The two-dimensional function is decomposed into a continuous array of grating-like functions having different periodicities and phase angles. Each (f_x, f_y) point in the Fourier space corresponds to an elementary "plane wave" type of function in object space. This complex exponential function has lines of constant phase separated by $(x^2 + y^2)^{-1/2}$ and at an angle of $\tan^{-1}(f_x, f_y)$ with the x-axis.

We have introduced the concept of *spatial frequencies*. This concept needs elaboration. Let us assume that we have a general data array consisting of measurements spaced equally along some relevant coordinate system. The length of any array can be denoted by $2^k = L$. One way of thinking of the array is to fit it into a box of length L. Spatial frequencies are basically those sine and cosine waves that will exactly fit in the box. For each spatial frequency, the Fourier domain contains one real and one imaginary point corresponding, respectively, to a cosine and a sine wave originating at the first point of the data array. The highest spatial frequency will be determined by the number of points in the array. The spatial frequencies depend only on the box size and number of points for their frequency definition. The recorded data define only the magnitude of each spatial frequency required to replicate the original data array upon their addition.

The Fourier transform operation simply provides the means to determine the magnitude of each spatial frequency present in a data set. From this spatial frequency view of Fourier transforms, it is reasonably apparent that any manipulations in Fourier space could be described as alterations in either the number or magnitude of spatial frequencies that describe the data.

We need to determine the resolution and contrast of an optical system as a function of input resolution and contrast. The MTF provides that information. If the MTF of a system is known, one can compute the image from an input object.

Contrast is defined as

$$C = (I_{\max} - I_{\min})/(I_{\max} + I_{\min}) \tag{2.21}$$

where I stands for intensity of light in the image or object. The MTF is a measure of how much image (output) contrast varies with object (input) contrast.

One can measure the contrast of the input and of the output for a given spatial frequency and take its ratio (output over input). This is the value of the MTF at that frequency. The maximum value of the MTF is 1 (no loss of contrast), and the minimum is 0 (total loss of contrast). The MTF is a relative, not absolute, measure.

The MTF is the series of values of the ratio of output to input contrasts for all relevant frequencies. This function is best determined by computing the Fourier transform of the image (output) intensity function and taking its modulus. The Fourier transform in this case takes the spatial distribution of light intensity in the image and converts it to spatial frequency information (the amplitudes of spatial sine waves of increasing frequencies in the image). This provides a curve that declines with increasing spatial frequency. The MTF curve is normalized to one at zero spatial frequency.

Image Development Using Ray Tracing Considerations

Let us consider the formation of an optical image P of a source O by the use of a lens (Figure 2.6) (6). The incident wave is slightly inclined to the axis of lens 2 such that its κ-vector, $(2\pi/\lambda)(\alpha\mathbf{i} + \beta\mathbf{j})$, has direction cosines α and β with respect to the axis x and y. The wave front passing through the pole of the lens will continue in the same direction, so that κ defines the angular position of the image point. The lens causes all parts of the wave front over the aperture of the lens to be co-added in the focal plane at a point P with coordinates

$$x\mathbf{i} + y\mathbf{j} = (\alpha\mathbf{i} + \beta\mathbf{j})f = P \tag{2.22}$$

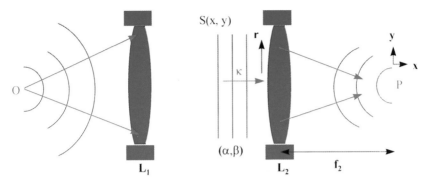

*Figure 2.6. The formation of an optical image P of a source O by the use of a lens; α ± β are the direction cosines of the plane wave incident on lens 2. S(x, y) is the signal, κ is the density distribution, **r** is the radius of the lens and f₂ is the focal of the lens. (Reproduced with permission from ref 6. Copyright 1991 Oxford Science Publications.)*

where f is the focal length of the lens. Thus the coherent superposition of contributions from the lens leads to an amplitude at P given by

$$\mathbf{E}(\kappa) = \mathbf{E}_0 e^{i\kappa f} \iint_{\text{aperture}} S(\mathbf{r}) e^{-k r} \, d\mathbf{r} \tag{2.23}$$

where $S(\mathbf{r})$ represents the amplitude and phase distribution across the aperture on the incident side, and the integration is performed over the aperture of the lens. This is equivalent to a Fourier transformation in which a lens weighting function $g(\mathbf{r})$ is unity inside the aperture and zero outside:

$$g(\mathbf{r}) = 1 \text{ inside aperture} \tag{2.24}$$

$$g(\mathbf{r}) = 0 \text{ outside aperture} \tag{2.25}$$

so

$$\mathbf{E}(\kappa) = \mathbf{E}_0 e^{i\kappa f} \iint_{\text{aperture}} S(\mathbf{r}) g(\mathbf{r}) e^{-ik\mathbf{r}} \, d\mathbf{r} \tag{2.26}$$

Assuming for the moment that $S(\mathbf{r})$ is uniform (for the sole purpose of illustrating a point), except for the term in front of the integral, $\mathbf{E}(\kappa)$ is just the Fourier transform of $g(\mathbf{r})$. This relationship shows that for a finite lens aperture one obtains a function that is simply the two-dimensional spatial FT of the aperture.

Suppose now that we have a real object with a density distribution $O(\kappa)$. The first lens converts the density function of light emerging from the object into a conjugate signal $S(\mathbf{r})$ given by

$$S(\mathbf{r}) = \iint O(\kappa)e^{-i\mathbf{kr}}\, d\kappa \qquad (2.27)$$

Nonideality of the first lens is accounted for by multiplying $S(\mathbf{r})$ by an appropriate spatial function $g_1(\mathbf{r})$. The second lens then reconstructs the image via the inverse transformation

$$I(\kappa) = \iint S(\mathbf{r})\, g_1(\mathbf{r})\, g_2(\mathbf{r})\, e^{-i\kappa\mathbf{r}}\, d\mathbf{r} \qquad (2.28)$$

The net effect of two lenses is given by $g(\mathbf{r}) = g_1(\mathbf{r})g_2(\mathbf{r})$.

The mathematics are complicated (although elegant), but the process has physical implications and can be explained rather simply using Figure 2.7 (4).

The specimen is illuminated from behind, and the light reaching the first lens is an intensity profile that corresponds to the light-transmitting properties of the specimen. The first lens in this system is situated exactly

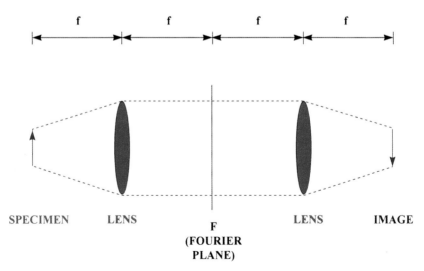

Figure 2.7. An optical system for imaging the Fourier transform of an object. (Reproduced with permission from ref 4. Copyright 1986 Plenum.)

at its focal length from the specimen plane. Consequently, light rays leaving the back of the lens travel parallel to each other. The second lens focuses these parallel rays and produces the inverted image seen at the image plane. The image at point F (Fourier plane) is the Fourier transform of the intensity profile of the specimen. The second lens also produces a Fourier transform. However, the image that this lens "sees" is the Fourier transform produced by the first lens. This second Fourier operation forms the inverted image of the specimen found in the image plane. Thus two Fourier transforms performed sequentially on the same function will reproduce the original function inverted. If the information represented in the Fourier transform can be preserved during its transmission through the microscope, the image formed at the image plane is an exact replication of the original specimen. However, no optical system is perfect, and some aberration of the image always occurs.

An ideal lens images an ideal Fourier transform of the intensity function of the specimen. This ideal Fourier transform extends to infinity, with the highest spatial frequencies (representing the smallest resolvable features of the specimen) being found farthest from the optical axis. The aperture of the microscope completely blocks all light from the ideal Fourier transform that is beyond the diameter of the aperture, so the aperture has the effect of removing the highest spatial frequencies (smallest objects) from the transform. When the image of the specimen is reformed by the second Fourier operation, the smallest features in the image will be missing, and resolution will be reduced. The wider the aperture function of the microscope, the higher the frequencies allowed to pass from the ideal Fourier transform, and the better the resolution of the microscope. This effect is shown in Figure 2.8 (*4*).

In optical microscopy magnification is achieved by appropriately adjusting the relative separation of the object and image planes for their respective lenses so that the wave vector κ translates into differing spatial separations. The PSF due to the finite lens apertures introduces a convolution broadening that corresponds to an image spatial separation of order $f_2\lambda/d$, where d/f_2 is the maximum angular aperture subtended by the lens.

In the Fourier sense, the resolution of optical microscopy is determined by the highest object spatial frequency that can be accommodated. This limiting Fourier component can be considered as a sort of sinusoidal grating in the object. When this grating is sufficiently fine that the first-order maxima occur at angles greater than the angle θ_{max} subtended by the lens, then the limit is reached. This is the *Abbe criterion* and corresponds to a spatial period $\lambda/\sin\theta_{max}$. Thus the resolution is given by

$$r \cong \lambda/\sin\theta_{max} \cong 2\pi/\kappa_{max} \qquad (2.29)$$

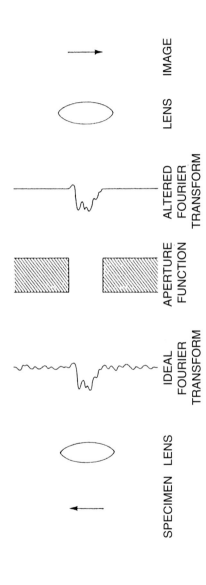

Figure 2.8. Model of a microscope imaging system using Fourier transforms. (Reproduced with permission from ref 4. Copyright 1986 Plenum.)

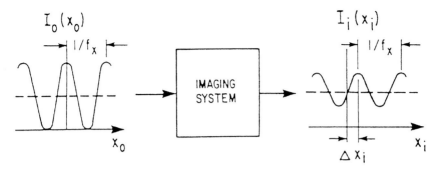

Figure 2.9. Imaging a sinusoidal test object. (Reproduced with permission from ref 4. Copyright 1986 Plenum.)

A more precise analysis including the refractive index yields

$$r = 0.61\lambda/n \sin \theta_{max} \qquad (2.30)$$

which was previously derived from light ray considerations.

Imaging of Simple Systems

Let us begin with the simplest of examples. Consider the imaging (without magnification) of an input object that is a sinusoidal wave (Figure 2.9) (*4*).

The peak-to-peak distance along the sinusoid is its spatial period. The reciprocal of the period is the spatial frequency, f_x, expressed in units of cycles per unit length. Three things can be observed experimentally about the observed image. First, the spatial frequency of the image is the same as that of the object. Second, the contrast or modulation is typically less in the image than in the object. Third, the object is centered about x_o but the image is displaced by an amount Δx. When we image several objects with different spatial frequencies, the image modulation and position shift will vary as a function of spatial frequency.

The MTF expresses the way the imaging process alters the contrast of the sinusoidal object, as a function of spatial frequency. The displacement Δx is equivalent to a phase shift $\Delta\theta$ of the sine function.

The MTF is determined primarily by two things for a spectroscopic microscope: the NA_{obj} and the aberrations in the optical elements. The MTF of a diffraction-limited microscope with circular aperture is shown in Figure 2.10 (*7*).

The typical MTF of a scanning imaging system may be assumed to be Gaussian:

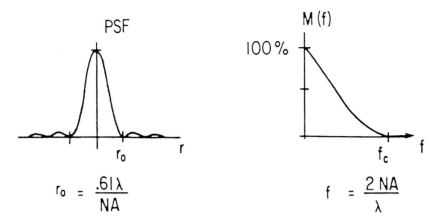

$$r_0 = \frac{.61\lambda}{NA} \qquad\qquad f = \frac{2\,NA}{\lambda}$$

Figure 2.10. Modulation transfer function of a diffraction-limited microscope with circular aperture and the corresponding point-spread function (PSF). (Reproduced with permission from ref 7. Copyright 1986 Plenum.)

$$\text{MTF}(f_s) = \exp\left(-\pi^2\sigma^2 f_x^2\right) \qquad (2.31)$$

where f_x is the spatial frequency and σ is the shape parameter. For example, for a circular aperture, σ is its actual diameter.

Image contrast is highest for low spatial frequencies, and decreases to zero as spatial frequency increases to f_c. The frequency, f_c, at which contrast goes to zero is called the *cutoff frequency*, and is related to system parameters by the formula

$$f_c = 2NA_{obj}/\lambda \qquad (2.32)$$

which expresses the well-known fact that resolution increases with larger NAs and shorter wavelengths.

Optical System Design

The ideal optical system is a collection of hypothetical optical components free from any aberrations that images a point source on the focal plane as a dimensionless, abstract point with an intensity. For a real, low-aberration system, the image of the point source has a real dimension, and inspection of the shape of the optical ray distribution at the image plane reveals a spread of the Gaussian distribution resulting from the presence of optical aberrations. Integration of the distributions over the image area yields an optical efficiency. The results can be presented as two- or three-dimensional graphs of the spatial distributions of the collected rays at the focal plane of the image.

There are currently on the market a number of software programs that allow one to do optical system design. Optical ray tracing programs allow the calculation and presentation of spatial ray distributions and of optical collection efficiency. These programs allow different optical elements to be designed with a variety of surfaces such as lenses, mirrors, apertures, and irises. The refraction, reflection, and propagation of light as a function of different media and wavelengths can be taken into account. Each of the generated rays is traced through the optical train. A sequence of ray segments is transmitted through a succession of optical surfaces, where each change of direction obeys the appropriate law of reflection or refraction at each surface.

References

1. Born, M.; Wolf, E. *Principles of Optics;* Pergamon: New York, 1980.
2. Heavens, O. S. *Optical Properties of Solid Films;* Butterworth: London, 1955.
3. Banerjee, P. P.; Poon, T.-C. *Principles of Applied Optics;* Richard D. Irwin and Aksen Associates: Boston, 1991.
4. Inoué, S. *Video Microscopy;* Plenum: New York, 1986.
5. Macovski, A. *Medical Imaging Systems;* Prentice Hall: New York, 1983; p 12.
6. Callaghan, P. T. *Principles of Nuclear Magnetic Resonance Microscopy;* Oxford Science: New York, 1991.
7. Hansen, E. W. In *Video Microscopy;* Inoué, S., Ed.; Plenum: New York, 1986; p 467.

3
Acquisition of Images

First get your facts; then you can distort them at your leisure.

—Mark Twain

Methods of Obtaining Spectroscopic Images

Characterization and mapping of real, complex samples can be best achieved by measurements of the samples with minimal perturbation to preserve information about sample components and their spatial distribution. The success of optical imaging depends on the acquisition of reproducible images of high fidelity with a minimum of blurring. Optical microscopy can often be carried out without the samples being "fixed," that is, sectioned or stained. However, as with high-resolution techniques such as electron microscopy, optical techniques can be enhanced by staining or labeling the specimen. Spectroscopic techniques, such as IR and Raman, on the other hand, have inherent "stains" or labels that constitute their vibrational frequencies, and these unique frequency markers serve as contrast mechanisms for the specimens. Spectroscopic image acquisition is critically constrained by the performance of the optics, nature of the sampling technique, and the response of the photon-detection system.

There are basically three methods for acquiring digital images (Figure 3.1): projection, scanning, and tomography. In *projection imaging*, the radiation from the specimen is transmitted through a lens onto a uniformly spaced, discrete array of detectors. In *scanning imaging*, the detector (or source) or the sample is moved so that different positions in the sample are rastered in sequence. In *tomography*, the detected signal obtained from different projections contains the spatial image information, and the image must be reconstructed mathematically from the projected signals.

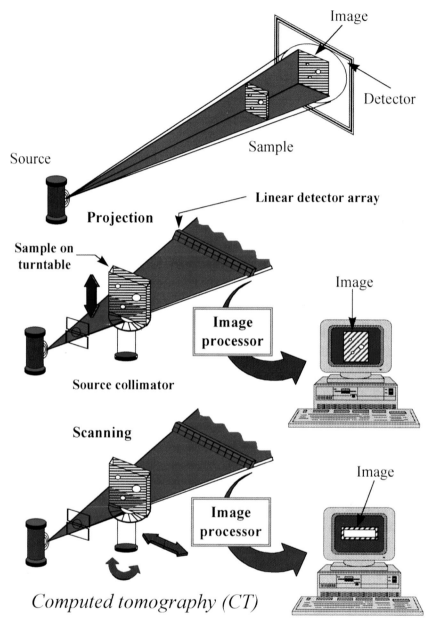

Figure 3.1. Diagram of the three types of imaging: projection, scanning, and tomography.

For projection methods, the image detector is often a color TV camera or a charge-coupled device (CCD) detector that produces a digitized signal. The spectroscopic information for projection methods can be generated by optical filtering. In the projection mode, the sample is uniformly illuminated by the rapid scanning of a broad beam over the sample area.

Scanning methods may be of two types. One type involves scanning of the sample systematically through a stationary field of view defined by the collection optics and detector. Stage or object scanning images are obtained by actually moving the specimen, sweeping it past the optical axis, scanning a line in the x-direction, moving an increment in y, and then performing another x-scan. In the second type, the source (or detector) is scanned over a stationary sample. Laser scanning microscopes were originally developed to direct an extraordinary amount of coherent light on a specified spatial region within a specimen. Such instruments use vibrating mirrors, and rotatory or other devices, to direct the laser light across the specimen in the xy plane.

There are some advantages to each of these scanning approaches. Point-by-point stage scanning uses a fixed beam and eliminates any lateral chromatic aberration that is inherent in beam-scanning methods. Thus, the signal intensity and resolution are independent of location. Also, with stage scanning one can scan the entire sample with the same magnification, whereas with a beam-scanning system, the data are collected strictly from the field of view visible to the system. This can be important for large samples or when searching the system for rare events. The major disadvantage of stage scanning is speed because the image is built up point by point. Beam scanning operates much faster.

For spectroscopic imaging using the scanning method, one must focus the beam to a small diameter that is scanned over the sample. Scanning requires hardware that can move the focused beam (or the detector) or the sample stage rapidly in a predefined pattern with good position repeatability and zero backlash. The signal must be acquired at each *voxel* (minimum detectable volume element) location without rectilinear distortion.

Attempts to increase sensitivity in scanned systems require getting more signal to the detector. This in turn means either increasing the optical element size or slowing the scanning speed so the dwell time (the time any one part of the sample is being imaged onto the detector) is increased. Either of these methods has undesirable consequences. Larger optical systems mean higher cost, and slower scan speeds increase measurement times.

For tomographic methods such as medical imaging, one obtains signals from two-dimensional slices and three-dimensional volumes of the sample, and the image is reconstructed using projection–reconstruction methods. The projection–reconstruction methods of imaging require that the experiment be repeated by rotating the sample about an axis or electronically rotating the acquisition signal. To obtain the image, the experiment is repeated until the rotation angle, θ, has been varied from $0°$ to $180°$. The image is then reconstructed by inverting the signal as a function of angle.

Factors Influencing Image Acquisition

Three primary factors affect the accuracy of the acquired image: field-to-field positioning, focusing, and illumination. It is sometimes difficult to ascertain which of these factors contributes to a faulty or highly blurred image.

When the measurements diverge between images over several spatial fields rather than randomly, this points to focusing as the possible source of the discrepancies because instability in illumination would be expected to occur randomly. Field-to-field positioning inaccuracies are expected to be cumulative (i.e., constantly increasing) and so can be detected by systematic shifts in the images taken from the same field of view. Upon reversal of this scanning process, substantial differences will be observed if field-to-field positioning is the problem.

Imaging Instrumentation

Illumination

The fidelity and resolution of microscope images depend on the performance of the illumination systems and the performance of the imaging optics. The illumination can be *episcopic* (reflected), *diascopic* (transmitted), or *oblique* (off-axis), or a combination of these, depending on the need. In addition, there are various illumination techniques.

Bright field is an illumination technique that provides flat, even illumination of the field of view. With bright-field illumination, the light irradiates the specimen from behind, leading to a diffracted or refracted image of the objective on a bright background. With bright-field illumination, the image clearly shows where objects are located, their aggregation, and the color and grain of the sample.

Dark field lights the specimen surface from an oblique angle so surface distortions can be observed. With dark-field illumination, one generates a highlighted image of the specimen on a dark background. A hollow cone of light with an angle greater than the acceptance angle of the objective is used. In this way, only light diffracted or scattered from a specimen at the apex of the light cone can enter the objective.

Critical illumination uses a small extended light source that is directly imaged by a condenser lens onto the object plane of the microscope. *Kohler illumination* uses a field lens to form a real image of the light source at the primary focus of the condenser lens.

Opaque objects must be viewed by light reflected off them. *Vertical illumination* uses the light as it passes through or around the objective to reach the specimen. This requires a beam splitter or prism to direct light down the optical axis of the objective. For *epi-illumination,* an annular beam splitter is used to form a hollow tube of light that passes through a doughnut-shaped condenser encircling the objective. In this lighting scheme the objective is bypassed altogether.

For highly transparent samples, phase-contrast microscopy is used. *Phase contrast* takes advantage of the unequal transmittance of light through a structure based on the different densities of the structure's components. An annular aperture is placed at the primary focus of the condenser of a conventional bright-field microscope. The condenser converts the hollow tube of light passing through this aperture into a hollow cone of light focused on the specimen. This light forms an image of the annular aperture at the back focal plane. Light is also diffracted by the specimen. The diffracted and undiffracted wave fronts can be selectively modified on the basis of their phases. A $90°$ (quarter-wave) phase shift simulates the diffraction effect of a partially absorbing medium. Consequently, a highly transparent object appears with the contrast of partially absorbing objects.

Polarized light and *differential interference contrast* are techniques that use polarizing materials or optical prisms (or both) to examine birefringent samples in terms of their orientation.

Sources for Microspectrometry

The ideal source for microspectrometry is one with:

- high energy in the appropriate frequency range;
- coherence with parallel collimation; and
- good beam stability in position and intensity.

Unfortunately, such ideal sources do not exist. Sources based on blackbody radiation in the IR, or lasers in the visible and near-IR, are the practically available sources. Tunable narrow-line-width lasers have potential as sources but are generally costly and complex to operate.

Beam divergence is a consequence of interference among the light waves propagating from an aperture. The smaller the aperture, measured in wavelengths of light transmitted through it, the less effectively the light waves interfere destructively and the greater the divergence.

Lasers are particularly useful as sources because of their high output power and wide range of wavelengths available. In addition, a laser has

- long coherence length,
- low beam divergence,
- good beam-pointing stability, and
- high-output polarization.

The wave nature of light limits the distance that a laser beam can propagate with a given spot size and irradiance. The *Rayleigh range* of a laser is, by definition, the propagation distance over which the irradiance decreases by a factor of 2. The Rayleigh range, Z_r, is given by

$$Z_r = \pi\omega^2 n/\lambda \qquad (3.1)$$

where ω is the radial spot size of the focused beam, n is the index of refraction, and λ is the wavelength of the light. In microscopic measurements, the laser is tightly focused to achieve high irradiance, reducing the Rayleigh range and thus reducing the distance over which high irradiance is maintained. For example, the Rayleigh range of gas lasers such as He–Ne is typically 20–30 cm. Solid-state and semiconductor lasers have shorter coherence lengths, approaching only a few millimeters. The minimum possible beam divergence is diffraction-limited, and lasers approach this theoretical limit. Lasers also have good beam-pointing stability with little angular drift. A He–Ne laser can achieve a beam-pointing stability of less than 0.03 mrad, which means that at 100 m from the laser the beam will move less than 3 mm.

However, the coherence of the laser beam leads to a loss of resolution because incoherent illumination provides greater interference effects and higher contrast. To obtain the maximum resolution, it is necessary to destroy the coherence of the laser illumination (*1*). It is possible to use diffusers or optical fibers to modify the laser beam.

Spectrometers for Acquiring Images

Spectrometers for acquiring images need to provide (*2*)

- high signal-to-noise ratio,
- high spectral frequency dispersion, and
- high spatial resolving power.

The primary optical property of an imaging spectrometer is its *optical conductance* at a given practical resolving power (*3*). The signal (the flux of radiation emitted by one line of the sample and transported through the spectrometer to the detector) is proportional to this property. The radiant flux ϕ (power) transported by any optical system from the source to the detector is described by

$$\phi = LG\tau \tag{3.2}$$

Here L is the *radiance* (power $\cdot$ solid angle^{-1} $\cdot$ area^{-1}) of the light source, G is the *optical conductance* (solid angle area), and τ is the *transmission factor* of the system.

This optical conductance of an optical element can be expressed in more useful spectral properties: L_v, the radiance per wavenumber, G_v, the optical conductance per wavenumber, and Δv^2, the bandwidth (*4*). The radiant flux can then be expressed in terms of the optical conductance by

$$\phi = L_v G_v \Delta v^2 \tau \tag{3.3}$$

The optical conductance is given by the area, F, of an element from which radiation is transported or received. An area, F, being irradiated from a solid angle, Ω, assumed to be a cone with the half-angle α, has an optical conductance given by

$$G = F\phi = F \cdot 2\pi(1 - \cos\alpha) = F \cdot 4\pi \sin^2(\alpha/2) \tag{3.4}$$

For a correctly designed optical instrument, the optical conductance between any two apertures is invariant. The radiance observed for any cross section of the beam of an optical instrument is invariant, except for losses by reflection, absorption, or scattering (which are taken into account by the transmission factor, τ). The effective optical conductance of an optical instrument is determined by that component, which cannot be enlarged for theoretical or technical reasons. In the case of microscopes, it is usually the objective lens.

The optical conductance of two apertures, F_1 and F_2, at a distance of a_{12} (for $a_{12}^2 > F_1, F_2$) is given to a good approximation by

$$G = (F_1 F_2)/a_{12}^2 \tag{3.5}$$

The optical conductance of a grating spectrometer, G_g, operating with a resolving power R is given by (*2*)

$$G_g = F_1 h/Rf \tag{3.6}$$

where F_1 is the area of the effective beam at the grating, R is the resolving power, h is the slit height, and f is the focal length of the collimator mirror.

The optical conductance of an interferometer, G_i, is given by (2)

$$G_i = 2F_1\pi/R \tag{3.7}$$

When the grating and the interferometers are operating with the same resolving power, the same area, F_1, and the same focal length, then the ratio of the optical conductances is given by

$$G_g/G_i = h/2\pi f \tag{3.8}$$

where h/f is in the range of about $1/10$ to $1/100$, so the optical conductance of the interferometer is larger by a factor of 60–600 than that of the grating instrument having the same effective beam area and operating with the same resolution. This is called the *Jacquinot* or *throughput advantage*.

The resolving power of a Michelson interferometer is determined by the maximum displacement of the moving mirror, Δl, multiplied by its wavenumber, v:

$$R_o = 2\Delta l \cdot v \tag{3.9}$$

To benefit from this resolving power for interferometers, one must use the maximum radius of the *Jacquinot aperture*, r

$$r = f \ (2/R_o)^{1/2} \tag{3.10}$$

where f is the focal length of the collimator mirror or lens producing its image. For Michelson interferometers, the Jacquinot aperture is equivalent to the entrance slit of a grating spectrometer. This aperture is necessary to reduce the beam to those rays that interfere to provide the necessary spectral resolution (4). For a Bruker spectrometer IFS 66/FRA, it has been reported that the diameter of the Jacquinot stop for a lens with a focal length of 33 mm at a resolving power of 2100 is calculated to be 2.0 mm (4), so the spectrometer can analyze radiation only emitted from a spot of this diameter. However, a smaller spot size is tolerated from a nearly parallel laser beam down to a few micrometers.

Interferometers also have the *multiplex advantage* compared to monochromators, which allows the simultaneous detection of all the spectral frequencies. For spectroscopic imaging, this advantage is considerable, particularly when a large number of frequencies are required to characterize the sample.

Digital Sampling of Images

Spectral Frequency Sampling

The sampling theorem for Fourier transform (FT) spectroscopy states that the sine wave with the highest frequency in a given spectrum, from whatever source, must be sampled at a rate of at least two points per cycle to be represented accurately. This sampling frequency is known as the *Nyquist* frequency. When a sine wave of a higher frequency than the Nyquist frequency exists, it will be folded back into the spectrum at a frequency exactly as much lower than the Nyquist frequency as the frequency of the sine wave is higher than the Nyquist frequency. This phenomenon is referred to as *aliasing*.

The sampling rate in FT spectroscopy is determined not only by the desired resolution, but by the frequency width to be observed. A 1000-Hz spectrum will be obtained only if the data system samples at 2000 Hz or once every 500 μs. Sampling faster will place the desired peaks in a cluster at one end of the spectrum; it will not increase the number of data points per peak or the resolution. The spectral resolution can only be increased by sampling for longer times at the same rate so that more points make up the block to be transformed. Further, it is not possible to examine only part of a spectrum in FT spectroscopy, since a smaller sampling frequency will only cause the data outside the "window" to fold back into the spectrum, leading to very confusing results.

In FTIR, the resolution need only be about 1/5 the frequency width at half-height of the bands. In this case the resolution of the spectrum may actually be degraded by taking more data points, since more noise is being included in the spectrum than would have been present in a smaller number of data points. Therefore, it is advisable to simply add a block of zeros to the acquired data and transform the data with this block of zeros included. This process, called *zero filling*, is mathematically justified as long as the block of zeros does not exceed the length of the original sampled data block. In practice, data are acquired in blocks of 2^n (n is a positive integer) of data points, because these are easiest to transform. Zero filling is then usually up to 2^{n+1} points.

Spatial Image Sampling

The spatial distribution of objects in a sample is unpredictable and must be treated as a random process. Consequently, an imaging system must be designed for a broad range of spatial distributions in the presence of noise.

To obtain an image that is representative of the sample, one must sample the image at a discrete number of points in the *xy* plane and store the values for computational purposes. If one partitions the gray scale into a set of discrete cells at regular intervals, then we have to specify only the number of pixels or grid size.

The *Shannon sampling theorem* gives us guidance in this area. In general, edges in an image introduce spatial frequencies. The sharper the edge, the higher the spatial frequency. Low spatial frequencies correspond to the absence of edges (i.e., regions of approximately uniform gray level). The basic idea behind the Shannon sampling theorem rests on the correspondence between abrupt changes in the intensity of an image and high spatial frequencies in its transform. If an image's transform contains no high frequencies, then the image contains no abrupt gray-level transitions, so the image need not be quantized very finely.

Let the spatial function $g(x,y)$ be a random process that represents any signal or noise component in the sample. Let us assume that the system is *isoplanatic*, and we shall examine an isoplanatic patch, A, that is rectangular and centered at $x = 0$ and $y = 0$. Let there be M voxels in the x direction and N voxels in the y direction in A. Let the corresponding sampling intervals be (X,Y). So x and y are limited by

$$-XM/2 < x < XM/2 \qquad -YN/2 < y < YN/2 \qquad (3.11)$$

and the solid angle subtended by A is given by $|A| = XYMN$.

Let the Fourier transform of this function, $g(f_x,f_y)$, be the corresponding frequency spectrum confined to the imaging bandpass frequency limit, B. The sampling intervals require that $f_x \leq XM/2$ and $f_y \leq YN/2$. It should be noted that the sampling intervals $(1/XM, 1/YN)$ in the frequency domain are not directly related to the sampling intervals (X,Y) in the spatial domain, except through the somewhat arbitrarily defined solid angle $|A|$ of the isoplanatic patch A.

Let us assume that the autocorrelation of $g(x,y)$ is given by

$$\Psi(x,y) = \Psi(r) = \sigma^2 \exp(-r/d) \qquad (3.12)$$

where $r^2 = x^2 + y^2$. The parameter d is the mean distance between edges of the morphological structures in the samples. We are interested in the mean distance d relative to the sampling interval of the imaging system. For example, when $d = 1$, the mean spatial detail is equal to the sampling interval.

To characterize the signal frequency spectrum $S(f_x,f_y)$, it is convenient to make the following two definitions: first, let W be the bandpass of the imaging system; ultimately, this bandpass is limited by the diffraction limit of the objective lens (i.e., $(f_x^2, f_y^4 < [2 \sin \alpha/\lambda]^2)$, where $\sin \alpha$ is the lens numerical aperture). Second, let B be the sampling bandwidth with corner points given by $f_x = \pm 1/2X$ and $f_y = \pm 1/2Y$, and sides parallel to the frequency coordinates (f_x, f_y).

An image $g(x, y)$ is band limited if its Fourier transform, $G(f_x, f_y)$, is zero whenever either $|f_x|$ or $|f_y|$ is greater than some number W called the bandwidth. The Shannon sampling theorem states that a band-limited function can be reconstructed exactly from image samples taken a non-zero distance apart, and specifies how the reconstruction is done.

Consider the one-dimensional case, where the image function g has a Fourier transform, G, which is band limited to bandwidth W. We expand $G(f_x)$ in the region $-W < f_x < W$ by a Fourier series,

$$G(f_x) = \sum_m c_m \exp[-2\pi f_x(m/2W)] \qquad (3.13)$$

where the coefficients c_m of the expansion are given by

$$c_m = \frac{1}{2^w} W2 \int_{-W} G(f_x)[-2\pi i f_x(m/2W)]df_x \qquad (3.14)$$

Now G is identically zero outside its bandwidth, so in the inverse transform we can replace the infinite limits with $\pm W$. Hence, the integral is precisely the inverse transform $g(x)$ evaluated at the point $x = m/2$. Therefore,

$$c_m = 1/2W\, g(m/2W) \qquad (3.15)$$

Substituting,

$$G(f_x) = 1/2\ W \sum g(m/2W)\ \exp[-2\pi i f_x(m/2W)] \qquad (3.16)$$

for $-W < f_x < W$ and $F(f_x) = 0$ otherwise.

From this equation, we can see that the image function g is completely determined by samples taken finitely far apart, since its transform depends only on values of g taken at intervals of $1/2W$. The inverse transform yields

$$g(x) = \sum g(m/2W)\ \text{sinc}[2W(x - m/2W)] \qquad (3.16)$$

where sinc $x = \sin x/x$.

This result is one statement of the Shannon sampling theorem. It indicates that one can reproduce g by centering a sinc function every $W/2$ units along the x axis, scaling each sinc function by multiplying it by the value of g at its center point, and adding all the scaled sincs. The greater the bandwidth W, the smaller the sampling interval $W/2$ required.

Tests of Repeatability

Samples should be analyzed to determine the image repeatability. This can be accomplished by measuring spectral features over a number of fields of view on each specimen and repeating the steps a second time. The two sets of measurements are then compared statistically. The statistical method used to test the differences can be the standard error of the difference between the two means. This method, also called the t-test, is based upon the average or means of the measurements over the fields of view, the standard deviations of the measurements, and the number of observations. In this method, the calculated t value is compared with published values that indicate the level of significance of the difference between the measurement means. For these tests, the 95% level of significance may be selected. The t table states that for 10 observations, the calculated t values must fall between $\pm$ 2.101 to establish with a 95% confidence that the two measurement means are not significantly different.

Detectors for Spectral Imaging

An optical system images the object and a portion of its background onto the detector. Imaging detectors typically include an $m \times n$ detector array. If the amount of the radiation originating from the object point and incident on the conjugate point on the detector is sufficiently large, then the object point generates a signal that differentiates it from the background. In such a system, an optical signal impinges on the detector and produces a current, $I(t)$, that flows through a load resistor, R_L, and produces a voltage, $v(t)$, according to the following relationship:

$$v(t) = I(t)R_L = [S(t) * h(t)]RR_L \tag{3.18}$$

where $S(t)$ is the intensity envelope of the optical signal, $h(t)$ is the impulse response function of the detector, R is the responsivity specified in units of amperes per watt, and * denotes convolution. Responsivity is the ratio of the photosignal to the radiation power incident on the detector. The higher the responsivity, the more sensitive the detector.

The quantity of light falling on a detector can be defined in three ways. It can be expressed as a number of photons at a specific wavelength (such as 100 photons/s at 800 nm). It can be given as the radiant power on the detector at a specific wavelength (40 pW at 400 nm), or it can be expressed as the amount of luminous flux falling on the detector (10 microlumens). The ideal detector has:

- high sensitivity,
- linear response,
- absolute accuracy,
- uniform and broad frequency response,
- fast response,
- a configuration that is spatially distributed as a two-dimensional array,
- low cost,
- robustness and stability to dynamic changes in system parameters,
- insensitivity to ambient condition fluctuations, and
- small size.

The quality of an image depends on the signal-to-noise ratio (S/N), and a limiting factor is often the response of the detector. Radiant sensitivity of a detector is the amount of electrical current that is produced by a given amount of radiant flux and is described in units of amperes per watt (A/W) at a specific wavelength. Spectral response is a term used to state the useful spectral range over which a detector will operate. The response can be characterized by a rise time, which is the time required for a detector output to rise from 10% to 90% of peak amplitude when illuminated by a short light pulse.

The power incident on the detector represents noise if it does not originate at the conjugate object point. Noise increases the amount of signal radiation required for object detection or recognition. The amount of noise of a detector at the output can be represented by a figure of merit, the *noise equivalent power* (NEP), which is the light power in watts of sinusoidally modulated light incident on a detector that produces an S/N of 1 for a bandwidth of 1 Hz (5):

$$\text{NEP} = P/(S/N) = P/(V_s/V_n) = P/(I_s/I_n) \tag{3.19}$$

where P is power, V_s and I_s are the signal voltage and current, and V_n and I_n are the noise voltage and current. For the best near-IR detectors (Ge or InGaAs), reported NEP values are about 10^{-15} W Hz^{-1}. The reciprocal of the NEP is defined as the *detectivity* (D^*), and is a measure of the minimum detectable incident light power for the given conditions. NEP and D^* are dependent on the temperature of the background, the detector temperature, area and field of view, gain and bandwidth of the detection system, and wavelength and modulation frequency of the incident light.

There are basically two classes of photodetectors: *quantum detectors* and *thermal detectors*. The principal distinction between the two types of detectors is in their *spectral response*. Spectral ressponse measures the responsivity of a photon detector at various wavelengths. Thermal detectors have a flat spectral response over a wide range of wavelengths. Quantum detectors, on the other hand, display a responsivity that increases with wavelength until the characteristic cutoff point, where responsivity drops to zero. The spectral response of quantum detectors arises because photon energy is inversely proportional to wavelength, and responsivity measures the electrical output of a detector at a given optical power. Responsivity therefore increases with wavelength because the output of a quantum detector is proportional to the number of photons it receives, and 1 W of 1-μm light contains twice as many photons as 1 W of 0.5-μm light. For thermal detectors, the output depends on the temperature shift induced by the absorbed energy, not the number of photons.

Quantum detectors involve direct quantum interactions between light and matter. In these detectors, incoming photons bounce electrons from the valence to the conducting band across an energy gap. Photon detectors are quantum detectors based on photoemission, photoconductive, or photovoltaic effects. Some quantum detectors use pure, undoped materials like mercury cadmium telluride (HgCdTe), termed intrinsic semiconductors. Extrinsic semiconductors use doped silicon or germanium. These systems function in the same manner as photodiodes in the optical and UV ranges, but the problem with IR radiation is that the band gap becomes smaller as the wavelength of light becomes longer. For IR detection, the composition of the primary ternary compounds must be adjusted to fine-tune the band gap. HgCdTe has a very broad range, going from 2 to 30 μm, but this compound is difficult to produce uniformly. This difficult production problem makes HgCdTe expensive for detectors. Developments in material-growth techniques will be required before HgCdTe can be used in array detectors. Array detectors from HgCdTe have considerable pixel-to-pixel variation. In addition, HgCdTe must be cooled to liquid-nitrogen temperature to yield high efficiency and low-noise-level performance. Recently, a new II-V ternary material, indium thallium arsenide (InTlAs) has been proposed for medium-wavelength detector arrays to replace HgCdTe because of the higher mechanical strength, greater uniformity, and lower cost.

Another possibility for long-wavelength IR detection is a focal-plane array of quantum-well IR photodetectors (QWIPs) (6). A quantum well is deep and narrow and its energy states are quantized. The potential depth and width can be adjusted so that there are only two states: the ground state near the well bottom, and a first excited state near the well top. A photon striking the well excites a ground-state electron to the first excited state, where an externally applied voltage sweeps it out, producing a photocurrent. Only photons having energies corresponding to the energy separating the two states are absorbed, resulting in a detector with a sharp absorption spectrum. A quantum well can be obtained by sandwiching a layer of GaAs between two layers of $Al_xGa_{1-x}As$. The GaAs layer controls the well width, and the Al composition, x, governs the potential depth. This system can be modified to enable detection of light at wavelengths longer than 6 μm. Stacking of the detector layers further improves the photon absorption.

Thermal detectors sense light by responding to the temperature increase of the entire detector element (thermoelectric effect, resistive bolometer effect, and the pyroelectric detector) that results when matter absorbs light. Energy resistive bolometers depend upon a change in resistance of a material due to a temperature rise when radiation is absorbed. The thermoelectric effect is used in the thermocouple and generates an output voltage (Seebeck effect) when the temperature of the junction between two metals changes with respect to the temperature of a reference junction. These two effects are found in metals and semiconductors. The pyroelectric effect is the generation of an external current in the absence of electrical bias by a time-dependent temperature change.

Thermal detectors can be traced back to 1800, when William Herschel discovered IR radiation using a prism and a mercury thermometer. The thermal couple was discovered in 1821 by the Russian-born German physicist Johann Seebeck, who found that if the ends of two dissimilar metal wires are joined together to form a loop, an electric current flows around the loop when one junction is kept at a different temperature than the other. The internal voltage responsible for current flow in a thermocouple is directly proportional to the temperature difference between the two junctions, where the constant of proportionality is called the *Seebeck coefficient*.

In conventional photon detectors, the absorption of IR radiation causes electronic interband transitions (intrinsic detectors) or transitions between impurity ground and excited states (extrinsic detectors), which lead to a photocurrent. Photon detectors are used as homogeneous photoconductor devices (i.e., Schottky barrier or *p–n* junction devices). The dependence of the absorption constant on the frequency determines the response of a photon detector. These devices are inherently faster than thermal devices. For the mid-IR region, narrow-gap semiconductors like HgCdTe, InSb, InGaAs, and lead salts are the materials of choice. To reduce thermal noise, these devices are cooled to 80 K.

The detector always sees a raw, unprocessed optical signal that contains both signal and noise. There are a variety of types of noise in detectors. The most fundamental source of noise is *photon noise*, which originates from the random arrival of photons at the detector surface. Because of this randomness, the number of photons collected within a given time interval varies around some statistical average, and the variance depends on the light source. Coherent laser light approximates a Poisson distribution, whereas incoherent blackbody radiation follows a broader geometrical distribution called Bose–Einstein. Thermal light sources are inherently noisier than lasers.

Background noise originates from sources other than the one of interest. Limiting the detector's field of view is one way to reduce background noise. *Dark current* originates within the detector and is a property of the detector material and its temperature. It is defined as the current produced by the detector when it is not exposed to any external radiation source. This randomly fluctuating current usually arises from the thermal excitations of charge carriers within the detector and can be attenuated by cooling the detector. When the magnitude of the source signal falls below that of the dark current, a signal can no longer be detected. *Johnson (Nyquist) noise* is the random signal generated by the thermal motion of electrons in the detector and its circuitry. This noise, which is proportional to the absolute temperature, exists even when no signal voltage is applied to the device. As the temperature is decreased, the thermal motion also decreases; hence, Johnson noise can be reduced. *Shot (Schottky) noise* arises whenever current flows through a detector or circuit. It originates from the random arrival of charge carriers at any point within the device. Shot noise generated in the detectors is affected in magnitude by the presence of the arriving radiation. *Gain noise* arises from random fluctuations in the internal gain of photodetectors.

Modulation noise, flicker noise, contact noise, or $1/f$ *noise* exists only when current is flowing through the device. This noise is due to traps that capture and release charge carriers randomly, causing random fluctuations in the current itself. The traps are created by contaminants or crystal defects at semiconductor junctions. Lower temperatures contribute to shallower traps and less trapping. As the name implies, $1/f$ noise decreases with increasing frequency.

Detectors are usually made with very small sensitive areas, typically 1 mm^2. This small detector target can pose a problem with light from an extended image because the source radiance (power per unit area per solid angle) must be conserved throughout any optical system, assuming no losses due to absorption, scattering, or obstruction. Thus, for a source of area a_0 and angular extent θ_0 (i.e., the maximum angle of divergence with respect to the optical axis), the area and angular extent anywhere in the optical path are given by (7)

$$n_1^2 a_1 \sin^2 \theta_1 = n_0^2 a_0 \sin^2 \theta_0 \qquad (3.20)$$

where n_1 and n_0 are the refractive indices of the respective media. Note that the conserved quantity is often given in the paraxial approximation as $n^2 a \theta^2$, but this is not appropriate for large angles. To reduce the source to an area as small as possible to fill the detector area, the smaller a_1 (the detector area), the larger θ_1 must become. By observation that the largest θ_1 could possibly be is $\pi/2$, the maximum concentration ratio, C_{max}, is given by

$$C_{max} = (a_1/a_0) = (n_1/n_0 \sin^2\theta_0) \qquad (3.21)$$

where a_1 is the smallest possible value for a given a_0 and θ_0. Due to blurring aberrations, it is not possible to obtain C_{max}. The irradiance (power per unit area) is inversely proportional to the square of the f *number* (the focal length divided by the aperture, which is a direct measure of the light-gathering power of the optical system). Thus, even a perfect $f/1$ system provides a factor of 4 less light than C_{max}, and a real system is expected to do much worse.

The recorded amplitude of the signal must be linearly proportional to the light intensity on the detector during acquisition if the spectrum is to have the proper spectral intensity. The nonlinear responses of detectors can result from too much photon flux, improper constant current biasing, or preamplifier saturation. Nonlinearity produces poor spectrophotometric accuracy, irrespective of the cause.

Single-Channel Detectors for the UV–Visible Region

Because the UV and visible spectral regions are generally free of interfering background radiation noise, it is possible to achieve very low intensity light detection with a suitable very low noise detector. In this frequency region, the primary detector is the *photomultiplier tube* (PMT). A phototube consists of an evacuated glass envelope containing a surface coated with an active metal. Incident radiation causes emission of electrons from this surface by the photoelectric effect; these electrons are collected by a positively charged plate, termed a dynode. The plate current is proportional to the intensity of the radiation. The photomultiplier effect arises if the electrons emitted from the cathode of a phototube are accelerated by a large potential and then allowed to strike another active surface; it is possible to get multiple emission of electrons from the second surface for each of the original electrons. These secondary electrons may, in turn, be accelerated, and upon striking the next surface, give rise to a larger number of electrons. By combining 10 or 12 such amplification stages, it is possible to get stable amplification of the original signal by a factor of 1 million or more.

PMTs are used because of their large internal gain and low noise readout. The noise arises mainly through the dark current due to the random thermal excitation of electrons. The PMT has a D value of about

10^{16} at room temperature. By cooling of the PMT, the dark current can be made negligible, so then $D \approx 10^{19}$, which corresponds to about 1 false photon per second. For comparison, the detectivity of the dark-adapted eye is about 10 to 100 photons/s (8). For detecting low-intensity UV and visible light, nearly ideal performance (i.e., limited only by light fluctuations) can be achieved by using a cooled PMT with a high quantum efficiency for the spectral region of interest.

The disadvantages of these devices include the following: the peak quantum efficiency of PMTs rarely exceeds 20%, only one wavelength can be monitored at a time, and they are difficult to use in a multichannel configuration. PMTs can also be damaged by exposure to high light levels. The cost of PMTs is high, and they are mechanically fragile. Their spectral sensitivity is limited; typically, PMTs operate from about 190 to 900 nm, although some can be manufactured to operate at up to 1100 nm.

Single-Channel Detectors for the IR Region

The IR spectral region, on the other hand, is plagued by background radiation, which limits the minimum detectable power. Thus, IR detectors usually do not have to be noiseless to be useful, but must introduce less noise than that from the background. An IR detector operating in this condition is said to be a background-limited photodetector. This performance level is achievable by many modern IR detectors, though such detectors often require operation at very low temperatures. Because D is dependent on the square root of both the detector area and the detection bandwidth, the comparison of IR detectors independent of these factors is facilitated with a figure of merit, D^*, which has been defined by

$$D^* = D(a_d \Delta f)^{1/2} \ [(\text{cm Hz}^{1/2})/\text{W}] \qquad (3.22)$$

where a_d is the detector area and Δf is the bandwidth. In general, the wavelength and modulation frequency used to determine D^* must be specified, and the background temperature is taken to be 290–300 K. The field of view is chosen to bring the background noise down to the detector level while allowing adequate light to be brought to the detector. In practice, the field of view is usually limited by a cold shield, with a circular aperture defining a conical field of view with cone angle equal to 2ϕ. For a detector that does not add any noise of its own, the dependence of D^* is given by

$$D^* = (1/\sin \phi)D(90°) \qquad (3.23)$$

where $D(90°)$ is the full hemispherical field of view. For real detectors, limiting the field of view will increase $D*$ until the detector noise equals the noise in the background radiation. There are two classes of mid-IR detectors, namely, thermal and photon detectors.

Thermal detectors, including thermocouples, bolometers, and pyro-electrics, produce an output signal proportional to the increase in temperature of the sensing element induced by exposure to light. The speed of response of thermal detectors is governed by their thermal time constant (typically in the millisecond range). In normal operation in the IR region, the radiant power striking the detector is about 10^{-7} W, corresponding to a temperature rise of the detector junction of about 0.001 to 0.01 °C. The *thermocouple* is a heat-sensitive junction of dissimilar metals, which develops an electromotive force that changes with temperature. Thermocouples have a constant response independent of wavelength, except at longer wavelengths (beyond 30 μm), and are usually operated in the vacuum to increase sensitivity and decrease noise. The sensitivity of thermocouples is roughly proportional to the inverse of the receiver area and heat capacity. Assuming background-limited conditions for a background temperature of 300 K and a hemispherical field of view, the theoretical $D*$ for ideal thermal detectors (i.e., for which all incident energy is converted into signal) is 1.8×10^{10} and is independent of wavelength (9).

Photon detectors, such as photodiodes and photoconductors, convert photons directly to mobile charge carriers, which are then measured as a current or voltage. The speed of response is dictated by the free-carrier lifetime (typically in the millisecond range). Semiconductor devices are used in which electrons are raised (by absorption of a photon from a nonconducting energy state or valence band) into a metastable conducting state. For photoconductors, the maximum $D*$ is dependent on wavelength with a minimum of about 4×10^{10} at about 10 μm (10). One reason for the superior $D*$ for photoconductors is that they have a limited spectral range, which makes them less sensitive to background noise. For most thermal detectors, $D* \approx 10^8–10^9$, while the photon detectors can have $D*$ values within a factor of 2 of the ideal value.

Optical Array Detectors

Two-dimensional array detectors have many more elements than single channel detectors, and if the incident radiation can be dispersed properly they have the multiplex advantages over single detector systems. Such detectors can eliminate the need for a scanning mechanism as the image is directly focused on the array detector. A high $D*$ for a single pixel of the array detector is not enough to ensure a high-performance imaging array.

Pixel-to-pixel nonuniformity produces a spatial fixed-pattern noise, which has to minimized as well. The spatial resolution of an array detector depends on the size of the elements and the number of elements in the array; and spatial resolution down to 5 μm have been achieved in the near-IR region.

Photodiode Array Detectors

Silicon photodiodes have diverse applications as detectors. A standard silicon detector has useful sensitivity from approximately 400 nm up to 1100 nm, whereas the UV diode detects from about 200 nm to 1100 nm. These detectors can measure optical power from only a few picowatts up to milliwatts with reproducible sensitivity and rise times as fast as 10 ps.

The silicon photodiode has a simple structure. The most common devices are p–n junction semiconductors made from silicon that has been doped with an appropriate impurity, usually boron. The diode has three distinct regions: the electron-rich n-type material, the neutral depletion layer, and the positively charged p-type material (this positive charge is due to the lack of electrons, a condition usually described as the presence of "holes").

The discrepancy in charge between the p-type and n-type layers creates an electrical field between them. Any photon carrying enough energy may be absorbed by the diode, causing the separation of electron-hole pairs. The electric field then causes a migration of this charge. It is this migration that the user measures as photocurrent.

Charge-Coupled Device Detectors

A relatively new type of detector is the CCD, which is an optical array detector based on silicon metal oxide semiconductor technology (*11*). The CCD imaging array itself has no inherent gain. Every incoming photon generates a single free electron in the pixel it impacts, with a probability ranging from 0% to 70%. This probability is known as the quantum efficiency and is wavelength-dependent. The number of electrons is determined by the light level in the image and the exposure (integration) time.

Use of the CCD as a spectroscopic detector is based on collecting and storing photon-induced charge on a continuous silicon substrate divided into individual elements (pixels) by a series of electrodes that are used to manipulate the charge. The correspondence between object pixels and CCD pixels is determined by the magnification of the optical system. Exposure of this two-dimensional imaging area (called the parallel register) to light leads to charge separation at the n–p junction, and an image of charge accumulates, localized by potential wells established by electrodes

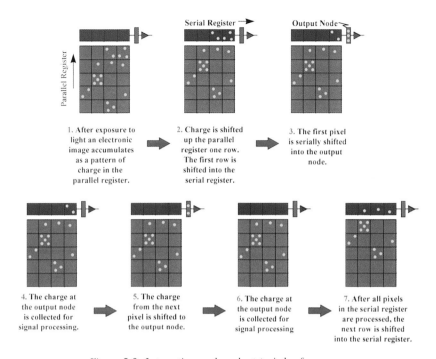

Figure 3.2. Integration and readout periods of an array.

on the detector surface. This charge image can be propagated, row by row, across the CCD by a series of potentials applied to the electrodes. The charge is then transferred to a second, one-dimensional array, called the serial register. During the readout, the serial register is emptied into the output node of the device and refilled with the next row of charge (Figure 3.2) (*12*). This process continues until the entire array is emptied of charge. This readout process introduces statistical variations known as read noise. The readout speed is usually limited by the speed of the analog-to-digital converter.

A number of parameters can be controlled to optimize performance, including exposure (integration) time, operating temperature, readout speed, binning, frame rate, and gain. Lengthening the exposure time will increase the signal. However, longer exposure time will also increase the dark current per pixel. For a constant image, the signal increases linearly with time, whereas the noise increases according to the square root of time. To minimize read noise, it is preferable to have one long exposure rather than several images of shorter exposure time.

Once each image is stored, it can be manipulated. One of the most important corrections is the flat-field correction, which allows images to be corrected for response nonuniformities due to inherent illumination variations in the image or because of anomalous pixel response.

The inherent signal-to-noise ratio of the image stored in these pixels is determined by the shot noise, which is the natural statistical variation in the number of photons arriving at each pixel.

The advantages of CCDs as detectors are (*11*)

1. a single device may consist of more than 250,000 independent detector elements,

2. high quantum efficiency,

3. low noise levels,

4. a spectral range that can extend from the near-IR well into the far-UV, and

5. a large dynamic range.

CCD detectors have a unique capability, which distinguishes them from other detectors. This is the readout mode, where charge from more than one detector element is combined on a chip before being read out. This process, called *binning*, involves moving the charge from a number of neighboring detector elements (e.g., a single column or row of the CCD chip) into a single "bin," and then transferring the charge from the bin to the output node (*13*).

CCDs do have some disadvantages as compared to detection with single-channel methods (*14*):

1. The frequency range is small.

2. The spectral length is limited by the dimensions of the array.

3. The density of measurement points, in terms of wavenumbers, is restricted by the dimensions of a single detector element.

4. Each detector element has its own sensitivity and noise character.

5. Cosmic rays generate undesired large spikes in CCD-detected spectra.

CCD detectors are particularly useful in the near-IR and are considered a viable alternative to FT Raman spectroscopy (*see* Chapter 9). The NEP, defined in eq 3.19 of CCDs is several orders of magnitude lower than for the far-visible/near-IR region detectors. In addition, because the CCD is an integrating device, the longer it integrates the better the NEP.

Spatial Positioning and Focusing

At the heart of every imaging system is a micromanipulator or micro-positioning device that controls the spatial position of the sample, the source, or both. Precision manipulators are integrated from highly precise linear- and/or circular-motion axes. The images captured by the systems are assumed to be orthogonal grids, with equal intervoxel spacing between the two-dimensional axis representing the sample. With stepper motors, one counts the pulses in the x and y directions. Geometrical distortions give rise to spatial blurring due to imperfections or maladjustments of the beam-positioning system or optical focusing errors.

A computer controlled x–y stage has several important parameters: range of motion, smallest increment of movement (0.1 μm may be necessary), repeatability of movement (no hysteresis or backlash), and ability to hold a position.

Two types of accuracy are required of the micropositioner. *Path accuracy* is the ability to move the focus spot along a specified path. Path accuracy is often less because of computer-control limitations than the mechanical specifications would lead one to believe. *Positioning accuracy* is the ability to statically locate the focal spot at a desired point. Positioning accuracy and resolution of these micromanipulator systems are often measured in micrometers or tens of micrometers. Errors in the movements of the positioning system lead to nonlinear effects and an effective increase in the spot size on the sample.

For spectroscopic imaging, lens-based beam manipulators introduce chromatic aberrations, since each wavelength transmitted through a lens-based optical system has its own index of refraction. In addition, lens-based optical beam pointing devices absorb some of the light energy and expand, enlarging the spot size. Mirror-based micromanipulators do not suffer these chromatic aberration problems.

Resolution

Resolution is the ability to show detail. The human eye, without any lens, can distinguish objects 0.11 mm (110 μm) apart. This limitation of the eye results from the spacing of the rods and cones in the eye. To see anything smaller, one needs a microscope.

The relative sizes of objects can be compared. The size of a small pinhead is 800 μm, a human egg is 100 μm, a human chromosome is 2.5 μm, a compact disk bit is 1 μm, a virus is 0.05 μm, and a large molecule is only 0.007 μm.

To see a l-μm particle, it must be magnified, without blurring, up to the 100-μm size that the eye can see. Objects to be examined by the eye can only be brought to the near point of the eye and remain in focus. The *near point* is the position at which if an object is brought closer, the image will blur because the retina can no longer compensate. For a person with normal eyesight, the near point is approximately 25 cm. A magnifying lens placed at the near point of the eye extends the eye's natural focal point by forming a virtual image that the eye can focus on. The focus of the magnifier makes the object positioned in front of the lens appear larger. The lens magnification, M, is given by the ratio of the focal length, f, of the lens and the near-point distance (25 cm)

$$M = 25/f + 1 \qquad (3.24)$$

Thus, the shorter the focal length the higher the magnification. Obviously, a maximum magnification of 25 is all that can be achieved. When the focal length decreases, so does the size of the lens, so there is a limit to the power of a single lens based on light-gathering capability.

To achieve higher magnification, one must use a combination of lenses. A compound microscope consists of two lenses, aligned along their optical axes. The first lens, called the *objective*, forms a real image of the object, while the second lens, called the *eyepiece*, magnifies the image for the eye. The total magnification of both lenses, M_T, is given by

$$M_T = -25x'/f_o f_E + 1 \qquad (3.25)$$

where x' is the distance of the real image from the back focus of the objective lens, and f_O and f_E represent the focal lengths of the objective and eyepiece, respectively. With higher performance objectives, one can achieve focal lengths as short as 1.5 mm and magnifications as high as 100. If paired with a $25\times$ eyepiece, a $100\times$ objective can produce a total magnification of $2500\times$. However, at these high magnifications, lens aberrations become significant.

The ultimate purpose of microanalytical techniques is to magnify the spectral information of a microsample to the size sufficient for interpretation. The direct correspondence between the sample and the images means we can think of the image as a map of the spectral responses of the sample at the resolution of the probe. For sufficiently low probe resolution, only general compositional patterns and large inhomogeneities will be discernible. For higher resolution, specific objects can be detected, and at even higher resolution, the clusters or aggregates comprising these objects can be observed. The ability to deduce the object from the image is not limited to features that are clearly resolved by the probe. The image is controlled by the local geometric patterns, which are localized spatially on a much finer scale than the resolution of the incident probe.

Spatial Resolution

The *probe resolution* is simply a length scale defined by the distance at which two objects image as two separate peaks. Moving the objects closer together results in a single image feature whose shape cannot often be predicted. The resolutions of the spectroscopic probes are in the multimicrometer range; this is significantly inferior to the resolutions obtained by nonspectroscopic probes such as electron microscopy. However, the spectroscopic probe information brings out chemical and structural features not available without spectroscopic contrast.

Spectroscopic measurements involve photons, and the spatial resolution is ultimately limited by the spatial coherence of the photons used. The *spatial coherence* characterizes the distance below which the interference of the harmonic signal is constructive. The wave vector given by $\mathbf{K}$ = $\mathbf{k}_f - \mathbf{k}_i$ is reciprocally related to the radiation wavelength according to

$$\mathbf{K} = 2\pi/\lambda \ (\text{length}^{-1}) \tag{3.26}$$

so the associated wavelength controls the spatial limitations of a given spectroscopic experiment. On this basis, the resolution in the IR is between 10 and 100 times less than for a visible system. For spectroscopic techniques such as IR, the wavelengths vary from 10^{-3} to 10^{-6} m, so the spatial dimensions probed are beyond the molecular distances, although the IR spectroscopic information is at the molecular level.

Pixel Resolution

In images, dimension measurements are limited to 1 pixel. For a given field of view, the *pixel resolution* is determined by the dimensions of the field of view divided by the number of pixels measured. Only the total resolution is influenced by the width of the pixels. Unfortunately, for spectroscopic imaging, this statement of resolution is wrong most of the time.

In the first place, the pixel spacing is equidistant only with respect to geometric spacing; it is not equidistant in wavelength or in wavenumbers. If the spectrometer is linear in wavenumber or wavelength, then the actual resolution systematically changes with the pixel resolution.

Spectral Line Width

The inherent spectral line width of the imaging system either in wavelength or frequency influences the resolution of the images. The instrumental line width contributes to the spatial uncertainty or "fuzziness" in the image. If the spectral line width is less than the pixel resolution (which is virtually always the case spectroscopically), the effect of line width on resolution is minimal.

Factors Limiting Resolution

Limitation of Diffraction

The theoretical performance of a spectroscopic microscope is mainly determined by its optical properties (numerical aperture, magnification power, and focal length). The resolution depends greatly on the angle, α, over which the objective lens of the microscope collects light, the so-called numerical aperture (NA). Because the effects of diffraction establish the ultimate limits to resolution, a perfectly aberration-free instrument is termed diffraction limited. Using optical considerations, the transverse resolution (within the x–y image plane of a wide-field microscope) is given by (15)

$$r = 1.22\lambda_0/2NA_{obj} \qquad (3.27)$$

where λ_0 is the wavelength of light in air, and NA_{obj} is the numerical aperture of the objective lens, given by

$$NA_{obj} = n \sin \theta \qquad (3.28)$$

where n is the refractive index of the medium between the specimen and the objective lens, and θ is the half-cone angle of light captured by the objective lens. For an optical microscope, λ is the wavelength of visible light (about 0.5 μm). This means that an optical microscope with a 1.4 NA objective can, at best, resolve spatial details of about 0.22 μm. One can go up to magnifications of about 1000× with an optical microscope.

The horizontal resolution of each pixel is ideally equal to half of the diffraction-limited resolution. Higher magnification will cause no increase of resolution because diffraction, not pixel size, is the limiting factor. Higher magnification will, however, lower sensitivity because less power will fall onto each pixel. The depth of field (DOF) or longitudinal resolution is determined from basic optical principles. The DOF is given by (16):

$$DOF = 2\lambda/NA^2 \qquad (3.29)$$

so the longitudinal resolution will be 0.51 μm. Note that the longitudinal resolution is different by a factor of 2 from the transverse resolution.

Aberrations

In theory, the resolution given by the equations just presented is ideal, but the actual resolution is less. Optical aberrations such as spherical aberrations, coma, astigmatism, and chromatic aberration exist in real optical systems, and all of these aberrations reduce both transverse and longitudinal resolution.

Brightness of an Image

The NA of the microspectroscopic system is important, as it determines the amount of energy gathered by the system, and this ultimately determines the signal-to-noise ratio (S/N). The "brightness" of an image in a microscope is given by

$$\text{brightness} = NA^2/\text{magnification}^2 \qquad (3.30)$$

Ultimately, the *brightness* of an image is determined by the intensity of illumination and the change in light introduced by the sample.

Minimum Detectable Volume

For microscopic spectroscopic techniques, there is another important factor which determines the spatial resolution, and that is the minimum detectable volume element (voxel) required to obtain a value of S/N that yields interpretable spectra. This minimum "useful" voxel depends on the nature of the sample, performance of the spectrometer, experimental time available, and S/N requirements of the image in the application. For any sample, a preliminary evaluation of S/N must be made in terms of this minimum volume. Some criteria for making such a determination will be given later.

Field of View

In microscopy, increased resolution is only obtained at the expense of the field of view (FOV). As the resolution goes up, the FOV must decrease. This decrease in FOV is sometimes a problem, as the FOV may not represent the character of the sample if it is too small.

Noise

Noise limits the fidelity of an image but has no effect on resolution; it only places a limit on the total experimental time necessary to acquire an image with a specific S/N. The S/N can be increased by averaging a number of acquisitions, since the signal adds coherently while the noise adds randomly. The S/N therefore increases with the square root of the number of acquisitions. However, this increase in S/N is accomplished at a considerable cost in experimental time. To record an image with twice the resolution at the same sensitivity level as for a previous image, four times the number of acquisitions are necessary. This relationship is for one imaging dimension; therefore, if the resolution is to be increased by a factor of 2 in two dimensions, the experimental time must be lengthened by a factor of 16.

Stray Light

Stray light in a microscope often limits the resolution by allowing signal to appear in the image that is not in the FOV of the sample. This is particularly true for IR systems, which have inherently long wavelengths and a central obstruction in the optical system. Much of the stray light arises from diffraction at the apertures. As indicated earlier, the energy from a point is not imaged ideally as a point, but rather as a diffraction pattern or Airy disk. Thus, there is a central bright spot followed by a succession of dark and bright rings. The bright rings are called *lobes* or *pods*. For an optical imaging system, roughly 85% of the energy is in what is called the central maximum of the pattern. Messerschmidt (*17*) calculates that with a numerical aperture of 0.5 and a wavelength of 20 μm for an IR microscope, the first dark ring occurs 24 μm out from the specimen, the second at 44 μm, the third at 64 μm, and the fourth at 84 μm. This results in the entire sample being contaminated by stray light from outside the specimen. If the area of interest is less than about 50 μm, there will be a stray light component in the image.

In addition, photons from out-of-field view sources may be deflected into the beam. These photons do not carry information about the object when they are incident on the detector plane—they represent noise. Microscopes have traditionally incorporated two special apertures to improve image quality: a glare stop and a field stop. The glare stop defines the largest bundle of rays that is transmitted through the optical system. The best location for a glare stop is at the last image of the diffracting optical component. The field stop is used to limit the field size. The field stop is usually placed at the last intermediate image plane before the detector.

Spectroscopic Considerations

For mapping, the spectroscopic performance must be characterized by

- high optical speed,
- accuracy, and
- reproducibility.

These factors place additional stress on the instrumental requirements of the imaging spectrometers.

Optical Speed

The most important component of any spectral instrument is the energy flux through the instrument, because energy flux is the principal component in the combined signal and noise. Another important component is the optical speed in an optical system, which defines the system's light-gathering power or, in the case of a photometer, its energy throughput (*18*).

A way of expressing optical speed is the *f number*, which is essentially the ratio of the focal length, *f*, of the mirrors or lenses to their aperture. A photometer with mirrors having 500-mm focal lengths and 50-mm diameters is an $f/10$ system. An FTIR instrument is usually an $f/10$ instrument.

If radiation emanates from a point source equally in all directions, then the *f* number defines the portion of that sphere of radiation that is accepted by the optical system. The larger the *f* number, the smaller the area of sphere accepted, and vice versa. Since we are dealing with areas, the differences in energy throughput for optical elements vary as the square of the ratio of the *f* numbers.

Signal-to-Noise Ratio

The amount of signal present in a photometric system above the background noise (*S/N*) is the single most important parameter affecting imaging performance. This parameter determines overall accuracy and minimum detectable concentrations. The signal is a result of the radiation produced by the source, the energy throughput (*f* number), and the quality of the optics. Most noise comes from source fluctuations and random electronic noise in the detector and amplifying circuits.

Sampling Methods

There are three general sampling approaches to spatially resolved spectroscopy of samples:

- microsectioning,
- microbeam methods, and
- localized spectroscopic methods.

The oldest and most general sampling method involves the *microsectioning* of the sample itself (*19–21*). The microsections of the sample are individually examined, and the spectroscopic results are given spatial labels based on the specific microsection of the sample examined. The basic method involves confining the sample in an appropriately small area and condensing the beam dimensions to maximize the energy passing through the sample. This ultramicrosampling method requires skillful microscopic sectioning techniques, such as microtoming, but has the advantage of being generally useful for most samples.

The second method for obtaining spatially resolved spectra is to examine the macrosample using *microbeam techniques,* which can position the beam at desired portions of the sample. Microbeam techniques rely on a focused beam to probe a given sample volume, and the beam or the sample is rastered systematically across a geometrical grid to obtain the map. For laser-based systems, like Raman spectroscopy, this technique is particularly useful as the beam can be focused to its diffraction limit (1 μm) and the sample itself can be moved systematically on a stage.

Finally, there are *localized spectroscopic methods,* which use selected excitation methods to isolate a particular microregion of interest (ROI), from which the spectroscopic results are obtained (*22*). The concept here is to isolate the desired spatial region and arrange to spectroscopically excite only that domain, leaving the remainder of the sample unaffected. For optical methods, special microsampling techniques such as attenuated total reflectance (ATR) and grazing angle incidence are used to probe a given region of the sample. With resonance techniques like NMR and EPR, selective excitation methods are often used to isolate the ROI using, for example, surface coil techniques.

Optical slicing might be termed optical microtoming, and it involves the use of confocal microscopic techniques that allow one to observe optical sections in the axial (thickness) dimension produced by successive small changes in focus. The confocal effect on axial resolution appears as a sectioning effect because the detector behind the confocal pinhole detects the light only from a thin x–y region in the neighborhood of the focal plane. These image stacks can be employed to visualize the three-dimensional properties of the sample.

Microspectroscopic Measurements

Any spectroscopic signal that varies in different areas of a sample can, in principle, be used to make a map or image. Spatial scanning greatly simplifies the generation of an image. It is assumed for simplicity that we have a parallel collimated source. Distortions will be considered later. This beam is partially absorbed and scattered by the sample, while the remaining transmitted energy is observed by the detector.

We also assume that the scattered light does not hit the detector and that the absorbers in the light path are nonluminous (i.e., no thermal emission or fluorescence). Let I_0 be the number of monochromatic photons impinging on the sample. The number of photons absorbed from the beam, ΔI, will be proportional to the number of photons available for absorption, I_0, and the number of absorbing molecules per cubic centimeter in the differential "slice" (db) of the sample. The number of absorbed photons is equal to the product of the number of absorbing molecules per cubic centimeter, which is proportional to the concentration of absorbing molecules, c, and to db, so

$$\Delta I = -aI_0 c\,db \tag{3.31}$$

where a is the absorptivity whose magnitude depends on the choice of units for c and b. So the number of photons absorbed is proportional to the number of incident photons, the path length, and the absorptivity of the sample.

If we start with I_0 intensity, and after a thickness, b, have I intensity, we can write the differential form of eq 3.31 so we have the integral relationship

$$\int dI/I_0 = -ac \int db \tag{3.32}$$

After solving this equation, we have the classical absorptivity relationship, and expressing it in log base 10, which is the usual method,

$$\log_{10}[I/I_0] = -abc \tag{3.33}$$

which is the Beer–Lambert law of absorption spectroscopy used for quantitative analysis of solutions.

For mapping purposes, we represent the sample as a two-dimensional matrix with x pixels in one dimension and y pixels in the other. Thus the intensity measured at frequency ν through thickness z ($z = b$) at a given (x, y) position is given by

$$I(x, y, \nu) = I_0 \exp[-a(x, y, z, \nu)c(x, y, z)b] \tag{3.34}$$

where $a(x, y, z, \nu)$ is the absorption coefficient at the voxel (x, y, z) at frequency ν, and $c(x, y, z)$ is the concentration of the analyte at the voxel (x, y, z).

Microspectroscopic measurements sample an isolated volume, which can be the total sample or some subvolume element (a voxel) of the sample.

Relationship Between Macroscopic and Microscopic Measurements

In general practice, spectroscopic measurements implicitly assume that the sample is homogeneous (i.e., that there are no spatial differences in the composition and structure of the components). At the microscopic level, multicomponent samples may have an inhomogeneous distribution of the components. For M components A, B, C, ... M, the voxels can contain different amounts of the components as distributed in the microvolumes, which, in turn, can have different sizes and shapes (i.e., different states of aggregation). The total volume, V_{tot}, is the sum of the voxel volumes, V_{vox}:

$$V_{tot} = \sum (V_{vox})_i \qquad (3.35)$$

The maps or images of the sample are obtained by scanning the iX_i voxels in some fashion. The experimental volume of a voxel is given by

$$V_{vox} = (FOV_x)(FOV_y)(FOV_z)/n_x n_y n_z \qquad (3.36)$$

where FOV is the field of view in the corresponding spatial direction and n is the number of encoding levels applied along the respective x, y, and z directions. For most spectroscopic techniques, FOV_z corresponds to the thickness, and $n_z = 1$.

The V_{vox} have microsubvolumes, V_m, made up of the relative contributions of the components:

$$V_{vox} = \sum V_m \quad \text{for} \quad m = A, B, C, ... M \qquad (3.37)$$

Let us describe the effect of inhomogeneities on the spectroscopic results in terms of specific "unit cells" of volume V_m characteristic of each component m. The composition dispersions in the sample will be determined by the number and type of these "unit cells" for the components in the measured voxel. Obviously, the geometric size and shape of the inhomogeneities are a result of the arrangements or aggregations of these unit cells in the available volume.

Table 3.1. Various Techniques of Spectral Imaging

	Detector		
Illumination	Single-channel Photomultiplier Tube	Multichannel, One-dimensional Photodiode Array	Multichannel, Two-dimensional Charge-coupled Device
Point	1 wavelength, 1 point	n wavelengths, 1 point	n wavelengths, 1 point
Linear		1 wavelength, 1 profile	n wavelengths, 1 profile
Global			1 wavelength, 1 image

The number of unit cells of component m in the voxel i is given by

$$n_{mi} = (V_{vox})_i / V_m \qquad (3.38)$$

Variation in n_{mi} leads to the differences in $(V_{vox})_i$, yielding different results for different voxels. The spatially resolved spectroscopic measurements detect these differences in the voxels and generate a map or image of the sample inhomogeneities.

Summary

From the information presented in this chapter, it can be deduced that the imaging technique depends on the nature of the illumination, either a point or single-frequency source, a linear or profile source, or a multi-frequency global source. The source of illumination, combined with the nature of the detector, that is, whether the detector is single-channel or multichannel (one- or two-dimensional), determines the type of imaging experiment that can be performed. A matrix of these combinations of sources and detectors determines the various techniques of spectral imaging and is shown in Table 3.1 (*23*).

References

1. Pallister, D. M.; Morris, M. D. *Appl. Spectrosc.* **1994**, *48*, 1277.
2. Schrader, B.; Keller, S. *Abstracts of Papers*, 8th International Conference on Fourier Transform, 1991; p 30; SPIE 1575.
3. Hansen, G. *Optik* **1950**, *6*, 337.
4. Schrader, B.; Baranovic, G.; Keller, S.; Sawatzki, J. *Fresenius' Z. Anal. Chem.* **1994**, *349*, 4.
5. Poehler, T. O. In *Physical Optics and Light Measurements;* Prentice Hall: Englewood Cliffs, NJ, 1988, pp 89–94.
6. Gunapala, S. D.; Bandara, K. M. S. V. *Phys. Thin Films* **1995**, *21*, 113.

7. Buontempo, J. T.; Rice, S. A. *Appl. Spectrosc.* **1992**, *46*, 725.
8. Stockmann, F. *Appl. Phys.* **1975**, *7*, 1.
9. Putley, E. H. *Phys. Technol.* **1973**, *4*, 202.
10. Putley, E. H. *J. Sci. Instrum.* **1966**, *43*, 857.
11. Pelletier, M. J. *Appl. Spectrosc.* **1990**, *44*, 1699.
12. Pemberton, J. E.; Sobocinski, R. L.; Bryant, M. A.; Carter, D. A. *Spectroscopy* **1990**, *5*, 26.
13. Deckert, V.; Kiefer, W. *Appl. Spectrosc.* **1992**, *46*, 32.
14. Clymer, B. D.; DeVore, T. B.; Jagadeesh, J.; Tomei, L. D. *Appl. Opt.* **1991**, *30*, 5056.
15. Kingslake, R. *Optical System Design;* Academic: New York, 1983.
16. Smith, W. J. *Modern Optical Engineering: The Design of Optical Systems;* McGraw-Hill: New York, 1966.
17. Messerschmidt, R. G. In *The Design, Sample Handling, and Applications of Infrared Microscopes;* Rousch, P. B., Ed.; ASTM STP 949; American Society for Testing and Materials: Philadelphia, PA, 1987; pp 12–26.
18. Banerjee, P. P.; Poon, T-C. *Principles of Applied Optics;* Richard D. Irwin and Aksen Associates: Miami FL, 1991.
19. Allen, T. J. *Vib. Spectrosc.* **1992**, *3*, 217.
20. Abbott, T. P.; Felker, F. C.; Kleiman, R. *Appl. Spectrosc.* **1993**, *47*, 180.
21. *The Design, Sample Handling, and Applications of Infrared Microscopes;* Rousch, P. B., Ed.; ASTM STP 949; American Society for Testing and Materials: Philadelphia, PA, 1987.
22. Koenig, J. L. *Spectroscopy of Polymers;* American Chemical Society: Washington, DC, 1992; pp 285–318.
23. Barbillat, J.; Delhaye, M.; Dhamelincourt, P. *Proc. 50th Annu. Meet. Electron Microsc. Soc. Am.* **1992**, *50*, 323.

4

Image Analysis

We have observed that images can be misleading as well as informative. Consequently, it is our view that the greater the number of independent ways in which a new type of imaging can be verified, the less likely will be the misinterpretation *or* overinterpretation *of the image (1).*

—P. J. Todd, R. T. Short, C. C. Grimm,
W. M. Holland, and S. P. Markey

Image processing can be said to begin when the radiance reflected or emitted by a scene (sample) strikes the optics of an image-gathering system, and to end when the acquired visual information has been recorded in the nervous system of human beings (2).

—F. O. Hack, S. John, and S. E. Reichenbach

If a picture is worth a thousand words, it's worth at least a million numbers (3).

—Larry Yeager

Image Analysis

A spectroscopic imaging system consists of three components: an image-capturing system, an image-processing and analysis system, and a visualization system. The spectral data obtained from imaging are analyzed to produce a variety of response factors that are spatially expressed in terms of the individual data-grid coordinates. The data collection associated with the spectroscopic mapping contains all of the information required to deconvolute the contributions of the different species in the sample. Using a computer, one can analyze the entire spectral range at each positional coordinate or, for a given frequency, develop a spatial image demonstrating the spatial distribution of an object-specific spectral feature.

Microscopic mapping experiments generate the spectral absorbances at a given spatial position and frequency. One-dimensional plots of histograms can be made of intensity versus pixel position where only one dimension is required, and stack plots can be used to encode a time dimension. The two-dimensional maps can be represented by color maps, gray-level maps, axonometric plots, or contour plots. A chemical functional group image is obtained by plotting a specific functional group absorbance in a spatial (x,y) fashion. Three-dimensional images can be constructed *sectional images* that are stacks of two-dimensional (x,y) maps in the z direction or by *projections* using reconstruction techniques. The presentation of sectional images requires only the recall of the data points and placing them in the proper framework, while projection images require a large amount of computation as one must solve the trigonometric problem of locating each point in three-dimensional space relative to all sampling intervals. A volume-rendering technique can be used to generate different three-dimensional views. The preferred method of presentation is determined by the level of interpretation required for the analysis.

We are interested in scientific visualization of the spectroscopic images in terms that we can understand, and the dimensions of such visualizations can be as many as needed to make the description of the multidimensional system accurate. *Stereology* is the study of projections and sections of multidimensional objects into spaces of lower dimension, and *image analysis* is the study of shapes and structures in two-dimensional images. Images in two or three dimensions can be easily presented on a screen. Adding color yields a fourth dimension. Time scaling adds a fifth.

Image analysis is generally a process of simplification: the complicated image is converted into a simpler, more interpretable, form by some sequence of mathematical steps. It is reasonable to regard a materials spectroscopic image as a spatial arrangement of two- or three-dimensional spectroscopically distinguishable objects superimposed on a background. Image analysis is the counting, measuring, and classifying of these features in an image, assuming that the distinguishable objects tend to be spectrally homogeneous and separable from other objects and the background.

Recognizing and classifying features for subsequent analysis is generally known as *image segmentation*. Image segmentation is generally based on gray-level and spectroscopic differences (characteristic frequencies) between classes of objects. Gray-image segmentation makes the fundamental assumption that each class of objects is represented by a unique spectroscopic frequency with different brightness (absorbance) values.

Of course, the spatial resolution of the experimental image relative to the objects is important. If the resolution is small with respect to the objects, then it will be possible to identify the size and shape of each object as well as the gray levels within it. Thus the objects are many times larger than the pixel resolution, and the image of each object will contain many pixels. Therefore adjacent pixels will tend to have the same brightness, and hence be spatially correlated. When the object is smaller than the pixel resolution, each pixel value represents a brightness for a combination of objects and background, and the adjacent-pixel correlation will be reduced. In this case, adjacent points in the image are more similar spectrally on average, and the mean distance at which this *spatial autocorrelation* disappears is related to the size, spacing, and shape of the objects in the image. Because covariance describes the behavior of the function at one point relative to its behavior at another point, it can also be regarded as an auto-covariance

However, often the gray levels do not uniquely describe object classes, and the methods of *mathematical morphology* can be used to segment the image based on a range of parameters (*4*). Using mathematical morphological techniques, segmentations can be made according to texture. Regions of different graininess are transformed to regions of unique gray levels. Normal thresholding and measurement of these regions can be performed as if they were individual features.

In any mathematical treatment of the digital images, we have made two assumptions: *spatial stationarity*, which assumes that the parameters of the underlying function of which the image is a part do not vary with spatial position, and *ergodicity*, which assumes that the spatial statistics taken in the field of view are unbiased estimates. Spectroscopic images are not stationary in the broad sense due to scan angle effects, for example, but can generally be regarded as stationary in small increments.

Image analysis can be a simple matter if the image resolution and fidelity are high, or complicated if such is not the case. Image analysis can involve a number of processes (*4*): image processing, image rectification, image enhancement, image restoration, and image reconstruction.

Image Processing

The visual world, according to Marr and Hildreth (*5*), is constructed mostly of contours, creases, scratches, marks, shadows, and shading. Image-processing techniques are used to enhance the image for interpretation.

Image processing is concerned with the mathematical manipulation of the digital imaging data (*4*). It involves the conversion of raw signals into display images that can be evaluated. Gray-image operators are the image-processing operators that use a gray-scale image as input and, through mathematical calculations, output another gray-scale image. These functions are divided into two broad types: calculations performed on a single pixel within an image without regard to surrounding pixels (point operations), and calculations that use neighboring pixels (neighborhood operations) to transform the value of a pixel.

Single-Pixel Operations

The simplest image-processing algorithms involve point or pixel processes where each pixel is treated independently of its neighbor. The general point-processing equation can be written as

$$O(x,y) = M[I(x,y)] \qquad (4.1)$$

where $O(x,y)$ is the output image array, $I(x,y)$ is the input image array, M is the function that maps input to output brightness, and x and y refer to the point coordinates of the input and output images. All of the pixels in the input image are normally processed by the function M to provide an output image.

Assuming 0 is pure black and 255 is pure white, brightness can be increased in the output image by adding a constant value B to every pixel in the image:

$$O(x,y) = [I(x,y)] + B \qquad (4.2)$$

Similarly, the contrast can be increased or decreased by multiplying each pixel by a constant value C:

$$O(x,y) = [I(x,y)] \times C \qquad (4.3)$$

Other simple mathematical processes can also be performed in this manner to modify the images in a desirable manner.

Neighboring-Pixel Operations

To vary the spatial information contained in an image, the gray-level properties of adjacent pixels must be taken into account. Gray-image group operators are typically designed to create images that more clearly reveal features by removing random variations in individual pixel brightness. These variations (spatial noise) when removed make edges of features visually more distinct by sharpening using edge-enhancement methods. Reducing variability in brightness across the image often introduced by scan angle can be accomplished by using a shading correction. Increasing contrast in the image to reveal subtle shifts in feature brightness by stretching the scale is often useful visually.

The higher the spatial frequency in a given region of the image, the more pixels in that region can vary in brightness and the more detail the region contains. By lowering the brightness range in a given range, that is, by lowering the spatial frequency, the contrast is lowered and the details of the image are blurred. This blurring is reduced by averaging the pixels in an area to determine the brightness value at any given point. Conversely, one can increase the visibility of the image details by increasing the local contrast by accentuating the brightness changes among adjacent pixels. This is accomplished by *image convolution*. A masking function is used with coefficients that are chosen to perform the desired convolution. The output image is obtained by calculating the brightness value of each pixel in a 3 × 3 array multiplied by its masking coefficient, and the sum of the products is the brightness value of the output pixel. The convolution for each pixel requires nine additions and as many multiplications just for a 3 × 3 mask or kernel. An image of 512 × 512 pixels contains 262,144 separate points and requires more than 2.3 million mathematical operations per convolution.

These gray-scale functions to enhance contrast can be categorized as spatial-domain operations or frequency-domain operations. They are also referred to as filters, transforms, or convolutions.

Image Rectification

Image rectification is concerned with spatial transformations that can remove positional distortions and permit images to be properly registered with respect to each other. *Registering* is the translation–rotation alignment process by which two images of like geometries and of the same set of objects are positioned coincident with respect to each other so that corresponding elements of the same area appear in the same place on the registered images. Registration allows the same sample area to be measured under different imaging conditions and compared. In this manner, the corresponding gray scales of the two images can be compared. Different techniques have been suggested including using the center point and the brightest point for registration (*4*). Image rectification is an important process as it allows the study of the mechanism of the imaging process itself. Image artifacts can be identified, and ultimately distortions can be minimized.

Image correlation (Figure 4.1) involves the comparison of a reference

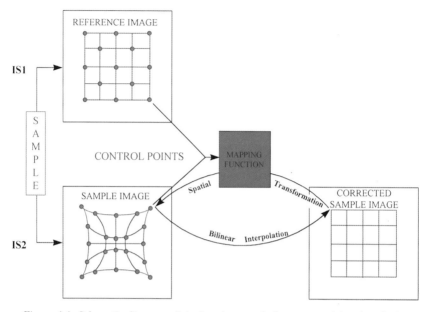

Figure 4.1. Schematic diagram of the imaging correlation process. After the selection of control points from both the reference and sample images, a mapping function is obtained and applied to define the spatial transformation from the sample to the reference image. The corrected sample image is then filled, pixel by pixel, by finding the corresponding pixel in the sample image. Gray-level values are assigned by using bilinear interpolation. (Reproduced from ref 6. Copyright 1987 American Chemical Society.)

and a sample image generated using identical imaging systems (6). The first step in the registration process is the selection of control points from both the reference and sample images. Then a second-order polynomial is used to compute the mapping function. From this function, the corresponding pixel location in the sample image is found to fill every pixel location in the corrected image. Each pixel in the output image can be mapped back to a corresponding pixel in the sample image. Finally, the gray-level value of each of the pixels in the corrected image is duplicated by interpolation from the corresponding gray level in the sample image. Interpolation is achieved by nearest-neighbor pixel resampling (6).

Image Enhancement

Image enhancement techniques are used to process a given image so that the result is more interpretable than the original image for utilization in specific applications. The problem with image enhancement is the existence of noise. In the language of communication theory, we say that the digital image function $I(x,y,z)$ is the sum of two functions: an "ideal" signal $s(x,y,z)$ and a pure noise image $n(x,y,z)$. The problem is that we really want to work with $s(x,y,z)$ but we only have $I(x,y,z)$ and must find some way of dealing with $n(x,y,z)$.

Enhancement techniques are basically procedures that are designed to manipulate images in order to take advantage of the human visual system. Examples are the improvement of resolution, removal of high- and low-frequency noise by filtering, suppression of background, and smoothing of white noise.

In materials research, some prior knowledge about the spatial structure of objects in the image is often available. Such images are termed *discrete-element images*. The knowledge of the arrangement of the objects can be used to improve the S/N of an image since the image obtained is a projection of many such objects, in which the individual substructures overlap. The improvement in the signal-to-noise ratio depends on the number of objects in a given pixel element and the degree of overlap relative to the resolution of the probe (7).

Thresholding results in a separate image called a binary image, which is composed of only two pixel-brightness values. Ideally, the objects to be measured will have a different range of gray-scale (brightness) values than the remainder of the image so the pixels within the image can be separated into two gray-scale groups, one representing the features to be measured and another representing the remainder of the image.

An important set of binary operators are based on the *dilation* and *erosion* functions. In dilation, objects are enlarged by additional pixels around their perimeters; erosion performs exactly the opposite by removing layers of pixels from the boundary of an object. Opening is a two-step process in which a layer of pixels is removed from the perimeter of each object represented in the binary image and then a layer is added to the perimeter of each remaining object. The net effect is that thin objects can be entirely removed from the image, while objects with larger area-to-perimeter ratios will remain and retain nearly the original dimensions. Another binary operation that is useful is *object deletion*, which allows objects to be selectively removed from the binary image based on specific spectral characteristics of the object.

Feature measurements and field measurements are two additional useful enhancement functions. *Feature measurements* determine each individual object in a field of view and produce a unique set of measurements. Feature or object measurements commonly provided are area, perimeter, length, and width. *Field measurements* produce information for an entire field of view and do not provide information for individual objects within the field. Software exists to measure the total number and average size, for example, of the objects in the field of view.

Image Restorations

Image restorations are methods that attempt to reconstruct or recover a degraded image by using some prior knowledge of the degradation phenomenon. In light microscopy, image restoration refers to the recovery of in-focus components from the measured bright-field images using computer-processing techniques. These techniques invert the imaging process of a microscope mathematically. Thus restoration methods are based on modeling the distortion process and applying the inverse process in order to recover the original image. Image restorations can help by removing distortions in images produced by optical distortions, or "blurring," caused by the lens. If the imaging process is linear, the restoration is also referred to as *deconvolution*. The Fourier analysis of the distorted image is, mathematically, the product of the Fourier transform of the undistorted image multiplied by the transfer function of the blurring factor. The correct image can, therefore, be recovered by a division process if one establishes the nature of the transfer function.

Image Reconstructions

Image reconstructions are mathematical methods involved in the reconstruction of a three-dimensional image from its projections (*8*). The three-dimensional image is reconstructed using a matrix inversion of the projection data. For complex images, this task is formidable. Iterative and direct reconstruction methods have been in existence for many years.

In tomographic reconstruction, the object is represented by a three-dimensional array of cubic voxels, and the individual projection sets become two-dimensional arrays. Thus, ideally, one sets up an array of cubic voxels, collects projection data from series of views in three dimensions, and solves for the density of each voxel either algebraically or by back projection. Practically, however, one defines one plane at a time as an array of square pixels, collects a series of views in two dimensions, solves for the densities in that plane, and then proceeds to the next plane.

Image Understanding

Image understanding (*8*) is a method that generates a more global understanding of what the image means and how the features detected in the image are related to each other. For example, it is possible that image analysis can be used to enhance an image, allowing the detection of certain geometric features such as edges, and image understanding looks at the relationship between the edges. It is desirable to develop algorithms and procedures for automatic evaluation of images. In this context, image algorithms describe complete computational tasks that evaluate images to get a desired final result.

One important approach to image understanding and interpretation is the use of fractals. Regular geometric shapes like circles and rectangles can be described in precise terms using solid geometry and precise integer dimensions with respect to such things as a radius or diagonal. Fractal mathematics makes it possible to describe complex objects, especially random objects, in terms of noninteger dimensions. Usually, the representation is in terms of an exponential dependence of one dimension of the described object.

Imagine a one-dimensional line lying on a two-dimensional surface. Then imagine the line becoming more and more curved and kinky. To do so, more and more length of line will be required, and the surface area will be increasingly covered by the line. The fractal approach reflects this increase in fine details as a dimensionality that increases from its initial value of one for the straight line toward the surface's two-dimensions when the line completely covers the surface. The fractal dimensions, D, of the kinky string then will fall in the interval

$$1 < D < 2 \qquad (4.4)$$

When a two-dimensional surface becomes more and more corrugated and the corrugation increasingly detailed, the surface fills more and more space, and the fractal model states that the surface dimensionality gradually increases from two to three

$$2 < D < 3 \qquad (4.5)$$

When the whole volume is tightly packed with a highly irregular surface, the dimensions of the surface reach that of the volume, that is, three. Conversely, when a solid becomes more and more filled with defects, the fractal dimensionality of the solid gradually decreases from the original three dimensions toward the two dimensions of a flat surface, the dimensions of the surfaces of the defects (*9*).

One general characteristic of fractals is that they are the end result of physical processes that modify shape through local action such as diffusion, aggregation, and turbulent flow. Fractal geometry allows us to represent shapes and surfaces by computations from the image data (*10–12*).

The most profound property of fractals is self-similarity, which means that a fractal has no characteristic length scale. If you expand a piece of a fractal, it looks the same as the original picture (i.e., the same as the whole fractal). Every piece of a fractal looks like the whole. There is no fixed-length scale in the system.

Natural objects are random, and a piece will not look exactly like the whole. But if you calculate a correlation function that tells you how the object is put together geometrically, you'll see the same behavior in a small piece of the object as in a much larger piece. This is the essence of fractal geometry.

Suppose that we want to measure the area of an object, or its surface. We do this by taking a measuring instrument of size L, determining that n measurements with this instrument will generate the area to be measured, and applying the formula

$$M = nL \tag{4.6}$$

where M is the metric property to be measured and L is the topological dimension of the measuring instrument. If the surface is rough, there will be features smaller than the size of the measuring tool that will be missed, whatever the size of the measuring tool selected. Therefore, the length of such a curve or area will depend not only on the roughness but also on the length of the measurement tool itself. To obtain a consistent measurement of the rough surface, we must include fractional dimensions. The fractional power in our measurement compensates in effect for the length or area lost because of details smaller than L. The unique fractional power that yields consistent estimates of a set's metric properties is called that set's fractal dimension.

However, real images and surfaces cannot be true mathematical fractals because the latter are defined to exist at all scales. Physical surfaces, in contrast, have an overall size that places an upper limit on the range of applicable scales. A lower limit is set by the size of the surface's constituent particles. Fractals, in common with all mathematical abstractions, can only approximate physical objects over a range of physical measurements.

Computational Techniques for Image Enhancement

Measurement Considerations

The usefulness of the results of imaging measurements depends heavily on two analytical properties: accuracy and representativeness. Accuracy is the difference between the average of a series of measurements and the "true" value given by a standard or a series of standards. Accuracy is a measure of the consistency between the results obtained and the actual concentration of the analyte in a particular sample. Inaccuracy, termed bias or systematic uncertainty, is the nonrandom component of measurement error. Representativeness refers to consistency between the results and the analyzed samples in relation to the analytical problem.

Reproducibility is the closeness of agreement of multiple measurements under the same conditions of measurement, measuring instrument, and environment, but with repeated readings taken by multiple operators. Precision is the variation of a value over a series of repeated measurements. Precision in standardization is required to evaluate the stability, performance, and adequacy of the selected reference or measurement system. Tolerance is the minimum acceptable combination of systematic and random errors resulting from the measurement system.

Repeatability is the closeness of agreement of multiple measurements under the same conditions of measurement, observer, measuring instrument, and environment. It is equal to the standard deviation. The standard deviation σ is an estimate of the precision of a measurement attributable to random uncertainty. The standard deviation is calculated by taking the square root of the variance estimate. The variance estimate σ^2 is given by

$$\sigma^2 = \sum_{i=1}^{n} (x_i - X)^2/(n - 1) \qquad (4.7)$$

where X is the data set mean, x_i is an individual data point in the set, and n is the number of elements in the data set.

Conventionally, a detection limit (c_L) is experimentally defined as the analyte concentration that yields a net analyte signal (x_A) equal to k times the standard deviation (σ_B) of the background (x_B):

$$c_L = k\sigma_B/[x_A/c_0] \qquad (4.8)$$

where c_0 is the concentration yielding a net analyte signal x_A. The right side of the equation is the quotient of the net signal at the detection limit $k\sigma_B$ and the sensitivity $[x_A/c_0]$. The value of $k = 3$ is recommended (*13*).

Standards

The entire imaging-measurement system, including instrumentation, methodology, human factors, and the conditions of measurement, can be systematically verified by the use of proper standards. Accurate standards are required if there is to be confidence in the analytical results. In developing a method, accurate standards are used to test the predictability of the relationship between the measured parameter and the property of interest. Standards are selected because they possess a measurable property that correlates quantitatively with the property of interest, such as spectral absorption with concentration. The standard's measurable property must scale in a predictable fashion over the entire calibration range.

Commonly, standard samples that have been certified for their bulk compositions are used as calibration samples in mapping experiments. In this case, it is often implicitly assumed that the standard is homogeneous (i.e., that there are no spatial differences in the concentration of the constituents). However, the composition of the sampled subvolume is not necessarily representative of the overall composition of the sample. In practice, it is frequently observed that the concentrations are considerably different, and standard inhomogeneity can contribute significantly to the overall uncertainty in the results.

The error due to heterogeneity is expected to increase as the size of the area of analysis decreases. It has been suggested that the heterogeneity as measured with a very small area of analysis is one order of magnitude lower than expected from the heterogeneity as measured with a very large area of analysis (*13*). The relative standard deviation of intensities measured for different pixels of a sample is proportional to the inverse square root of the pixel volume (*13*).

All standards are part of a hierarchical order in which each class of standards is subordinate to and derived from standards conforming to higher levels of accuracy and precision. Standard reference materials or reference standards are the highest level of laboratory standards and are calibrated with respect to a primary standard. Certified standards are reference materials for which the parameters of interest are certified to be within specified limits of accuracy and precision by a laboratory like the National Institute of Standards and Technology. Practically, one usually uses working standards that are calibrated using reference materials available to the spectroscopists.

Calibration

> *Calibration is the mathematical and statistical process of extracting information, usually analyte concentration, from the instrument signal (13).*

> —K. S. Booksh and B. R. Kowaiski

In any instrumental procedure designed to make quantitative measurements, the first step involves performing a calibration. The calibration is performed by measuring the analyte concentration for a variety of samples using a well-defined experimental procedure. The calibration establishes an empirical model relating the instrument response to the concentration. This empirical calibration model is used with the instrument measurement of an unknown sample to estimate the analyte concentration for which the calibration was developed.

Linear Models

In the simplest case, the relationship is a simple linear regression between a single measurement (y) from a single wavelength of the spectrometer, and the level of analyte (c) is given by

$$y = a + bc \qquad (4.9)$$

where a and b are constants. In this case, the instrumental response and independent measurements are used to construct a model (i.e., measure a and b). This model is then used to predict the concentrations of analyte made solely on the instrumental responses at the specific wavelength. In the ideal case, the system obeys theoretical models such as the Beer–Lambert law, where $a = 0$.

In the case of univariate calibration (referred to as the classical method) the statistical model is

$$y_i = b_i c_i + e_i \qquad (4.10)$$

The experimental measurement error in y_i is given by e_i. In this case, calibration is accomplished by least-squares regression of the experimental measurements on a set of standard samples with concentrations bridging the domain of interest. In the peak-height method, for example, the height of a characteristic peak of a component is the spectral information for building the method. This method is extremely simple and useful except when a single characteristic peak cannot be isolated due to the presence of overlapping bands and interferences.

In the inverse method, the implied statistical model is

$$c_i = b_2 y_i + e_i \qquad (4.11)$$

where e_i is the measurement error associated with the reference samples c_i. Both of these methods are widely used but suffer primarily from the inability to detect the presence of interferences and from low precision.

The most common problems with the linear model calibration graphs are (*14*):

- curvature over the whole range, that is, a deviation from the straight line model in the high concentration range;

- a background correction that is difficult to make properly;
- the lack of a way of determining when outliers exist; and
- low precision (due to only a single measurement).

Multivariant Linear Calibration

Multivariant linear calibration (MLC) uses multiple instrumental measurements on a series of standards. Multiple measurements (at different frequencies, e.g.) allow:

- higher precision,
- the determination of multiple components in a system, and
- the detection of the presence of interferences.

For multiple linear regressions (MLR), a single analyte is measured for n instrumental measurements (i.e., frequencies that are specific for the analyte).

$$y_{ij} = b_j c_i + e_{ij} \quad \text{for} \quad j = 1, 2, \ldots, n \qquad (4.12)$$

where e_{ij} is the measurement error association with the jth measurement on the ith specimen.

For multiple-component systems obeying Beer's law, the equation becomes

$$A_i = \sum_{j=1} a_{ij} b c_j + e_{ij} \qquad (4.13)$$

where A_i is the absorbance of a multicomponent sample at a frequency i, a_{ij} is the absorptivity of component j at frequency i, b is the pathlength, and c_j is the concentration of component j. One can combine the path length and absorptivity terms ($k = a * b$)

$$A_i = \sum_{j=1} k_{ij} c_j + e_{ij} \qquad (4.14)$$

In matrix language

$$\mathbf{A} = \mathbf{KC} + \mathbf{e} \qquad (4.15)$$

where $\mathbf{A}$ is an $n \times m$ matrix constructed from the measured absorbance spectra of the m mixtures at n frequencies, and $\mathbf{C}$ is the $l \times m$ concentration matrix corresponding to the concentrations of each of the l components in the m known references. This method, known as the **K**-matrix method by spectroscopists, minimizes spectral errors.

Classical least squares is used for analysis of spectroscopic measurements when the model is well-understood and linear (e.g., Beer's law) (*14–16*). The first condition for using classical least squares for calibration is that the concentrations of all species giving a nonzero response must be included in the calibration step. In classical least squares, the pure-component spectra (at unit concentration) are used in building the calibration model. Using the full spectral information, the precision of the measurements is improved considerably.

The inverse relationship can be written as

$$x' = a_0 + (a_1 y_1) + (a_2 y_2) + (a_3 y_3) + \ldots + (a_n y_n) \qquad (4.16)$$

where y_n may be the instrumental response at different frequencies. In matrix language, this becomes

$$\mathbf{c} = \mathbf{Pb} + \mathbf{e} \qquad (4.17)$$

where **c** is a vector of the concentrations of all the samples, **P** is the instrument response function matrix (a collection of spectra), **b** is the vector of the model parameters, and **e** is the matrix of the concentration residuals.

The inverse relationship is known as the **P**-matrix method. The inverse method is a frequency-limited calibration method, that is, the number of frequencies included in the analysis cannot exceed the number of calibration samples used in the calibration, or in the case of polymers, the number of independently varying sources of spectral variation in the data. This inverse method minimizes the sum of the squared concentration errors rather than the spectral absorbances. This method has found wide application in near-IR spectroscopy.

Nonlinear Multivariate Methods

The primary advantage of multivariate analysis over univariate techniques is that it properly treats correlated behavior between two or more variables in the statistical modeling process. For this reason, and others, the use of principal component analysis for spectroscopic problems is becoming of common use.

In full spectrum (all measured frequencies) methods such as principal component analysis (PCA), and partial least squares (PLS), the number of measurements is unlimited. Continuum regression methods such as PLS and PCA use a very large number of instrumental measurements simultaneously to develop an empirical model that relates the measured spectra of the calibration samples to concentrations or properties of the samples. These methods use factor analysis to extract information from the spectra to build the model by partitioning the matrix $\mathbf{S}$ (containing the spectral information for the set of samples) into two smaller matrices called the loadings (or factors) matrix and the scores matrix. The factors are chosen in such a way that they describe as precisely as possible the underlying variations in $\mathbf{S}$. These factors lie in the same space as the columns of $\mathbf{S}$ and can be used as a new basis of the matrix. The scores matrix is then a new representation of $\mathbf{S}$ (new spectral intensities) on this basis. It is these new spectral intensities that are used in the calibration step in PCA and PLC to build the model.

With PCA, the principal factors extracted from the covariance matrix of $\mathbf{S}$ are used as a new basis for the scores matrix. The relation between this matrix and the concentration matrix is then estimated using multivariate linear regression.

With PLS, the underlying factors in both $\mathbf{S}$ and $\mathbf{C}$ are estimated simultaneously. Furthermore, the columns of $\mathbf{C}$ are used to estimate the factors for $\mathbf{S}$, and at the same time, the columns of $\mathbf{S}$ are used to estimate the factors for $\mathbf{C}$. Thus, the advantage of PLS over PCA is that in PLS the information about concentrations is incorporated in the factors.

In general, the assumed model for these methods is

$$x_i = b_0 + (b_1 t_{i1}) + (b_2 t_{i2}) + \ldots (b_n t_{ik}) \tag{4.18}$$

where t_{ik} is the kth score associated with the ith sample. Each score consists of a linear combination of the original measurements, that is

$$t_{ik} = (\gamma_{i1} y_{i1}) + (\gamma_{i2} y_{i2}) + \ldots (\gamma_{iq} y_{iq}) \tag{4.19}$$

Data from the calibration set are used to obtain the set of coefficients $\{\gamma_{ik}\}$, the model parameters $\{b_k\}$, and the model size (17).

If absorbance is measured at k wavelengths for m mixtures containing n analytes ($m > n$), then one has an equation system that can be expressed in matrix form as

$$\mathbf{A} = \mathbf{CK} + \mathbf{E} \tag{4.20}$$

where $\mathbf{A}(m,k)$ is the absorbance matrix, $\mathbf{K}(n,k)$ the absorptivity coefficient matrix, $\mathbf{C}(m,n)$ the concentration matrix, and $\mathbf{E}(m,k)$ the residuals matrix (18). In PCA calibration, the absorbance matrix is broken down into the product of the scores matrix, $\mathbf{T}(m,a)$, by the loadings matrix, $\mathbf{P}(a,k)$:

$$\mathbf{A} = \mathbf{TP} + \mathbf{F} \tag{4.21}$$

where $\mathbf{F}(m,k)$ is the residuals matrix and a is the number of principal components. Each factor in the model can be thought of as an abstract phenomenon that is a source of variation in the spectra of the calibration samples. The scores for a single factor can be thought of as the "intensities" of the corresponding abstract phenomenon for each sample, and the loadings for a single factor can be thought of as the "spectral signature" of the corresponding abstract phenomenon. Therefore, the scores for each factor can be used to determine whether the spectroscopy is sensitive to known trends in the samples, and the loadings can be used to better understand how the spectrum is influenced by different properties that vary in the samples. Furthermore, the spectral residuals (in the matrix $\mathbf{F}$), which contain the spectral information for each sample that is not explained by the PCA model, can provide information about the nature of the random spectral information that is not useful for the determination of the property of interest (*19*).

The scores matrix is related to the concentration matrix by the regressor matrix, $\mathbf{B}$, which is determined by a least-squares regression. Thus,

$$\mathbf{C} = \mathbf{TB} \qquad (4.22)$$

$$\mathbf{B} = (\mathbf{T^tT})^{-1}\,\mathbf{T^tC} + \mathbf{G} \qquad (4.23)$$

where the superscript t denotes the transpose of the matrix, and $\mathbf{G}$ is the residuals matrix.

Successful application of PCA requires that the number of distinguishable components contributing to the spectra of the materials be less than the number of samples in the calibration set and preferably less than half this number (*20*). Components whose contributions to the spectra are so small that they are lost in the noise are not considered. Also, components that vary together (spectroscopically indistinguishable) are regarded as a single component.

It can be shown that the sum of the eigenvalues is equal to the total variance of the original data (*20*). In this case, it represents the total variance about the origin and equals the sum of the squares of all the data points. The individual eigenvalues represent the contribution of each factor to this variance. That is, the factor with the largest eigenvalue represents the spectrum that varies the most from zero. In most circumstances the major variations from zero will be well described by the mean of the spectra. The factor associated with the largest eigenvalue is almost always a constant multiple of the mean spectrum. The second factor represents the major orthogonal variations of the spectra from the first factor. The smallest eigenvalue will represent mostly noise along with minor components, or differences in components that are indistinguishable from the noise. By eliminating from the expansion those factors in which the noise contribution outweighs the information about the components in the mixture, the original data can be represented more concisely, without loss of information. This is the implicit noise reduction in the PCA method.

Central to the performance of the PCA method is the proper selection of the number of principal components to be used to construct the calibration model. Some of the commonly used criteria are the number of factors required to minimize the error of prediction (20), cross-validation methods (21), ratios of successive eigenvalues (22), the reduced eigenvalue ratio (23), and the F test (24, 25). Part of the difficulty in determining the number of components can be attributed to the sensitivity of the method to noise.

With the PCA method, quantifying an unknown mixture involves determining its scores vector:

$$a_x P^t = t_x \qquad (4.24)$$

$$c_x = t_x B \qquad (4.25)$$

The PCA method has found wide application to a variety of polymer systems.

Signal Processing

The processing of a signal $x(\nu)$ can be represented by the mathematical operation:

$$y(\nu) = T[x(\nu)] \qquad (4.26)$$

where T is the operator (system transformation) and $y(\nu)$ is the output or response function. The purpose of the transformation is to extract some quantity of interest from the signal $x(\nu)$.

Choice of a Signal-Processing Scheme

Error is inherent in experimental data. The universal goal of all signal-processing schemes is to allow the measurement of the signal in such a manner that the uncertainty introduced by the noise variance is minimized. The choice of a signal-processing scheme is generally dictated by the nature and magnitude of the signal, as well as the statistical behavior of the noise. Quite commonly, the signal is a single peak, symmetric or asymmetric, and the desired information is related to some signal parameter, such as its height, width, or area.

Signal processing is used to remove interferences, reduce noise, or extract information related to the desired variables.

Data Pretreatment

It is assumed that the property of interest (absorbance) is dependent on the quantities of various components in the material. It is necessary that the spectra be standardized so that they represent an equal amount of material for each sample. A number of methods are used to remove extraneous sources that contribute no useful information about the sample.

Normalization

Normalization is performed to eliminate systematic "size" differences between samples. A proper normalization should minimize the differences between replicated samples and preserve the correlation between the original analytical signals.

Normalization is employed when the sum of the intensities should be constant and comparison of samples reveals that it is not. If the weights of replicate samples vary, there will be differences in the absolute intensities or areas but not in the relative proportions of the peaks. Normalization of spectral data to a constant sum over all absorbances is a common practice that minimizes variations in peak height or areas because of differences in sample amount. Normalization removes this bias in feature magnitudes that may arise as a result of experimental variations (26).

Spectra are commonly normalized to the most intense peak or to a constant sum (27). However, when the spectra exhibit what is known as heteroscedastic noise, meaning that the absolute noise increases with increasing intensity, then normalization to a constant sum effectively converts the noise from peaks with high intensity into systematic variation. It is desirable to transform the data in such a manner that the heteroscedastic noise is transformed to homoscedasticity (the noise is distributed independently of signal intensity) prior to the use of any normalization procedure. When the noise is homoscedastic, normalization to a constant sum is appropriate. When the noise is heteroscedastic, with a relative standard deviation that is constant, then the log normalization is appropriate. When the heteroscedastic noise shows a decrease with signal size, the power normalization is appropriate. If the standard deviation is proportional to the mean of the signal, the log transform is proper. If the standard deviation of the noise is proportional to the root of the mean, the square root transform is proper.

Normalization is defined as the transformation

$$\mathbf{z}_i^{\mathbf{T}} = (1/N)_i \mathbf{x}_i^{\mathbf{T}} \tag{4.27}$$

where $\mathbf{z}_i^{\mathbf{T}}$ and $\mathbf{x}_i^{\mathbf{T}}$ represent the normalized and raw intensity vectors, respectively, of sample i, and superscript $\mathbf{T}$ implies transposition of a column vector into a row vector. For a weighted normalization, the scaler N_i is determined by

$$N_i = \mathbf{x}_i^{\mathbf{T}} \mathbf{w} \tag{4.28}$$

The vector $\mathbf{w}$ is a column vector of weight factors. Normalization to a constant sum is defined as $\mathbf{w} = 1$. Normalization to internal standards or selective normalization is performed by choosing the elements of $\mathbf{w}$ corresponding to the selected peaks as 1, and 0 for the others.

Autoscaling

Autoscaling, or subtracting the mean and dividing by the standard deviation of each feature, converts each feature in the data set to a mean of zero and a standard deviation of one. This common data transformation removes inadvertent weighting caused by different means or different magnitudes of variation in each dimension.

Mean-Centered Data

Spectra are mean-centered prior to regression to remove nonzero intercepts in the models.
   Mean centered values x_i^* are given by

$$x_i^m = x_i - x_{\mathrm{avg}} \tag{4.29}$$

where

$$x_{\mathrm{avg}} = (1/n)\Sigma x_i \tag{4.30}$$

This operation makes subsequent computations less sensitive to roundoff and overflow problems.

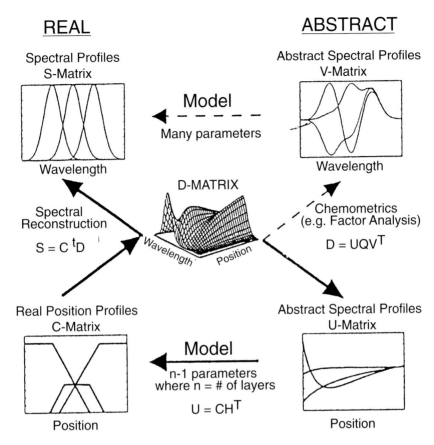

REAL ABSTRACT

Spectral Profiles
S-Matrix

Model
— — — — —
Many parameters

Abstract Spectral Profiles
V-Matrix

Wavelength

Wavelength

D-MATRIX

Spectral
Reconstruction

$S = C^{t}D$

Chemometrics
(e.g. Factor Analysis)

$D = UQV^{T}$

Real Position Profiles
C-Matrix

Model
n-1 parameters
where n = # of layers

$U = CH^{T}$

Abstract Spectral Profiles
U-Matrix

Position

Position

Figure 4.2. Graphical representation of the factor-analysis computational approach. (Reproduced with permission from ref 28. Copyright 1993 Society of Spectroscopy.)

Use of Factor Analysis for Image Enhancement

Computational techniques can lead to superior contrast because potential interferences from other species are effectively compensated. Sample quantification, discriminate analysis, and vector orthogonalization are examples of advanced numerical expressions of spectral response. As a result various chemometric techniques have been used to enhance the images. Factor analysis has been used to enhance the spatial resolution (*28*), and a graphical representation of the approach is given in Figure 4.2.

In this case, it is desired to isolate unique spectra of multilayer systems where the layers are less than 10 μm in thickness (which is below the physical resolution of the microscope). The observed spectra will, there-

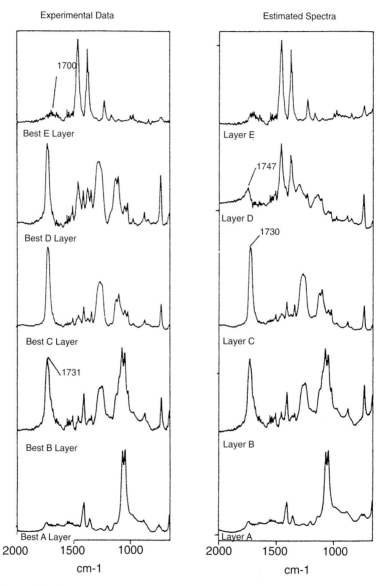

Figure 4.3. Spectral estimates compared with raw data from industrially produced multilayer polymeric film analysis. (Reproduced with permission from ref 28. Copyright 1993 Society of Spectroscopy.)

fore, be mixture spectra arising from contributions due to adjacent layers. A model was developed assuming that the area of the aperture that covers a specific layer is proportional to the contribution of that layer to the measured spectra. An abstract representation of the wavelength (**V**-matrix) and position space (**U**-matrix) is transformed using an appropriate model to a real representation of the position (**C**-matrix) and wavelength (**S**-matrix) spaces. Once the real position information is available it is possible to extract the spectral features of each layer. The method was evaluated by simulation using a structure that had two thick layers of polyethylene terephthalate (PET) and a vinylidene chloride containing copolymers (VCP) with a 5 μm ethylene–vinyl acetate copolymer (EVA) layer. The results are shown in Figure 4.3, which shows the estimated spectra for the 5 μm layer of EVA with the standard spectra. The degree of correlation with the pure EVA was 0.9979 for the factor analysis, but only 0.7391 with the raw data.

Image Enhancement using Interactive Self-Modeling Multivariate Analysis

Microimaging data have also been used to extract information about the pure components using a self-modeling mixture analysis termed

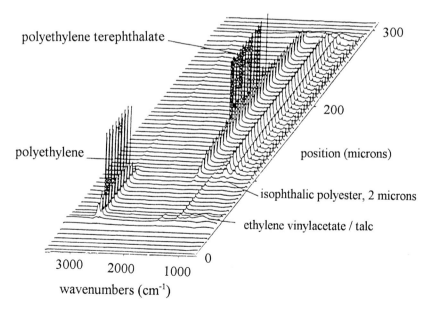

Figure 4.4. Three-dimensional plot of the spectral data collected vs. the position of the mask in micrometers. (Reproduced with permission from ref 29. Copyright 1994 Society of Spectroscopy.)

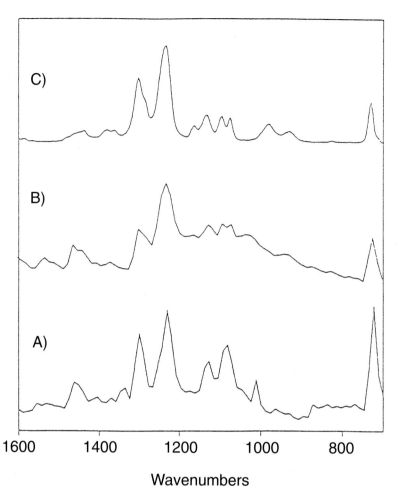

Wavenumbers

Figure 4.5. A comparison of (A) the calculated spectrum of the unknown component in the middle layer, (B) the spectrum obtained on minute amounts of the middle layer, and (C) the spectrum of a reference isophthalic polyester. Spectrum B has been baseline-corrected to allow an objective comparison with the other two. (Reproduced with permission from ref 29. Copyright 1994 Society of Spectroscopy.)

SIMPLISMA (simple-to-use, interactive self-modeling mixture analysis) (*29*). This algorithm is quite simple and is interactive because all of the intermediate steps in the calculation are displayed in the form of spectra. A polymer laminate of 240-μm thickness was used that had four layers ranging in thickness between 2 and 175 μm. The layers were identified as ethylene vinyl acetate with talc (layer 1), polyethylene (layer 2), and

PET (layer 4) (Figure 4.4). The 2–3-μm layer was not identifiable by normal spectroscopic methods. However, using SIMPLISMA it was determined that this layer was an isophthalic polyester (Figure 4.5).

Expectations for the Future

The large virtual memory on supercomputers has made possible the manipulation of images involving tens of thousands of spectra obtained at a large number of frequencies and spatial positions. Thus, it is possible to utilize the full spectrum in the image analysis rather than a selection of a specific frequency. This is particularly important for near-IR imaging. Such an image-analysis method has been described that utilizes parallel vector supercomputers and workstations (*30*). This new experimental clustering technique is termed the quantile bootstrap error-adjusted single-sample theory (BEST). In this method the spectra are represented by a *d*-dimensional hyperspace, where *d* is the number of wavelengths in the spectra. Each point is translated by an amount that corresponds to the magnitude of the signal observed. Thus similar samples produce similar spectra and project as "probability orbitals" or "clusters" in similar regions of hyperspace (*31*). The BEST system calculates the integral of a probability orbital in hyperspace by starting at the centroid of the orbital and working outward in all directions at a uniform rate. The distance between the center of the orbital and a sample spectrum is proportional to the concentration of the component responsible for the vector connecting the center to the sample points. The direction of the vector identifies the component.

Image processing software is available free of charge from the National Institutes of Health (*32*), and a number of commercial packages exist. Recently a number of monographs in this area have become available (*33–36*) and should be consulted for additional information in this important area of image processing.

References

1. Todd, P. J.; Short, R. T.; Grimm, C. C.; Holland, W. M.; Markey, S. P. *Anal. Chem.* **1992**, *64*, 1871.
2. Hack, F. O.; John, S.; Reichenbach, S. E. *SPIE* **1990**, 159.
3. Yeager, L.; *Visualization of Natural Phenomena.*
4. Serra, J. *Image Analysis and Mathematical Morphology;* Academic: London, 1982; 2 vols.
5. Marr, D.; Hildreth, E. *Proc. R. Soc. London* **1980**, *B207*, 187–217.
6. Turner, L. K.; Ling, Y.-C.; Bernius, M. T.; Morrison, G. H. *Anal. Chem.* **1987**, *59*, 2463.

7. Blumler, P.; Greferath, M.; Blumich, B.; Spiess, H. W. *J. Magn. Reson. Ser. A* **1993**, *103*, 142.
8. Morris, R. A. In *Microscopic and Spectroscopic Imaging of the Chemical State;* Morris, M. D., Ed.; Practical Spectroscopy Series 10; Marcel Dekker: New York, 1993.
9. Aharoni, S. M.; Edwards, S. F. *Adv. Polym. Sci.* **1994**, *18*, 38.
10. Mandelbrot, B. B. *Fractal: Form, Chance and Dimension;* W. H. Freeman: New York, 1977.
11. Feder, J. *Fractals;* Plenum: New York, 1988.
12. Mandelbrot, B. B. *The Fractal Geometry of Nature;* Freeman: New York, 1977.
13. Booksh, K. S.; Kowaiski, B. R. *Anal. Chem.* **1994**, *66*, 782A.
14. Blackburn, J. A. *Anal. Chem.* **1965**, *37*, 1000.
15. Antoon, M. K.; Koenig, J. H.; Koenig, J. L. *Appl. Spectrosc.* **1977**, *31*, 518.
16. Haaland, D. M.; Easterling, R. G.; Vopick, D. A. *Appl. Spectrosc.* **1977**, *39*, 73.
17. Stone, M.; Brooks, R. J. *J. R. Statistic. Soc., Ser. B* **1990**, *52*, 237.
18. Blanco, M.; Coello, J.; Iturriaga, H.; Maspoch, S.; Redon, M. *Appl. Spectrosc.* **1994**, *48*, 73.
19. Miller, C. E. *Appl. Spectrosc.* **1993**, *47*, 222.
20. Malinowski, E. R. *Factor Analysis in Chemistry;* Wiley-Interscience: New York, 1991; Chap. 10.
21. Wold, S. *Technometrics* **1978**, *20*, 397.
22. Rossi, T. M.; Warner, I. M. *Anal. Chem.* **1986**, *58*, 810.
23. Malinowski, E. R. *J. Chemom.* **1987**, *1*, 33.
24. Malinowski, E. R. *J. Chemom.* **1988**, *3*, 49.
25. Malinowski, E. R. *J. Chemom.* **1990**, *4*, 102.
26. Kvalheim, O. M.; Brakstad, F.; Liang, Y. *Anal. Chem.* **1993**, *66*, 43.
27. Reyment, R. *Chemom. Intell. Lab. Syst.* **1987**, *2*, 79.
28. Pell, R. J.; McKelvy, M. L.; Harthcock, M. A. *Appl. Spectrosc.* **1993**, *47*, 634.
29. Guilment, J.; Markel, S.; Windig, W. *Appl. Spectrosc.* **1994**, *48*, 320.
30. Cassis, L. A.; Lodder, R. A. *Anal. Chem.* **1993**, *65*, 1247.
31. Lodder, R. A.; Hieftje, G. M. *Appl. Spectrosc.* **1988**, *42*, 1512.
32. Rasband, W. *NIH IMAGE;* National Institutes of Health: Bethesda, MD, 1990.
33. Russ, J. C. *Computer Assisted Microscopy;* Plenum: New York, 1990.
34. Soumekh, M. *Fourier Array Imaging;* Prentice-Hall: Englewood Cliffs, NJ, 1994.
35. *Multidimensional Microscopy;* Cheng, P. C.; Lin, T. J. H.; Wu, W. L.; Wu, J. L., Eds.; Springer, New York, 1994,
36. Russ, J. C. *The Image Processing Handbook;* CRC: Boca Raton, 1990.

5
Experimental Imaging

Optical Effects

To be useful, an image should have a one-to-one correspondence between the object space that it represents and the spatial placement in the image. The exact correspondence is, however, disturbed by the presence of one or more degradation mechanisms:

- noise,
- aberrations, and
- astigmatism.

Two principal types of optical artifacts degrade images: *artifact contrast* and *misregistration*. *Artifact contrast* is a deviation from proportionality between signal intensity and concentration. *Misregistration* is the wrong spatial placement of the concentration value inside the analyzed volume (*1, 2*).

In spectroscopic imaging, noise can obscure the visibility of objects by degrading the signal-to-noise ratio (S/N) below the limit of 2.0, which the eye requires to discriminate objects. The noise in imaging can be reduced at the cost of longer imaging measurement time, higher source intensity, better detector efficiency, etc. At the minimum, the S/N of the spectroscopic image must be sufficient to visually detect two morphologically different phases of materials. This minimum S/N requirement depends on a variety of sample and optical parameters, some of which we will discuss here. Of course, the primary requirement is an object-specific optical signal from the sample that is detectable above the nonspecific background optical noise.

The lack of sharpness in spectroscopic images compared to optical images stems from basic physics. The problem is inherent in simultaneous multiwavelength imaging. Dispersion occurs when the light is broken out into its spectral components as it passes through the optics of the microscope. Different frequencies are imaged in different locations along the same optical path, both laterally and axially, depending on the chromatic function of the optics. In addition, focusing of the optical system will be specific for one wavelength of the light and defocused for others across the sample. Thus, the spectroscopic images have different brightness and contrast levels for the different wavelengths being used.

In addition to the desired sample-specific absorbance signal in the transmitted intensities, there is always additional attenuation or loss of signal due to optical effects, particularly light scattering and diffraction (3). The fundamental lower limit, below which background cannot be reduced, is determined by *Rayleigh scattering*. This is light scattering that is intrinsic to the material and is caused by microscopic differences in refractive index induced by density or concentration fluctuations that are inherent in the material. These fluctuations in density are caused by thermodynamic effects and cannot be reduced below certain limits.

Artifacts can also arise from distortions of the images due to nonoptical imaging parameters. These artifacts become more important as the resolution of the images increases. So the optical devices used should have minimal imperfections that cause distortion, loss of light, or other problems. It is also desirable to have compensating internal corrections for the aberrations where possible, for such factors as field curvature, astigmatism, coma, and spherical and axial chromatics. All of these factors lead to a loss of both resolution and signal.

The advent of computer processing allows the minimization of some of these artifacts by mapping a sample image to a reference image and by restoring the image when the degradation mechanism is known. However, suitable reference images are seldom available.

Refractive Index Dispersion

Matter slows down light. The *refractive index, n,* of a substance is a measure of the speed of light in that substance compared with the speed of light in a vacuum. The refractive indices of a material are determined by the polarizability of the molecules at optical frequencies. A material with a low refractive index has a low polarizability, that is, the dipole moment per unit volume induced by the electromagnetic field is minimized.

In a homogeneous medium (i.e., in a medium with a constant refractive index), light rays travel in straight lines. The optical path length of a wave in a material of thickness t is nt. The material contains nt/λ wavelengths for a path length of nt in a vacuum. The optical path difference,

Δ, due to the presence of this material is then $(n - 1)t$, and the phase difference produced is $(2\pi/\lambda)/\Delta$.

But in inhomogeneous media, the paths of the rays are considerably different, and they depend on the nature of the inhomogeneities and their respective refractive indices. In the simplest case, for an inhomogeneous medium with a continuous variation of refractive index, the trajectory of the ray becomes a continuous curve. For complex systems, the paths of the rays are essentially indeterminate. Dispersion in the refractive index is always present for inhomogeneous samples being examined by spectroscopic probes, and to date, the impact on the fidelity of the images has not been determined.

Optical Effects at Interfaces

Light passing obliquely from one transparent medium into another is *refracted* (bent) from its initial path. Considering that light consists of rays, which tend to travel in straight lines, the *angle of incidence* can be defined as the angle between the path of the light, as it approaches the interface, and the perpendicular to the interface at the point of incidence. The *angle of refraction* is the angle between the ray, after it has passed the interface, and the perpendicular to the interface. These angles are shown in Figure 5.1.

The angle of incidence, the angle of refraction, and the perpendicular to the surface are coplanar. As the ray passes from a rare medium (such as air) into a denser medium, it is bent toward the perpendicular. Hence, if it passes from the more dense into the less dense medium, it is bent away from the perpendicular. *All optical paths are reversible.* For a given wavelength of light, the ratio of the sine of the angle of incidence to the sine of the angle of refraction is constant and is the ratio of the refractive indices of the two media:

$$(\sin I)_1/(\sin r)_2 = n_2/n_1 \tag{5.1}$$

This is Snell's law.

When the beam of light passes from a dense to a rarer medium, the angle r will be greater than I. As angle I increases, and the ratio $\sin I/\sin r$ remains constant, the angle r must also increase and remain greater than I. If angle I is increased to the value where r becomes $90°$, the beam of light will no longer pass from the first medium to the second, but will travel through the first medium to the dividing surface and then pass along this surface, thus making a $90°$ angle, with the perpendicular to the surface. This is called the *critical ray*. If I is smaller than this particular value, light will pass through the second medium; if it is greater, all light

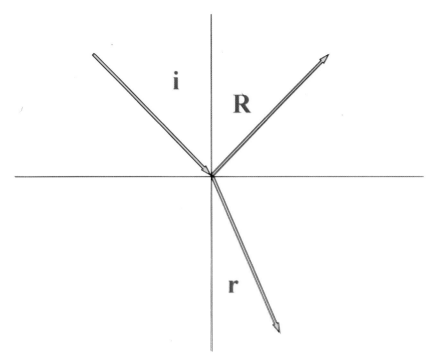

Figure 5.1. Refraction and reflection at an interface.

will be reflected from the surface back into the first medium. Total reflection can occur only when light passes from the denser to the rarer medium.

In addition, when the light beam strikes the interface between media that have different indices of refraction, a fraction of the light is reflected (Figure 5.1). The reflectance, R, depends on the ratio $\mu = n_1/n_2$ of the two media. Near normal incidence,

$$R = [(\mu - 1)/(\mu + 1)]^2 \qquad (5.2)$$

In addition, the reflected wave changes phase by 180° when the incident light travels from the low- to the high-index material. The phase change of a wave traveling in the other direction is 0. For an air–glass interface, μ is equal to the index n of glass, and R is about 4%.

In addition to the external reflection, there are also internal reflection losses at the boundary between air and the sample surface (Figure 5.2) (4). Specular reflectance and internal reflection at the inside of the sample's surface cause linearity problems, with the result that absorbance is

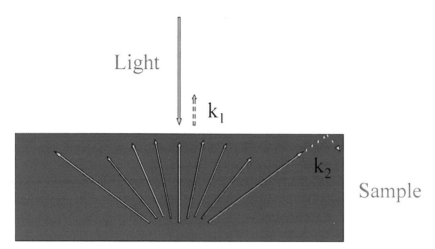

Figure 5.2. *The Saunderson correction factors for external and internal total reflection at the boundary between the sample and its immediate surroundings. Part of the collimated light beam striking the boundary is reflected specularly (k_1); the rest enters the sample and is scattered and partially absorbed by the chemical constituents. The unabsorbed, diffuse light emerging back out from the sample is partially reflected back into the sample because of the difference in refractive index between the sample and its surroundings. (Reproduced with permission from ref 4. Copyright 1985 Society of Spectroscopy.)*

not linearly related to concentration even at constant light-scattering levels. Saunderson (5) suggested a correction based on a corrected reflectance, R^*:

$$R^* = (R - k_1)/(1 - k_1 - k_2 + k_2R) \qquad (5.3)$$

where k_1 is the correction constant for external specular reflection, between 2 and 4% depending on the refractive index; and k_2 is the correction for internal reflection, between 40 and 60%. Where possible, this R^* should be used in place of R in reflection calculations.

Birefringence

Birefringence is the difference between the refractive index, Δn, of the extraordinary wave, n_e, and that of the ordinary wave, n_o. The linear birefringence is given by $\Delta n = |n_e - n_o|$, and is related to the linear retardance, δ, and the sample thickness, d, by

$$\Delta n(\lambda) = (\lambda/2\pi d)\delta(\lambda) \qquad (5.4)$$

A *birefringent* material is doubly refracting, that is, its index of refraction depends on the direction of the light traveling through it. Reflection of an electromagnetic wave at a plane surface changes not only its direction of propagation and its intensity but also its state of polarization. Primarily, this is because under the same angle of incidence reflection means different changes in amplitude as well as phase for electric vectors oscillating perpendicular and parallel to the plane of reflection. As a consequence, reflected radiation is generally found to be elliptically polarized, even though the incident radiation was linearly polarized.

Let us consider the case of uniaxial materials. When a plane wave is incident on a uniaxial crystal with its polarization at some angle to the optical axis of the medium, the light is decomposed into two secondary waves, traveling with speeds $\vartheta_o = c/n_o$ (ordinary) and $\vartheta_e = c/n_e$ (extraordinary), where n_o and n_e are the extraordinary and ordinary refractive indices for the wave traveling in the crystal at an average direction different from the principal axis of the medium, and c is the speed of light in a vacuum. If a narrow beam of unpolarized light falls on the crystal, two beams polarized orthogonally to each other will emerge. For this reason, the phenomenon is called *birefringence*.

Because of the anisotropy of the medium, when an incident light beam enters the medium with an incident angle, the extraordinary and ordinary waves inside the medium have refraction angles θ_e and θ_o, respectively. Consequently, the waves travel with light paths d_e and d_o. These differences in light paths cannot be separated easily from differences in concentration or thickness of the medium, thus complicating the interpretation of the results.

However, there are some imaging issues related to the propagation of polarized light through a medium with varying optical orientation. Along any particular light path, a change of polarization will occur. The composite effect of all the "domains" through which the light has passed is complex.

Provided that the number of domains is large and the effect of a single domain is small, it is reasonable to associate this polarization change with an effective average birefringence related to the average orientation, as discussed above. However, the resulting light beam contains contributions from a large number of light paths. Variations in effective birefringence from one path to another mean that no single polarization state characterizes the emerging light beam.

When the domain size is small, a given light path will sample many domains, and there will be little variation in effective birefringence from one path to another and, consequently, little "mixing" in the polarization state. Conversely, if relatively few domains are sampled, there will be substantial variations from one light path to another and a great deal of mixing of polarizations.

Light Scattering Effects in Inhomogeneous Media

Light scattering is a critical factor in the analysis of microscopic spectra and their corresponding images. When light travels in inhomogeneous media, it splits into three components arising from coherent, quasi-coherent, and incoherent effects. These three components are defined as *ballistic* (coherent), *snakelike* (quasi-coherent), and *diffuse* (incoherent).

The ballistic photons result from the coherent interference of light scattered in the forward direction, and they propagate nearly straight through the medium. The ballistic component is always present, and only its intensity is reduced by scattering (away from the forward direction).

Let us consider a simple example of an inhomogeneous medium, such as one with a continuous variation in refractive index. Photons follow the continuous variation of refractive index along the line of shortest optical path, so the ballistic component broadens. Some photons are scattered slightly off the straight line path and zigzag through the medium. This is the snakelike component.

Last, diffuse photons are present in both discrete and continuous media and are the most dominant component of the three. These photons undergo multiple scattering and follow a random walk through the medium. Diffuse photons lose all the signal information they carried on entering the medium and form noise at the image plane. The ballistic photons retain the signal information, and the snakelike photons retain some, and they both form the images on the image plane.

Light Scattering from Reflecting Surfaces

The reflection of light at a smooth surface coated with a reflective medium is a result of light scattering. The light wave incident on the reflecting surface drives the electric charges in a thin surface layer of the coating. Such charges, in the case of a smooth surface, are uniformly distributed. This causes the light waves, radiated by the charges, to cancel in all directions but the direction of specular reflection.

Defects, such as scratches, on the reflecting surface spoil the uniformity of charge distribution. In this case, the waves emitted by such charge distributions in directions other than that of the specular reflection do not cancel. The incident light is scattered in addition to being reflected. Such scattering from defects creates aberrations in the images.

Another aberration arising from scattering of microbeams from reflecting surfaces is called *coma*. Coma is an optical aberration in which the off-axis beams do not form a single focused spot but rather comet-shaped patterns. The focal length of a mirror is dependent on the zone of the mirror, as shown in Figure 5.3 (*3*).

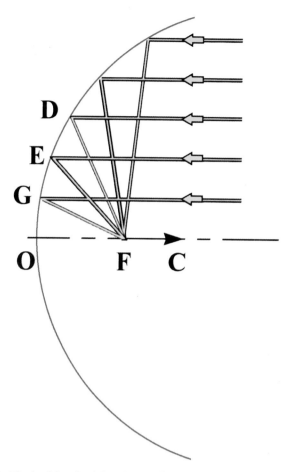

Figure 5.3. The focal length of the mirror is dependent on the zone of the mirror. For example, ray OF has a considerably different focal length than ray DF. (Reproduced with permission from ref 3. Copyright 1987 ASTM.)

Light Scattering from Particulates

When a light beam illuminates a sample containing particles with a variety of particulate sizes, light is scattered by particles in all directions, as shown in Figure 5.4 (5). When particles much larger than the wavelength of incident light are struck by the light, scattering occurs at very small angles, and the scattering intensity patterns consist of concentric annular rings of light and dark intensity. This type of scattering is known as classical *Fraunhofer diffraction*. As the particles become smaller, scattering angles are increased and scattering intensity patterns become very complex. The angle at which light is scattered is dependent upon the diameter of the cross-

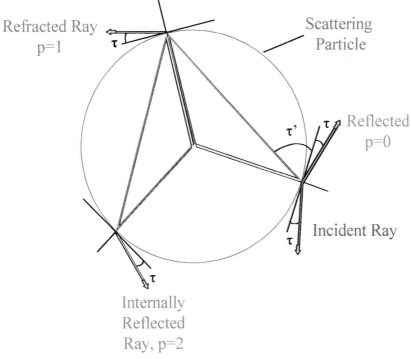

Figure 5.4. Geometric approximation of scattering from a spherical particle. (Reproduced with permission from ref 6. Copyright 1991 San Francisco Promotions.)

sectional area of the scattering particle. Spheres present the same cross-sectional area regardless of the orientation, whereas platelike and rod-shaped particles (with large aspect ratios) are capable of presenting a range of cross sections from the smallest to the largest dimension of the particle.

When a light wave impinges on a particle, the oscillating electric field of the light wave forces oscillations of the electric charges in the particle. The oscillating charges radiate in all directions. The radiated power equals the power removed from the incident beam. Thus, light scattering causes the attenuation of the light beam.

The light rays incident on the particle surface are partly reflected and partly transmitted. The transmitted rays are reflected repeatedly on the inner surface of the particle. At each incidence at the inner surfaces, a portion of the light is transmitted out. In a sense, each particle becomes a minute light source. Some of the scattered light reaches the detector and causes aberrations.

The scattered light power is a function of the angle of incidence of the light rays. Larger particles scatter at smaller angles, and smaller particles at greater angles. Irregularly shaped particles in the sample show a multiplicity of profiles to the incident light beam. Thus, light scattering from particles is a complex phenomenon.

In the process of light scattering by a particle that is smaller than about $1/10$ of the incident light wavelength, the charges in the particle oscillate in unison. This is because the entire particle at one time experiences essentially the same electric field due to the incident wave. As a result, nearly the same power is scattered in every direction. In this case, the angular pattern of the scattered light is approximately uniform. This pattern practically does not change with the particle size in this small range. The total power scattered in all directions by such a small particle is a very small fraction of the incident power. The scattered power increases as the power of the particle size and decreases as the fourth power of the wavelength (*1*).

If the particle size is comparable to or larger than the wavelength of the incident light, the electric charges in various portions of the particle are driven by the electric field in different sectors of the light wave. Because the oscillations of this field in different sectors of the incident wave differ in phase, so do the oscillations of the charges. So the paths from various portions of the particle are different.

In general, the larger the particle, the more light is scattered in the direction of propagation of the incident light than in other directions and the more complex is the angular pattern of the scattered light. The total power of light scattered in all directions by a large particle is nearly independent of the wavelength.

Stray Light

Stray light is defined as diffracted or scattered light reaching the detector that has a lower probability of being absorbed than the primary light. The presence of stray light leads to a nonlinearity in the measured absorbance with concentration. Light diffraction generates stray light and occurs at several optical components in the process of obtaining an image. Diffraction particularly occurs at the aperture as a result of decreasing the aperture size. This diffraction produces stray light, which degrades the quality of the image.

When the sample size is twice the diffraction limit for the imaging system, the effects of diffraction are minimal. In infrared microscopes, there is a small "secondary" mirror, which results in the loss of energy and generates additional stray light (*3*).

It is possible to estimate the effect of stray light on Beer's law (*8*). The total detected light in the absence of the sample, I_o, consists of two parts: primary light (F) and stray light ($1 - F$). The detector merely sums intensities; therefore,

$$I_0 = (1 - F)I_0 + FI_0 \tag{5.5}$$

When the incident light passes through the sample, the primary light is then absorbed as described by Beer's law, but the stray light is not diminished to the same extent, and the detector *cannot distinguish between the two*. The instrument calculates the absorbance from the total measured intensity, I:

$$I_o = (1 - F)I_o{\cdot}10^{-a_1 bc} + FI_o{\cdot}10^{-a_2 bc} \tag{5.6}$$

The absorbance, A, of the species at concentration c is then given by

$$A = -\log[(1 - F){\cdot}10^{-a_1 bc} + FI_o{\cdot}10^{-a_2 bc}] \tag{5.7}$$

The variables are a_1 and a_2, are the absorptivities of the primary and stray light. This equation is demonstrated in Figure 5.5 (*8*).

Because a_2 is defined to be the absorption coefficient for the stray light, it is smaller than a_1. Therefore, at low concentrations, the second term in the above equation becomes a constant. Therefore, at low concentrations the stray light effect is small. At low optical densities, the effect of stray light is also small. However, at high values, the absorbance error can be calculated as demonstrated in Figure 5.6 for an optical density of 1.0 (*9*).

Stray light is the ultimate nemesis of microscopists and spectroscopists, and every effort must be made to control the magnitude of the contribution due to stray light.

Contributions Due to Photon Noise

There are a number of sources of noise in spectroscopy, including:

- photon noise in the radiation source,
- photon noise from the instrument,
- size-independent detector noise,
- detector-size-dependent noise (D^* of the detector), and
- thermal photon noise from the detector housing.

Each of these types of noise will be considered in turn, but first a description of the effects of noise on imaging is presented.

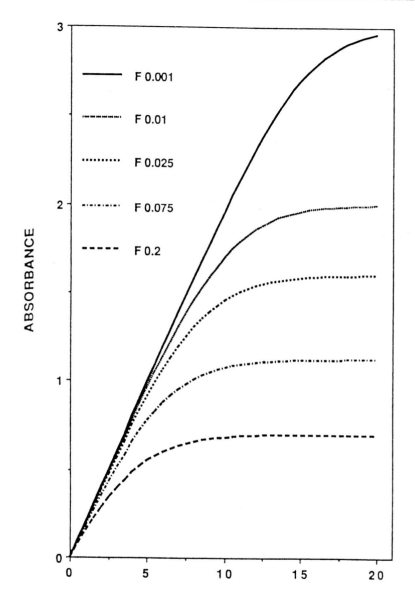

Figure 5.5. Effect of primary light intensity, F, on the working curve. The fraction of stray light is increased from 0.001 to 0.2. The absorptivity factors are $k_1 = 0.2$ and $k_2 = 0$. (Reproduced with permission from ref 8. Copyright 1990 Society of Spectroscopy.)

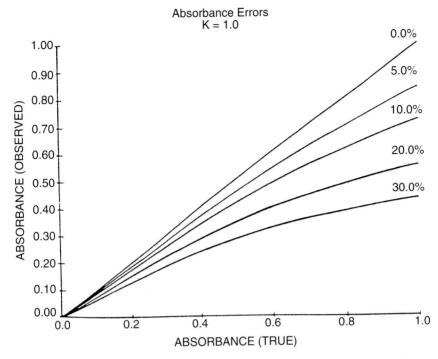

Figure 5.6. *Absorbance error for a band of optical density equal to 1.0 at varying stray light levels. (Reproduced with permission from reference 9. Copyright 1988 Marcel Dekker.)*

Effects of Noise in Imaging Measurements

Noise affects the ability to measure distances in images. Consider the problem of measuring a geometric edge. If the point spread function of the lens is radially symmetric and peaks at its center, symmetry shows that a geometric edge can be located by finding the half-intensity point of the edge response function. So we can estimate the effect of noise on edge location (Figure 5.7) (*10*).

In Figure 5.7, the solid curverepresents the idealized image of an edge with a baseline intensity of A_b and a peak intensity of A_p. The x's represent digitized data that are 1 pixel apart horizontally and include noise. The points labeled with subscripts 1 and 2 are the pixels closest to the half-intensity point, A_m, which is given by

$$A_m = (A_p - A_b)/2 \qquad (5.8)$$

To find the uncertainty, E_m, that results from the uncertainty in A_m, we differentiate the first equation and find that

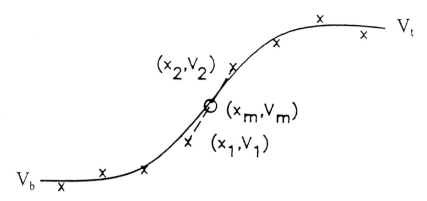

Figure 5.7. Image of an edge a few pixels wide. The solid curve is the exact image, and the crosses are the digitized data, including noise. The horizontal axis is distance, measured in pixels, and the vertical axis is intensity (or voltage from the video camera), measured in relative units. The term V_b is the intensity of the base line, and V_t is the intensity of the top line. Point (x_m, V_m) is the calculated half-intensity point, and points (x_1, V_1) and (x_2, V_2) are the closest pixels to the half-intensity point. (Reproduced with permission from ref 10. Copyright 1991 San Francisco Promotions.)

$$E_m = \delta A_m / (A_2 - A_1) \tag{5.9}$$

where δ means uncertainty and $x_2 - x_1 = 1$ pixel. We can find E_m by successively differentiating the expression with respect to A_2 and A_1, and then adding the resulting uncertainties in quadrature. We assume that the noise is the same for all intensities. If N points are averaged to find the baseline and the peak, we obtain

$$E_m = 1 / (S/N \sqrt{N}) \tag{5.10}$$

Similarly, we calculate that

$$E_2 = 1 / (2S/N) \tag{5.11}$$

and

$$E_1 = 3 / (2S/N) \tag{5.12}$$

so if we add these in quadrature, we find that

$$\delta x = \sqrt{[(1/N) + 2.5]} / S/N \tag{5.13}$$

If N is a small number (it is 3 in the above), the uncertainty (as a fraction of a pixel) of locating the half-intensity point is about twice the reciprocal of S/N. Therefore, a maximum in the S/N of the image is necessary for high-spatial-resolution measurements of objects, as well as the use of the maximum number of pixels. · '

Photon Noise in the Radiation Source

Radiation noise arises from the fluctuation in the amplitude of light itself, which results from the random nature of photoemission. This radiation noise is the ultimate limitation in light detection, but fortunately it is usually sufficiently low to be undetectable except in special types of measurements. Therefore, for microscopy and mapping, this source of noise can be neglected.

Photon Noise from Instrument Background

There is photon noise due to amplitude fluctuations in the background radiation emitted by objects in the instrument as a result of their nonzero temperature. This photon background noise contribution can be approximated as blackbody radiation whose spectral distribution is given by Planck's law with a maximum value at

$$\lambda(\mu m) = (2.897 \times 10^3) T_b \tag{5.14}$$

where T_b is the temperature of the background. If the background objects are at 295 K, the noise is at a maximum at a wavelength, λ, of 10 μm. Therefore, an examination of the spectra in this region will allow the detection of this instrumental photon noise. Because absorbance is $\log(I_o/I)$, the ratio should eliminate this noise in microscopy.

Sample Noise

Sample noise is present in the sample itself due to concentration fluctuations, but this level of fluctuation noise is quite small unless there is extensive turbulence such as diffusion or flow during the imaging process.

Detector Noise

The noise generated in the detectors can be divided into noise affected in magnitude by the presence of the arriving radiation, referred to as *shot noise*, and noise not so affected, referred to as *dark current*. The dark current is characteristic of the detector and its manufacturing history.

Electronic Noise

The noise generated by the electronics is amplified and sampled together with the signal for digital transmission but is independent of the magnitude of the arriving radiation. The minimum detectable current above the background for a typical electronic circuit is 1×10^{-13} C/s (*11*). This noise is generally insignificant in properly tuned systems.

Quantization Noise

The electrical signal must be quantized for digital processing. The *quantization noise* is a noise introduced by digital processing and is the basic limitation of digital systems in determining the true value of the signal, in the same fashion as random noise is the basic limitation of analog systems. The quantization error is inversely proportional to the square of the number of quantization levels, K. For $K = 8$, the root mean square magnitude of the quantization noise is 4.5×10^{-3} (*12*). This quantization noise can be reduced by using a larger number of quantization levels, that is, a larger word size.

Spatial Noise

Spectroscopic imaging typically deals with low-contrast signal variations on a large background. Both the low-contrast nature of the signal and the large magnitude of the background make spectroscopic imaging vulnerable to nonuniformities in the system. To overcome the image degradation caused by array nonuniformity, one uses computers to remove pixel-to-pixel variations. In ideal conditions, such techniques can remove array nonuniformity from spectral images. However, there is generally some residual nonuniformity in the images after correction, and we will refer to this residual nonuniformity as *spatial* noise (*13*). A random additive offset for each pixel is an example of a predominantly spatial noise source. Spatial noise arises when the noise is correlated over the period of the signal integration time.

Spatial noise is also known as *fixed pattern noise*. Spatial noise appears as a pattern that remains regardless of the image, and hence the name fixed pattern noise. *Temporal noise* appears as a salt-and-pepper pattern on the display, which changes with each image. Shot noise is an example of a predominantly temporal noise source. Temporal noise is uncorrelated, so the noise terms add in quadrature. Spatial noise is temporally correlated noise, so the noise terms add directly. The accumulated noise is the sum of the temporal and spatial noise. The spatial noise varies as a percentage of the mean signal, and the temporal noise varies as the square root of the mean signal, so the accumulated noise is determined by the signal integration time. The temporal noise is characterized by a power spectrum.

There are four known sources of spatial noise (*13*):

- nonuniform pixel nonlinearities,
- pixel $1/f$ noise,
- array $1/f$ noise, and
- spectral nonuniformities.

The pixel nonlinearities can be corrected for by proper calibration generating what could be termed a spatial instrument function. *Pixel* $1/f$ *noise* is the nonstationary noise associated with each pixel. When present, this noise causes each pixel to drift with respect to the other pixels on the array in a spatially uncorrelated fashion. *Array* $1/f$ *noise* is similar to pixel $1/f$ noise except that it arises from something external to all the pixels, which is unstable, and the effect of the instability is not the same for each pixel. The effect of array $1/f$ noise on images is to introduce a pattern in the image where the pattern itself changes little in time but the magnitude of the pattern is variable. Both types of $1/f$ noise are correctable for short time intervals.

Spectral nonuniformities are pixel-to-pixel variations in the spectral response of the detectors and can be corrected when the chromatic content of the calibration matches that of the image.

Single-element scanned systems are not subject to spatial noise if the drift occurs on a time scale greater than the signal integration time.

Sampling Anomalies

We are trying to characterize the sample and its spatial variations. However, the mere presence of the sample introduces optical effects that must be considered in interpreting the resulting images.

Sample Defocusing of the Beam

The sample itself defocuses the beam because of its inherent refractive index as well as the variations arising from differences in concentration. The focal shifts are demonstrated in Figure 5.8 (*14*), where the focal shifts associated with changes in refractive index are illustrated for the sample support, but it is intuitively obvious that a larger effect will arise from variations in refractive index of the sample.

In polymer systems, lens-shaped samples are often encountered, such

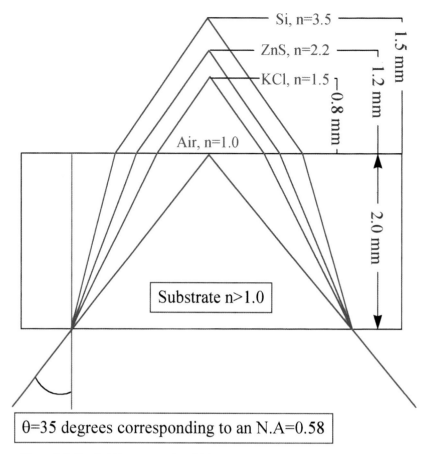

Figure 5.8. Focal shifts associated with changes in refractive index, illustrated for the sample support. (Reproduced with permission from reference 14. Copyright 1989 Society of Spectroscopy.)

as fibers, fabrics, and in fiber-reinforced composites. A fiber-shaped sample can act as a lens. When the energy impinges on the fiber, the beam is defocused in a fashion that depends on the portion of the fiber that the ray hits. If the ray is at the exact center of the fiber, no effect will be observed. However, at any other portions of the fiber the beam is defocused.

A further complication for a fiber-shaped sample is the range of thicknesses seen by the radiation. Some of the input radiation passes through the thin portions and some through the thicker portions, resulting in a substantial optical wedge effect (*15*). Hence, there is a distortion of the absorbance due to deviation from Beer's law. Flattening of the sample has been recommended as a method of reducing this effect (*16*).

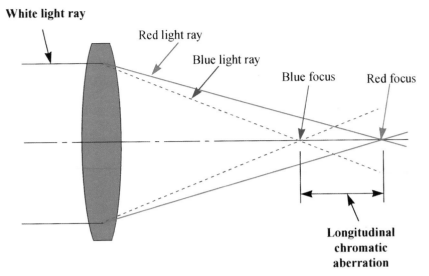

Figure 5.9. Illustration of chromatic aberration. The shorter wavelengths are bent more by a given lens than the longer wavelengths, resulting in a blurred focus. (Reproduced with permission from ref 3. Copyright 1987 ASTM.)

Sample Displacement of the Beam

If a sample has parallel faces and a refractive index greater than unity, the beam will be displaced. Quantitatively, the shift is equal to $(n - 1)b/n$, where n is the refractive index of the specimen and b is the path length in the sample (*14*). This effect is quite a problem in spectroscopic imaging, as the effect takes on the form of a *chromatic aberration*. That is, the magnitude of the shift depends on the wavelength (Figure 5.9).

Because the refractive index depends on the absorptivity for absorbing wavelengths, the amount of beam shift is a function of wavelength. This problem should be recognized, but few options exist for its correction.

Variations in Sample Thickness Relative to the Beam Dimensions

Quantitative absorbance measurements assume the validity of Beer's law, which assumes that the path length, b, through the sample over the impinging beam dimensions is constant. For microscopic samples, this may not be the case. The intensity is defined as the power per unit area, but the detector only measures the incident power after transmission through the sample. Substantial variations in the path length over the dimensions of the beam can lead to a nonlinear response of the measured absorbances because of changes in concentration (*15, 17*). This "wedgeness" or "path-length heterogeneity" effect is important for strongly absorbing substances such as polymers.

A similar report has been made concerning the error in attenuated total reflectance measurements introduced when the effective incident beam height is larger than the exposed adsorbate height. Errors as large as 75% have been calculated (*18*).

Imaging Acquisition Strategies

Brightness and Contrast

An initial decision must be made by the spectroscopist to assure that the acquired image has sufficient brightness and contrast to allow successful interpretation of the image. It is often difficult or impossible to decide which combination of intensity and contrast yields an image "truly" representative of the sample prior to an initial examination of the sample. Preliminary spectra will reveal the basic components of the sample, including the spectral nature of the components, thickness, presence of spectral interferences, and type and distribution of the background. The experience of the spectroscopist can be used to "guesstimate" the brightness and contrast levels required for interpretation and make a determination of the initial imaging conditions.

The approach must lead to the optimal S/N ratio per unit experimental time because we want to maximize the sensitivity to small chemical changes and have the maximum spectral dynamic range. The acquisition strategies can be tested by consideration of a *performance parameter*, P_P, defined by the ratio of the S/N ratio to the square root of the total measurement time, T_t:

$$P_p = (S/N)/\sqrt{T_t} \qquad (5.15)$$

It is desirable to obtain the highest possible performance parameter. There is also an obvious need for suppression of the background signal and maximum enhancement of the spectral signal.

For spectroscopic purposes, one can obtain a figure-of-merit for the technique on the basis of differences in concentration. If we assume that the two voxels have the same absorptivity coefficient but differ in concentration, then the figure-of-merit, Γ, is given by

$$\Gamma = \frac{\Delta S/N}{\Delta C/C} \qquad (5.16)$$

In other words, Γ is the relative concentration discrimination sensitivity for a given fractional change in concentration, ΔC, between the two voxels.

Determination of Pixel Resolution

One of the first requirements of spatially resolved spectroscopies is to ascertain the minimum pixel resolution required. The fundamental limitation to spatial resolution is determined by the available signal-to-noise ratio. The signal from each volume element decreases as the resolution is enhanced, leading to an imaging time dependent on the sixth power of the spatial resolution. The primary considerations for acquiring spectrally scanned maps are intensity and contrast at the highest possible spatial and frequency resolution. The strategy used must optimize these two factors in light of the experimental parameters available. Critical parameters for a determination of the spatial resolution are the speed of data acquisition and the instrument response (i.e., how fast the system responds to a change in the spectral intensity). The two major factors that affect the data acquisition rate are the scan rate and the number of scans required to assure spectra with a reasonable degree of fidelity.

The development of a voxel-sampling strategy is an important consideration in mapping. One must determine the required digital density of the spatial data recorded, so the number of voxels scanned should be carefully selected. The minimum size of the voxel is determined by the signal-to-noise ratio required to interpret the spectral data, and the maximum number of voxels measured is limited by the available experimental measurement time. The measurement time per voxel is a function of the magnitude of the S/N per scan. If a higher experimental S/N is required, necessitating signal averaging, the measurement time for mapping increases rapidly. In some cases, the S/N ratio can be enhanced by convoluting the mapping data with a smoothing filter.

Determination of Field of View

For microscopic investigations of a certain region of the object, the sensitive volume of the measurement must be clearly restricted and accurately positioned. When the signal is imaged onto the aperture by the condenser, the best or smallest image formed at the aperture is determined by the product of the diffraction-limited dimension and the condenser magnification. Contributions from adjoining spatial regions can lead to mixing of signals and spectral dilution. Extreme care must be taken to minimize these extraneous contributions. The spectrum arises from a well-defined region of the sample or field of view (FOV).

To obtain the spectrum from a sample without spectral interferences from outside sources, the distance between the individual components must be larger than the diffraction-limited resolution. Only 84% of the total energy is found in the first ring of the Airy disk. The remaining 16% of the energy may find its way into the image and contaminate the individual spectra. Sommer and Katon found that in the infrared, diffraction will spread light into an area larger than that of the sample for sample diameters of 20 μm or less (*19*).

The imaging aperture must be selected to accommodate the sample size. The minimum aperture size is the product of the sample size and objective magnification. For an infrared microscope, if an 8-μm-diameter sample is to be analyzed, the size of the aperture should be 15 times larger, or 120 μm. However, for 9.5-μm light and a numerical aperture of 0.58, the diameter of the diffraction-limited image formed at the sample plane is 20 μm. In this case, diffraction will spread light into an area larger than that of the sample for sample diameters less than 20 μm.

Use of Digital Computational Methods for Image Enhancement

The first steps in imaging are the acquisition and encoding of a digital image. Several means exist to accomplish these tasks. The usefulness of the data and the method of approach depend on what is known a priori about the image and the information that can be deduced from it. Before an image can be subjected to any digital computational operation, it must be sampled. The resolution of the sampling procedure is characterized by an instrument function of the apparatus with which the sampling is accomplished.

Important aspects of image data acquisition are

- sampling distance (i.e., sample spacing or pixel size),

- spectral band pass,

- level of any contaminative noise and its statistics,

- amplitude quantization range (i.e., the number of bits to which the data are resolved in amplitude),

- degree of uniformity with any geometric distortions,

- dynamic range,

- associated coefficients of linearity and saturation, and

- instrumental and environmental effects.

These factors must be taken into account in order to acquire and encode a proper image.

During the encoding process, it is possible to have *image contamination* by recording nonlinearities, recording noise, bit errors, missing information, and saturation. These types of image contamination have the character of being independent from pixel to pixel, and must be separated from the interesting detail in an image that is often correlated across several neighboring pixels.

A simple experimental check can be used to discover the aforementioned displacement errors. The experiment is repeated a second time with all spatial displacement errors reversed in the second sequence. If the spectra differ substantially, this clearly shows a displacement error. In this case, one obviously gains by spatial filtering.

Images are stored as quantized pixels in the computer memory, so image processing tends to rely on digital filters rather than on the classical analog filters. A *digital filter* is a discrete array of possibly complex numbers whose effect is to alter the spatial frequency content of an image during some processing operation. The filtering operation can be in either image space or Fourier space. Digital filters are direct or spectral if they are applied, respectively, in image space or Fourier space. Here, we will term the image to be filtered as the input image, and the filtered image as the output image. For a nonrecursive digital filter, each output pixel is a weighted sum of given pixels. For a recursive digital filter, each output pixel is a weighted sum of given pixels and previously calculated output pixels. We wish to form a refined image corresponding to what is termed the recoverable true image (*20–23*).

These are basic image encoding problems, and methods must be devised for overcoming or correcting for various image artifacts. For most imaging techniques, the artifacts are geometric distortion, ghost images, and optical field nonuniformity.

One of the golden rules of image reconstruction is to avoid processing data that exhibit any species of "glitches," that is, anything having the character of discontinuities.

Visualization of Mapping Results

Visualization of the images can take on many aspects, depending on the nature of the image and the detail of structural information expected. Volume definition via interactive thresholding or windowing allows one to reduce a complex image to a binary image quickly and easily. Interactive selection of individual volume clusters can be established by using edge location algorithms. Visualization of different topological classes can be accomplished by plotting maps of the characteristic spectral features.

Visual contrast of mixed components and their spatial concentrations can be enhanced by the assignment of different colors to the corresponding gray levels.

Once the experimental maps have been obtained, it is possible, with the computer, to display the data in several ways that aid in interpretation. Choosing an appropriate method for presenting the data is important.

All of the experimental data can be displayed in a single spectral window in which one axis is the frequency and the other axis the intensities, with each of the spectra offset on the third axis to give a three-dimensional display. The advantage of this mode of display is that all of the data can be viewed simultaneously The disadvantage is that no spatial location information is available.

To display experimental data so that it provides spatial information, each spectrum in the map is reduced to a single frequency intensity, and this intensity is plotted as a function of location. This reduction of a spectrum to a single frequency is called *profiling*.

Histograms are used in dealing with statistical data and distributions. A three-dimensional histogram shows a histogram for two variables.

Topographic or *contour plots* display lines of constant value as a function of the coordinate axis. Topographic plots are particularly useful for displaying intensity values.

A *carpet plot* is a derivative of a surface plot, which makes use of solid area fills in conjunction with three-dimensional curves.

Display of mapping data is subjective, and the choice of method depends on the spectroscopist's mind-set.

References

1. Kingslake, R. *Optical System Design;* Academic: New York, 1983.
2. Smith, W. J. *Modern Optical Engineering: The Design of Optical Systems;* McGraw-Hill: New York, 1966.
3. Messerschmidt, R. G. *The Design, Sample Handling, and Applications of Infrared Microscopes;* Roush, P. B., Ed.; American Society for Testing and Materials: Philadelphia, PA, 1987.
4. Geladi, P.; MacDougall, D.; Martins, H. *Appl. Spectrosc.* **1985,** *39,* 491.
5. Saunderson, J. L. *J. Opt. Soc. Am.* **1942,** *32,* 727.
6. Naqui, A.; Durst, F.; Liu, X. *Appl. Opt.* **1991,** *30,* 4949.
7. Jonaxz, M. *Am. Lab.* **1992,** *24.*
8. Black, S. S.; Robarge, W. P.; Boss, C. B. *Anal. Chem.* **1990,** *44,* 280.
9. Chase, B. In *Infrared Microspectroscopy;* Messerschmidt, R. G.; Harthcock, M. A., Eds.; Marcel Dekker: New York, 1988; p 97.
10. Mechels, S.; Young, M. *Appl. Opt.* **1991,** *30,* 2202.
11. Tripp, C. P.; Hair, M. L. *Appl. Spectrosc.* **1992,** *46,* 100.
12. Huck, F. O.; Park, S. K. *Appl. Opt.* **1975,** *14,* 2508.
13. Mooney, J. M. *Appl. Opt.* **1991,** *30,* 3324.

14. Katon, J. E.; Sommer, A. J.; Lang, L. *Appl. Spectrosc. Rev.* **1989**, *25*, 173.
15. Koenig, J. L. *Anal. Chem.* **1964**, *36*, 1045–1046.
16. Tungol, M. W.; Bartick, E. G.; Montaser, A. *Appl. Spectrosc.* **1993**, *47*, 1655.
17. Hirschfeld, T. *Anal. Chem.* **1979**, *51*, 495.
18. Free, M. L.; Miller, J. D. *Appl. Spectrosc.* **1994**, *48*, 891.
19. Sommer, A. E.; Katon, J. E. *Appl. Spectrosc.* **1991**, *45*, 1633.
20. Bates, R. H. T.; McDonnell, M. J. *Image Restoration and Reconstruction;* Claredon: Oxford, England, 1986.
21. Herman, G. T. *Image Reconstruction from Projections: The Fundamentals of Computerized Tomography;* Academic: New York, 1980.
22. Sekihara, K.; Matsui, S.; Kohno, H. *IEEE Transactions on Medical Imaging;* The Institute of Electrical and Electronics Engineers, Inc. New York, 1985; p 193.
23. Haacke, E. M.; Patrick, J. L.; Lenz, G. W.; Parrish, T. *Reviews of Magnetic Resonance in Medicine,* Pergamon, New York, 1986, Vol. 1, p. 123.

6
Optical Microscopy

Faith is a fine invention
When gentlemen can see,
But microscopes are prudent
In an emergency.

—Emily Dickinson (1860)

Optical Microscopy

Optical microscopy is a convenient microscopic technique for the study of materials (*1*). A number of optical configurations are available that together yield substantial information about the sample. It is relatively easy to produce specimens that are sufficiently thin and flat to be examined. Specimens for optical microscopy do not need to be conducting, and they do not have to be maintained in a vacuum. One can construct a modified stage so that controlled environmental conditions can be maintained. Also, visible light only rarely affects the chemical composition or molecular-weight distribution of the sample, so the technique is noninvasive and nondestructive, whereas some samples are beam-sensitive for some non-optical microscopic methods. For example, with high-resolution electron microscopy, organic systems are generally very sensitive to electron irradiation because the molecular chains become cross-linked or are decomposed (chain scission) in a short time upon electron irradiation (*2*). Because little or no sample processing is required after collecting the sample, one can be confident that the materials will reveal their phases, textures, and morphologies as generated by processing and fabrication (*3*). Thickness and local thickness variations are critical to most optical microscopic techniques. Most materials, particularly multiphase materials, have large variations in local thickness (*4*).

An optical microscope with an objective with numerical aperture

(NA) = 1.4 (the Abbe diffraction limit) can observe and study objects 200 nm wide with the 550 green light. The *optical microscope* relies on spatial variations at a single wavelength of the refractive index, absorption, and reflectivity in a specimen in order to produce the modulation in light intensity necessary to form the magnified image. The ability to identify the chemical nature of materials that comprise the specimen is very limited as *these optical properties are generally not material specific.*

Methods of Optical Microscopy

In *transmitted-light microscopy* (5), a collimated light beam passes directly through the sample and into the objective lens of the microscope. The magnified image is viewed through a microscope eyepiece. Morphological features that scatter light or give rise to optical density variations are visible. The sample must be thin enough that an appreciable fraction of the light is transmitted.

When specimens are opaque or excessively thick, *reflected-light microscopy* (6) may be used to examine variations in surface structure. Illumination is usually provided through the objective lens itself, but only secularly reflected light is allowed to reenter the objective. Contrast arises from variations in surface reflectivity.

Dark-field microscopy (7) may be carried out using either transmitted or reflected light. In contrast to bright-field illumination, directly transmitted or reflected light is prevented from entering the objective. In transmission, this is accomplished by placing a circular stop above the condenser to block the central beam. In reflection, special objective apertures are used to provide very oblique illumination of the sample. When viewed in dark field, objects that reflect or scatter light are visible against a dark background, and marked gains in contrast are often obtainable.

In *polarized-light microscopy* (8), the illuminating source is plane-polarized before it impinges on the sample, and a second polarizer is inserted in the reflected or transmitted beam. In most cases, the two polarizers are crossed at 90° to each other, and only those materials that cause a partial depolarization of the light are visible. The *retardation, R,* can be calculated from the analyzer angle, θ, thus:

$$R = (\theta/\pi)\lambda \qquad (6.1)$$

where θ is the analyzer angle at which the light intensity passing through the analyzer is the minimum. The initial direction ($\theta = 0$) of the analyzer is perpendicular to the polarizer. The birefringence can be calculated from the sample thickness, and the retardation value, *R,* as shown in eq. 6.2:

$$\Delta n = R/d \qquad (6.2)$$

where d is the sample thickness at the corresponding state. For polymers where an orientation function, f, is desired, it can be calculated as

$$f = \Delta n / \Delta n^\circ \qquad (6.3)$$

where Δn° is the birefringence of a perfectly oriented sample. Since Δn° is constant for a given sample, the orientation function is linearly proportional to birefringence. The *orientation function* calculated from the birefringence value represents the average orientation of all chains in the sample.

Phase-contrast microscopy (7) depends on the fact that differences the refractive indices of two transparent phases produce small phase shifts in the light exiting from each component. By the use of an annular ring condenser and a phase retarding absorption ring, these phase differences are converted to intensity variations in the observed image. The intensity variations are qualitatively proportional to optical path variations in the object.

In *interference microscopy* (9), monochromatic light from a single source is divided by the use of a beam splitter. One beam is reflected from or transmitted through the sample and then recombined with the reference beam. Contrast results from interference of the two beams. A pattern of bright and dark lines (fringes) results from an optical path difference between the reference beam and the sample beam. To convert a traditional interferometer into a scanning white light system, the interferometer is mounted inside a microscope objective. The microscope is fitted with a high-precision vertical scanning stage, and a charge-coupled device (CCD) camera is used to gather data. As the stage scans the sample vertically, a three-dimensional interferogram is collected and stored in a computer where it is mathematically transformed into a quantitative three-dimensional image. Scanning white light interferometry can be used to study problems in surface structure analysis.

Confocal Microscopy

Two-dimensional spatial images obtained by ordinary microscopy have a number of limitations. They offer images that distinguish between events over the x and y dimensions of the sample, but reveal no depth profile information. In confocal imaging systems (10), the illuminating pinholes imaged on the specimen and a moving mechanism scan across the specimen in a raster pattern. A scanning mirror sweeps the small spot while the computer collects x and y image data. The light emitted from the specimen is rescanned by the same mechanism and reimaged through the pinhole again. A piezoelectric crystal steps the sample in the z direction to accumulate a series of depth slices for three-dimensional reconstruction of the image.

Confocal microscopes (11–13) allow for improved axial resolution. Confocal microscopes offer a viewing capability with resolution that is better than the ordinary light microscopes. The system can collect image data for a series of optical slices by successive small change in focus such that a three-dimensional reconstruction of the sample is possible. It is also nondestructive and does not require a high-vacuum system. However, confocal designs are primarily limited to fluorescence and single-wavelength bright-field images and cannot currently provide polarizing, phase-contrast, or interference contrast images.

In a *confocal microscope,* excitation light is focused onto a small spot within the sample; light originating at the in-focus point is passed back through the same imaging optics and aperture [hence the term "confocal" for conjugate (interchangeable) foci] and imaged onto a detector. The small aperture improves the resolution and shortens the depth of focus by eliminating out-of-focus light. The small-pinhole aperture rejects light from other planes within the sample.

The problem of spherical aberration is of special importance in a confocal system. The main benefits of confocal microscopy are increased resolution and narrow depth of field. These factors allow the researcher to significantly increase apparent X,Y resolution, eliminate out-of-focus information to increase contrast, and create $X–Z$ scans. Spherical aberration limits these abilities as the depth into the specimen increases. A confocal reconstruction of an $X–Z$ slice of a perfectly round object will become deformed into an oval elongated object in the Z-axis.

Commercial confocal microscopes operate by reflectance or epifluorescence. A pair of scanning mirrors direct a diffuse laser beam in a raster pattern through an objective lens, which focuses it on the specimen. Reflected light passes back along the same light path and is "descanned" by the moving mirrors and then diverted by a beam splitter into a detector. On its return to the detector, light from the focal plane is focused through a pinhole, which blocks light from out-of-focus regions of the spectrum. The microscope stage is moved up and down by a stepping motor to collect images at different depths.

Confocal microscopes have a very high level of discrimination against light from outside the image plane and have shown themselves to be capable of providing high-quality images from significant depths below the surface of highly light-scattering materials. By coupling the confocal microscope with a spectrophotometric detection system, it is possible to construct wavelength-selective images of any thin slice through a semitransparent sample. By collecting and analyzing a series of such slices, it should be possible to reconstruct the pattern of absorption at any particular wavelength in three dimensions.

Confocal microscopes employ optical systems in which both the condenser and objective lenses are focused onto a single volume element of the specimen. Light from a point source is focused onto the image plane by the objective lens. Reflected light is gathered by the same lens and directed by a beam splitter onto a small-aperture confocal with the focal spot in the image plane. Light scattered or reflected from the planes above and below the image plane will not focus onto the aperture and will be selectively rejected. By scanning the incident beam or the sample, a two-dimensional image can be built up (*14*).

In essence, in confocal microscopy, the out-of-focus information is removed by real-time Fourier filtering, by means of a pinhole or slit or assembly of them. Because well over 90% of the intensity originates from above or below the plane of best focus, most of this light is rejected by the filter system. One consequence of this filtering is the low throughput of the system.

UV–Visible Spectroscopy

The first visible spectroscopy (*15,16*) can be traced back to M. A. de Dominis, who gave the first correct explanation of the rainbow. Sir Isaac Newton (1642–1727) discovered the phenomenon of the dispersion of white light as it passed through a prism into a visible spectrum that he identified as red, orange, yellow, green, blue, indigo, and violet (recall the classroom mnemonic Roy G. Biv).

Overview

For absorption spectra in the visible and UV region, the absorption consists of displacing an outer electron in the molecule. The spectrum is a function of the chromophores in the molecule rather than specific bonds. Electronic absorption is very strong with molar absorptivity values frequently up to 10,000, whereas in infrared they rarely exceed 1000.

The visible and UV absorption is a highly specific property of the molecular structure, and the frequency range within which energy can be absorbed is specifically dependent on the molecular structure of the absorbing material. The more mobile the electrons, the smaller the energy difference between the ground state and the excited electronic state and the lower the frequency of absorption (i.e., the longer the wavelength).

UV–visible spectrophotometry is possibly the most solidly established and widely used analytical technique (*17*). However, its limited selectivity and the lack of specific chromogenic moieties have frequently limited its application. However, improved instrumentation such as diode array spectrophotometers, and improvements in methodology such as new chemometric procedures for processing complex signals, will improve its utility for polymer systems

Fluorescence Imaging

When a molecule absorbs a photon of light, it is elevated to an unstable excited state and then it can release its excess energy by various pathways, one of which is *fluorescence* emission (*18, 19*). All fluorescent probes have chromophores that can be photochemically promoted to an excited state by irradiating with light. A fluorescent indicator molecule typically emits 10^4 to 10^5 photons before it is destroyed by photobleaching side reactions. Excitation wavelengths can be in the UV, blue, or green regions of the spectrum. The wavelength distribution of the outgoing photons from the fluorescent probes forms the emission spectrum, which peaks at longer wavelengths (lower energies) than the excitation spectrum and is characteristic of the particular fluorophore.

Several requirements must be met for the development of the optical arrangement for fluorescence detection. The optical system must have maximum collection efficiency over a wide wavelength range, and it must discriminate between background and analytical signals without significant sacrifice in the latter. There must be minimal losses of signal along the optical path due to reflection from different surfaces and apertures. Low optical aberrations are required for efficient spatial discrimination between the fluorescence and background.

The difference between the maxima of the absorption and emission spectra is called the *Stokes shift*. Because of this shift, optical filtration can be used to separate excitation light from the longer wavelength fluorescent emitted light. By selecting an appropriate fluorochrome, a sensitive and quantitative analysis can be made. Fluorescence is highly specific, extremely sensitive, and amenable to microscopic detection. By spectrally filtering the emission from the sample, the localization and concentration of the probe target can be mapped.

When fluorescent molecules are excited with plane-polarized light, they emit light in the same polarized plane, provided that the molecule remains stationary through the excited state. However, if the excited molecule rotates or tumbles during the excited state, then light is emitted in a plane different from the excitation plane. If the fluorescently labeled molecules are large (i.e., polymers), they move little during the excited state interval, and the emitted light remains highly polarized with respect to the excitation plane. If fluorescently labeled molecules are small, they rotate or tumble faster, and the resulting emitted light is depolarized relative to the excitation plane.

Fluorescence depolarization can be used to measure the mobility of fluorochromes. Freely rotating groups do not retain any memory of the orientation of the polarized excitation prior to emission. However, in the event of hindered rotation of the fluorochrome, emission of light will occur with some degree of correlation with the plane-polarized excitation. The extent of polarization, P, is given by

$$P = (I_{VV} - I_{VH})/(I_{VV} + I_{VH}) \qquad (6.4)$$

where I_{VV} is the emission intensity with vertically polarized excitation and vertically polarized emission and I_{VH} is the emission intensity with vertically polarized excitation and horizontally polarized emission. A label with no mobility will have a value of $P = 1$, and a totally mobile label will have a value of $P = 0$.

Molecules remain in the excited state for approximately 10^{-9} s before releasing their energy and returning to the ground state. The time delay between initial absorption and emission—the "fluorescence lifetime"—can be thought of as the amount of time the excited molecule has to interact with and be modified by other molecules in its immediate vicinity. These modifications can affect the wavelength, intensity, or polarization of the light that the molecule finally emits.

The advantage of fluorescence is that many lifetimes fall in the 1–20-ns range. This time scale coincides almost perfectly with the time scale of molecular interactions. "Time-resolved" fluorescence methods are used to resolve events that occur on this time scale. These measurements use a continuous beam of light as an excitation source, and the resulting fluorescence is observed on a time scale appropriate for the experiment, generally on the order of milliseconds to seconds. Confocal fluorescence microscopy has been used to observe in real time the dynamics and conformational changes in DNA molecules in solution labeled with a single fluorescent tag (*20*).

The selectivity and excellent detection limits of fluorescence spectroscopy have made it a very popular technique for the imaging. In fluorescence microscopy, the excitation is at one wavelength and the fluorescence is at a longer wavelength. It is generally assumed that the intensity is directly proportional to the concentration of the fluorophore, but error is introduced due to processes termed collisional quenching.

A standard *fluorescence microscope* is like a conventional microscope except its light source, usually a mercury or xenon arc lamp, produces UV, blue, and green light. This light is passed through a monochromator or interference filter to select the excitation wavelengths that induce fluorescence in the sample being examined. It is possible to detect a few thousand molecules of a strong fluorophore in a volume of 1 μm^3. The depth of field for a fluorescence microscope is only a few micrometers.

Fluorescence microscopy is limited by the physics and photochemistry that is associated with the production of fluorescent light. Due to the presence of oxygen in most specimens, fluorophores have a statistical lifetime, measured in the average number of times they can be excited before they become photolyzed from reactions with the singlet oxygen molecules.

Fluorescence imaging requires highly efficient excitation. For many important fluorochromes, this means that objective lenses must be quite transparent in the UV region. Many optical glasses are limited in this range, becoming virtually opaque below 360 nm.

For fluorescence microscopy, there must be no chromatic aberrations, either longitudinal or lateral, for rays of both wavelengths (absorption and emission) to be in the proper positions to pass through the entrance and exit pinholes of the scanning system.

A microflow fluorescence detector has been developed that lowers the limit of detection for a number of substances (*21*). This new system uses an optical isolation system that eliminates stray light interference. Two filters are used. One filter removes all laser light except the wavelength needed to make the material fluoresce; a second filter system eliminates that wavelength after fluorescence is excited. This way just the fluorescence is detected.

For example, the microflow detector can provide 10^{-11} M detection in 1-nL samples, equivalent to 10^{-20} moles detection. In comparison, standard absorption detectors can sense 10^{-5} M in 10-nL samples (10^{-13} moles), and commercial fluorescence detectors are adequate for 10^{-8} M in 500-nL samples (5×10^{-15} moles). With this device it is possible to analyze the chemical makeup of a single human red blood cell.

Scattered radiation (Rayleigh, Mie, or Raman) can be an interferent in measurements of sample fluorescence. The presence of scatter raises the detection limit of the fluorescence measurements.

Near-Field Scanning Optical Microscope

Diffraction limits the spatial resolution of conventional optical imaging instruments. In practice, this diffraction limit is approximately one-half the wavelength of light being used. Thus, for confocal laser imaging with green light ($\lambda = 500$ nm), resolution is limited to approximately 300 nm. *Near-field scanning optical microscopy* (*22, 23*) overcomes this boundary by scanning a subwavelength-sized light source very close to a sample and building up an optical image of the specimen pixel-by-pixel. The light source is an optical aperture (~25 nm in diameter) fabricated at the tapered apex of an aluminized optical fiber. Using force feedback, the tip of the probe maintains a constant separation from the sample (~5 nm). Thus, as the light emanates from the probe tip, it only illuminates a volume of the sample approximately equal to the aperture size. Any collected optical contrast, either transmitted through or reflected from the sample, originates from this small volume (*2*). Hence, resolution is limited by the size of the aperture and *not* by the wavelength of light. Serial-scan images with resolution of 13 nm using visible radiation have been reported (*24*). Recently, near-field scanning optical microscopy has demonstrated single-molecule sensitivity combined with nanometric spatial resolution (*25*).

Applications of Microscopy of Polymers

It would be foolhardy to attempt to even outline the numerous applications of optical microscopy and imaging to polymeric systems. Polymers of all fashion from crystals to films to fibers have been examined in one way or another using some or all of the different microscopic methods that have been described. The microscopic examination of fibers, for example, relies on visual comparison of characteristics such as color, diameter, cross-sectional shape, birefringence, refractive index, and fluorescence, and the generic class of a fiber can usually be made.

A number of books and general references have been written giving details of the various applications of light microscopy (*26–29*), and I will leave it to the interested party to examine these references. An excellent reference also exists for UV microscopy (*30*).

Video Microscopy

A major development in optical microscopy is the use of video-recording techniques to detect dynamic processes (*31*). It is simple to connect a video camera to the microscope and record processes having a time dependence. The scan rate is 30 scans per second. These video images of kinetic or thermally induced process can be recorded using a frame grabber and converted to digital information. The dynamic video images can be manipulated using ordinary image-analysis techniques and displayed as static images or video recordings as desired.

Video cameras detect the light-induced electrical charge built up on a photoreactive plate. The light passes through a positively charged, transparent window and falls on a photoreactive layer. The light changes the conductance, which is a function of the intensity of the light, and causes charge to be drawn from the charged window and stored in the photoreactive layer. An electron beam scans the layer and detects the stored charge, producing a voltage difference that produces a current that generates the image.

Video cameras can also be based on a CCD. One of the main advantages of CCD cameras is that there is no electron gun so they are lighter and cheaper.

Image detail is a function of an image's pixel population and gray scale resolution. An image's pixel population is the number of pixels that are used to form the image. The pixel population is given by the square of the number of pixels recorded (typically arranged as a square array with, e.g., 512 or 1024 in each dimension). The gray-scale resolution gives the number of possible shades of gray that can be recorded for each pixel in the image. A typical gray-scale resolution is 256 gray levels reflecting the 8-bit scale corresponding to 2^8. The product of an image's pixel population and gray-scale resolution reflects the maximum detail that can be recorded in the image.

Video cameras do not improve the wavelength-dependent optical resolution of 0.2 μm but yield an image consisting of 512 $\times$ 512 pixels. This means that the minimum field of view that can be usefully examined is 512 $\times$ 0.2 = 102.4 μm. A video magnification of about $\times 40$ can be achieved for the objective and projector lens combined.

References

1. Wilson, T.; Sheppard, C. *Theory and Practice of Scanning Optical Microscopy;* Academic: London, 1984.
2. Kopelman, R.; Tan, W. *Microscopic and Spectroscopic Imaging of the Chemical State;* Morris, M. D., Ed.; Practical Spectroscopy Series; Marcel Dekker: New York, 1993; Vol. 10, p 227.
3. Winchell, A. N.; Winchell, H. *The Microscopical Characters of Artificial Inorganic Solid Substances: Optical Properties of Artificial Minerals,* 3rd ed.; Academic: New York, 1964.
4. Hartshorne, N. H. *The Microscopy of Liquid Crystals;* Microscopy Publications: Chicago, IL, 1974.
5. Abramowitz, M. *Contrast Methods in Microscopy, Transmitted Light;* Olympus Corporation: Lake Success, New York, 1990; Vol. 2.
6. Abramowitz, M. *Reflected Light Microscopy;* Olympus Corporation: Lake Success, New York, 1990; Vol. 3.
7. Pluta, M. *Advanced Light Microscopy: Specialized Methods;* Elsevier: New York, 1989; Vol. 2.
8. McCrone, W. C.; McCrone, L. B.; Delly, J. G. *Polarized Light Microscopy;* McCrone Research Institute: Chicago, IL, 1984.
9. Hariharan, P. *Basics of Interferometry;* Academic: New York, 1991.
10. Wilson, T.; Sheppard, C. *Theory and Practice of Scanning Optical Microscopy;* Academic: New York, 1984.
11. *Confocal Microscopy;* Wilson, T., Ed.; Academic: New York, 1992.

12. *Handbook of Biological Confocal Microscopy;* Pawley, J. B., Ed.; Plenum: New York, 1990.
13. Cheng, P. C.; Lin, T. H.; Wu, W. L.; Wu, J. L. *Multidimensional Microscopy;* Springer-Verlag: New York, 1994.
14. Inoué, S. In *Handbook of Biological Confocal Microscopy;* Pawley, J., Ed.; Plenum: New York, 1990; p12.
15. Demchenko, A. P. *Ultraviolet Spectroscopy of Proteins;* Springer-Verlag: New York, 1986.
16. Perkampus, H. H. *UV–Vis Spectroscopy and Its Applications;* Springer-Verlag: New York, 1992.
17. Brace, R. O. "Bibliography of Ultraviolet Applications"; Bulletin 7102, Bechman Instruments:, 1964.
18. Wolfbeis, O. S. *Fluorescence Spectroscopy;* Springer-Verlag: New York, 1992.
19. Lakowicz, J. R. *Principles of Fluorescence Spectroscopy;* Plenum: New York, 1983.
20. Nie, S.; Chiu, D. T.; Zare, R. N. *Science (Washington, D.C.)* **1994,** *266,* 1018.
21. Yeung, E. *R&D (Cahners)* **1992,** *34,* 19.
22. Lewis, A.; Lieberman, K. *Anal. Chem.* **1991,** *63,* 625A.
23. Lewis, A.; Lieberman, K. *Nature (London)* **1991,** *354,* 214.
24. Harris, T. D.; Trautman, J. K.; Mackilin, J.; Grober, R. D.; Hess, H. *Abstracts of Papers,* PITTCON '95, New Orleans, LA; Pittsburgh Conference: Pittsburgh, PA, 1995; Paper 368.
25. Dunn, R. C.; Xie, X. S. *Abstracts of Papers,* PITTCON '95, New Orleans, LA; Pittsburgh Conference: Pittsburgh, PA, 1995; Paper 365.
26. Sawayer, L. C.; Grubb, D. T. *Polymer Microscopy;* Chapman & Hall: London, 1987.
27. Hemsley, D. A. *Applied Polymer Light Microscopy;* Elsevier Applied Science: New York, 1989.
28. Hemsley, D. A. *The Light Microscopy of Synthetic Polymers;* Oxford University: Oxford, England, 1984.
29. Skirius, S. A. *Microscope* **1986,** *34,* 28.
30. Billingham, N. C.; Calvert, P. D. *Dev. Polym. Charact.* **1982**.
31. Inoué, S. *Video Microscopy;* Plenum: New York, 1986.

7

Infrared Microspectroscopic Imaging

Introduction

In visible microscopy and imaging, the spectroscopic response of the sample arises from differences in the refractive index resulting from variations in materials, textures, and orientation. Visible microscopy can be made chemically specific by using staining techniques that tag different materials. In the infrared spectral region, the specific vibrational absorptions of the sample can be considered natural stains that allow chemical identification. The fingerprint nature of the absorbances in the vibrational spectrum provides excellent chemical selectivity and high image contrast. Like visible microscopy, the infrared imaging technique is nondestructive and applicable to a wide range of materials. With the advent of reflection microscope probes, little if any sample preparation is required. Infrared microscopy has developed rapidly and is one of the valuable and versatile tools in the analytical laboratory (*1–7*). With the use of computer-controlled stages, infrared imaging has also developed into a useful technique in the infrared region (*8*).

Elementary Theory of Infrared Spectroscopy

An *infrared spectrum* results from the interactions of the vibrational motions of a material with electromagnetic radiation. These vibrational interactions can be described by a simple harmonic oscillator model. The elementary theory given in this section is supported by material in refs 9–12.

After separating the electronic contributions, each molecule has an internal vibrational energy, U, which can be expressed in terms of the coordinates and interbond forces between the atoms constituting the molecule. A nonlinear molecule consisting of N atoms has $3N - 6$ degrees of freedom. Thus, a set of $3N - 6$ generalized coordinates, G_i, can be found that completely describes the internal motions of this nonlinear molecule. The internal energy of the molecule can be written as

$$U = U(G_1, \ldots, G_b) \tag{7.1}$$

where $b = 3N - 6$.

For small oscillations of the atoms around the equilibrium position, U can be expressed as a Taylor series expansion:

$$U = U_0 + 1/2\Sigma (\partial^2 U/\partial G_i' \partial G_j') G_i' G_j' + \cdots \tag{7.2}$$

For all i and j, and I from 1 to b,

$$G_I' = G_i - G_{i0} \tag{7.3}$$

where G_{i0} are the generalized internal coordinates of the atoms in the equilibrium state. The first term, U_0, is the internal potential energy of the molecule in the equilibrium state. Because the molecule is near equilibrium, the first-derivative term of the Taylor series is zero. The second-derivative term, $\partial^2 U/\partial G_i' \partial G_j'$, can be interpreted as a matrix element of a matrix, $\mathbf{II}$. Matrix $\mathbf{II}$ is symmetric because

$$\partial^2 U/\partial G_i' \partial G_j' = \partial^2 U/\partial G_j' \partial G_i' \tag{7.4}$$

This symmetric matrix can be diagonalized with the use of a unitary transformation, $\mathbf{V}$. This introduces a new set of coordinates:

$$X_a = \Sigma_b V_{ab} G_a' \tag{7.5}$$

such that

$$\mathbf{V' II V} = \Omega \tag{7.6}$$

where X_a are the new $3N - 6$ generalized coordinates, and Ω is the diagonal matrix. We can write

$$U = U_0 + (\tfrac{1}{2})\Sigma X_a^2 \Omega_{aa} + \cdots \tag{7.7}$$

Thus, we have a system of weakly coupled harmonic oscillators. The second term represents the contribution to the potential energy from the fundamental collective vibrational bands, while the cubic and higher terms are responsible for combination, difference, and overtone bands.

The decomposition of coupled harmonic oscillators into a collection of independent oscillators is known as a normal mode expansion, and the independent oscillators are called *normal modes*. Normal modes are defined as modes of vibration where the respective atomic motions of the atoms are in "harmony"; that is, they all reach their maximum and minimum displacements at the same time. These normal modes can be expressed in terms of bond stretches and angle deformation (termed internal coordinates) and can be calculated by using a procedure called normal coordinate analysis (*13, 14*).

Selection Rules

A normal vibrational mode in a molecule may give rise to resonant absorption (or emission) of electromagnetic radiation only when the transition is induced by the interaction of the electric vector, **E,** of the incident beam with the electric dipole moment, μ_i, of the molecule. That is, the dynamic dipole moment of the ith normal mode, $\delta\mu_i/\delta x_i$ or μ_i, is nonzero. The intensity of the transition is proportional to the square of the transition dipole moment, that is, the matrix element of the electric dipole moment operator between the two quantized vibrational levels involved. For room temperature and below, it is assumed that only the lowest vibrational energy levels are occupied if the wavenumber of the normal mode is >500 cm^{-1}. For example, the first excited state of a mode with a vibrational wavenumber of 2000 cm^{-1} is populated with a probability of only ca. 10^{-5} at room temperature, and that of a mode at 200 cm^{-1} still only has a probability of 0.3.

A typical IR absorption spectrum of a molecule consists of a series of sharp lines, each corresponding to a transition between the vibrational ground state of the molecule ($\epsilon = 0$ for all *I*, where v is the vibrational quantum number) and the first ($v = 1$) excited state of the ith mode.

The number of optically active normal modes can be predicted from symmetry considerations alone. All molecules can be classified into groups according to nature of the symmetry elements (mirror planes, rotation axes, etc.). Each normal mode belongs to a particular symmetry species. Each transition moment integral has a clearly defined behavior with respect to products of these symmetry species. Consequently, the vanishing or nonvanishing of the integrals is the same for all transitions between states of two particular symmetry classes. Nonvanishing integrals occur when the excited state belongs to the same symmetry species as one component of the dipole moment. Thus, the optical selection rules of each class of vibrations are predetermined by its symmetry species, and the infrared activity of the normal modes can be determined solely from a knowledge of the symmetry of the molecule in question. The symmetry species of the normal mode also determines the dichroic behavior of the infrared absorption bands for oriented molecules (*14*).

Consequently, on the basis of a structural model of a molecule, the number and polarizations of the normal modes can be predicted. A comparison of observed spectra with predicted spectra can be used to test the validity of the structural model. The use of selection rules in the interpretation of vibrational spectra has historically proven to be an extremely valuable approach for distinguishing between different structural models of molecules. The structural information obtained from selection rules obtained relates to the symmetry of the molecule, and it thus complements the information on bond types obtained from the frequency data in the spectra (*13*).

Spectral Parameters

A spectrum is completely described when the following spectral parameters are specified:

- frequencies,
- number of constituent bands,
- band shapes,
- bandwidths, and
- amplitudes or intensities.

The frequencies measured are a function of the vibrational energy levels of the molecule and are primarily determined by the masses of the atoms and the intramolecular bonding. The number of constituent bands is determined by the number of atoms (N) in the molecule ($3N - 6$, where the three translations and rotations of the molecules contribute zero frequencies) less the number lost because of symmetry considerations. The shapes of the bands are determined by the instrument function and the relaxation mechanism of the molecule. Theoretically, band shapes are expected to be Lorentzian, but for polymers a variety of band shapes have been experimentally observed. The widths of the bands depend on the environment of the molecules. The amplitude is a function of the magnitude of the change in the bond dipole moment and is larger for greater changes. The intensities are determined by the concentration and path length.

An IR spectrum is ordinarily recorded in wavenumbers, $v(\text{cm}^{-1})$, which is the number of waves per centimeter. The relationship between ν and the wavelength, λ, is given by

$$\nu(\mathrm{cm}^{-1}) = 10^4/\lambda \ (\mu\mathrm{m}) \tag{7.8}$$

which can also be written as

$$\nu(\mathrm{cm}^{-1}) = 3 \times 10^{10} \ \mathrm{Hz} \tag{7.9}$$

The wavenumber scale is directly proportional to the energy and vibrational frequency of the absorbing unit. In wavenumbers, the relationship is given by

$$\Delta E_{\mathrm{vib}} = hc\nu(\mathrm{cm}^{-1}) \tag{7.10}$$

where ΔE_{vib} is the vibrational energy level separation, h is Planck's constant (6.62×10^{-27} erg s), and c is the speed of light (3×10^{10} cm/s). The fundamental IR region arbitrarily extends from 4000 cm^{-1} to approximately 300 cm^{-1}. The far-IR region extends from 300 to 10 cm^{-1}, but the low-IR source energy available makes this region generally inaccessible except with special instrumentation. These definitions of the different infrared regions, which were originally based on instrumental requirements of dispersion and detectors, are less important now with Fourier transform IR (FTIR) spectroscopy, but the terms are still in common use.

Infrared spectroscopy is useful for the recognition of the functional groups present in a sample as well as the identification of the molecular species (15–18). On this basis, IR spectroscopy has a large number of applications.

The vibrational energy levels can be calculated from first principles by using normal coordinate analysis, and as a result, some of the factors influencing spectra have been discovered. These factors include bond stiffness and atomic mass, as well as the geometry and interaction between neighboring chemical bonds (*11*). Shifts in frequency observed for variations in chemical structure of the functional groups can generally be rationalized in terms of changes in geometry or bonding.

Quantitative Absorbance Spectroscopy

For quantitative absorbance measurements, it is assumed that there is a linear concentration–response function (Beer's law), and this requires that

- the stray light is zero,
- the absorbers are assumed to be nonluminous (i.e., no thermal emission or fluorescence),
- the absorption width is assumed to be independent of the concentration of absorbers (no change in intermolecular interactions with concentration),
- the absorption coefficient is independent of the power of the incident radiation (no saturation),
- the spatial distribution of the absorbers across the light beam is uniform, and
- the spectral response of the detector is linear with light intensity.

These assumptions are seldom, if ever, entirely met by the experiment, and efforts must be made to either minimize the deviations from these assumptions or to design the experiment or the data processing to correct for nonadherence to the assumptions. Sources of nonlinear response can be attributed to

- physical sources, such as nonuniformity of the sample, particularly variations in thickness,
- rough surfaces and voids causing scattering and a nonlinear background,
- instrumental sources, including improper illumination, nonlinear response of amplifiers, etc., and
- chemical sources, including complex formation and interactions between species.

Unfortunately, infrared microscopic experiments are susceptible to all of these problems, and when quantitative results are required, extreme care must be taken in the design and execution of the mapping experiment.

Optical Considerations in the Infrared

The measurement of intensities at various radiant energies is basic to spectroscopy. When the radiation intensity transmitted through a sample is measured, absorption spectra are obtained. However, when the intensity of radiation reflected by the sample is measured, reflectance spectra are obtained. The intensity of reflected or transmitted radiation depends on the energy of the incident radiation and the surface's composition and morphology. When electromagnetic radiation is incident on a surface, four fundamental processes occur. The path taken by radiation as it meets a sample for transmission, reflection–absorption, diffuse reflection, and specular reflection is shown in Figure 7.1.

Consider a reflectance measurement on an inhomogeneous film. The reflected light consists of several components: front-surface reflection, back-surface reflection, and diffuse reflection. The measured reflectance, R_m, may be written

$$R_m = R_s + R_d + R_b \tag{7.11}$$

where R_s is the specular reflected component from the front surface, R_d is the diffuse reflected component, and R_b is the specular reflected component from the back surface.

If all of these phenomena are superimposed, then observed spectra are highly distorted. As a consequence, in any reflectance measurement a sampling method that is optimized for one of these components must be employed.

Infrared Sampling Techniques

Sampling for infrared microscopy of polymers is more demanding than for the visible light microscope because of the intrinsically strong infrared absorbance of polymers. The sampling technique selected depends on the nature of the specimen and the effort required to prepare the sample. The transmission technique is preferred for characterizing a material on the basis of the Lambert–Beer law. For IR, sample preparation for transmission requires extremely thin and uniform samples and has been a labor-intensive operation requiring some measure of skill and experience. Conventional dispersive IR spectroscopy of polymers is mainly based on transmission measurements, except for samples for which the preparation of a thin layer is problematic, inadequate, or prohibited. In these cases, external and internal reflection is used. In normal practice, except for microscopy, reflection is a more versatile technique because there is only one interface and the sample thickness is not critical. In addition, recent instrumental developments have made reflection techniques accessible for infrared microscopic investigations.

With the advent of FTIR and the higher energy throughput of these instruments, a number of additional sampling techniques have become useful, including diffuse reflectance, emission, and photoacoustic techniques. For microscopic measurements, several approaches to sampling

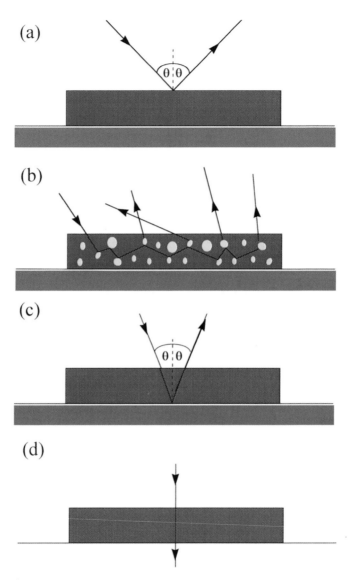

Figure 7.1. Path taken by radiation as it meets a sample: (a) specular reflection; (b) diffuse reflection; (c) reflection absorption; and (d) transmission. (Reproduced with permission from ref 22. Copyright 1992 Elsevier.)

have been reported (*19–21*). A particularly complete description of 11 different sampling techniques for IR microscopy focusing on paint samples, but useful for most polymer systems, has been published (*22*).

Ideally, the spectrum should be obtained on a sample unmodified by the sampling technique. However, this is not always practical, so the sample preparation procedure should generate little sample modification.

Transmission Sampling

Principles of the Method

Transmission spectroscopy involves the transmission of light through a uniform, thin polymer film. The transmission, T, is given by

$$T = I_t/I_i \qquad (7.12)$$

where I_i is the incident intensity and I_t is the transmitted intensity. The percent transmission is given by

$$\%T = (I_t/I_i)100 \qquad (7.13)$$

The absorbance, A, is defined by

$$A = \log(1/T) \qquad (7.14)$$

Assuming that the material obeys Lambert's Law, it follows that the absorbance will be proportional to the optical path length through the sample.

With transmission, the spectrum often contains large interference fringes arising from the multiple reflections of the light from the front and back surfaces of the sample. These interference fringes can create some difficulty in detecting absorbance bands. On the other hand, the thickness of the sample can be determined from the period of the fringe, λ, by

$$\lambda = 1/(2nD \cos \theta) \qquad (7.15)$$

where n is the refractive index of the film, D is the thickness, and θ is the angle of incidence of the infrared beam with the surface of the film.

Transmission Microsampling

Transmission microsampling is applicable to a wide variety of physically small samples, and sensitivity as low as 1 ng has been reported (*23*). Typically, to obtain the microspectra of samples of limited amounts of material, one reduces the sample to a small area, and reduces the beam dimensions with a beam condenser (8 ×), to maximize the energy passing through the sample, with apertures on the order of 10–500 μm. Small apertures can be obtained by using "pinholes" to mask off the light transmission through the sample.

For larger samples, microsamples can usually be obtained by slicing with a razor blade or knife, or by microtoming. If the sample is still too thick it can be placed in a hydraulic or hand press, a KBr pellet die, or a diamond anvil and pressed to a proper thickness for microtransmission measurements. Microdisks of Kbr (0.5–1.5 mm diameter) can be used to hold the samples.

Microdiamond optical cells can be used for compressing the sample and as sample holders for small samples, but they introduce polarization effects and interference bands due to the diamond (24). The low optical throughput of these microsampling devices also decreases the signal-to-noise ratio.

The sample must be optically thin for transmission measurements. Many small samples are not naturally optically thin. These samples may be compressed or microtomed, both of which may distort the resulting spectrum.

Microtomed cross sections can be obtained by hand-sectioning. Microtoming of laminates, thin films, and fibers is best performed on specimens whose thickness is less than 10 μm. Usually an area thin enough for transmission spectra can be found. These samples may be termed free-standing because there is no embedding or immersion medium (19). The free-standing samples may be attached to the microscope stage with tape or on adhesive paper holders.

For industrial fibers, the diameters of the fibers are often too large for good transmission measurements. However, improved spectra can be obtained by flattening the fibers with a roller-knife (25).

For some polymer samples, particularly biological specimens, it is an advantageous to look at structural features of solid samples immersed in a liquid medium (19). Samples can be embedded in paraffin or epoxy. Spectral subtraction can be used to minimize the spectral contributions of the embedding material. For protein studies of specimens like hair, D_2O is suitable, as the secondary structure of the protein can be determined.

Specular Reflectance

The perfect sample for specular or external reflectance spectroscopy (ERS) is one that is

- optically flat,
- optically thick, and
- homogeneous.

However, real samples never meet this ideal. The specular reflection spectra are superimposed on the absorption spectra in many real polymer samples such that the specular component appears in the resultant absorption spectra as anomalous dispersion (i.e., first-order derivative spectra). So the nature of the sample has to be carefully considered when specular reflectance is being considered. ERS requires a flat reflective surface, and the results are sensitive to the polarization of the incident beam as well as the angle of incidence. In addition, the orientation of the electric dipoles in the films is important to the selection rules and the intensities of the reflected beam.

Infrared reflection techniques are used for surface characterization because they provide highly detailed information about molecular structure, orientation on the surface, and intermolecular interactions; they are also nondestructive and relatively easy to use.

The External Reflection Method

Specular reflectance is the Fresnel reflection from the exterior surface of a material, without the beam penetrating the sample. Specular reflection obeys the law of reflection: the angle of reflection equals the angle of incidence. The spectra obtained from this type of interaction are markedly different from transmission spectra. The bands are derivative in shape, and peaks appear shifted. In order to compare specular reflectance data with absorbance spectra, the Kramers–Kronig transformation must be applied to convert the specular reflection data to absorbance-like spectra.

There are two possible polarizations of the incident beam, *s* and *p* polarization. The *p*-polarized light occurs when the electric vector of the incident light is parallel to the plane of incidence defined by the incident and reflected infrared beams off the metal surface. At a high (grazing) angle of incidence, the intensity of a reflected *p*-polarized IR light beam is enhanced at a metal surface, so that even submonolayer quantities of chemisorbed species can be observed.

Light polarized with its electric vector parallel to the surface (*s*-polarized) results in a low-intensity electric field at the film–metal interface at all incident angles, because the electric-field dipole and its image dipole are antiparallel and of equal magnitude, that is, they sum to zero field strength. This situation prevents absorption of *s*-polarized radiation by a surface-contained vibration.

In contrast, *p*-polarized light has a nonzero electric-field vector component perpendicular to the surface. This component is maximized at Brewster's angle, ϕ, which is given by (30)

$$\sin \phi \tan \phi = 2^{1/2}/(1.05 \times 10^{-6})(\nu\varepsilon\rho)^{1/2} \qquad (7.16)$$

where ν is the frequency of the light, ε is the dielectric constant of the ambient phase, and ρ is the resistivity of the metal. For high-conductivity metals, such as aluminum, silver, gold, copper, and platinum, the Brewster's angles are $>89.5°$. This means that optimum spectral sensitivity is achieved at large angles of incidence with p-polarized light (31). At present, angles less than 85° are most often used.

The surface-induced infrared selection rule states that only vibrating dipoles with a nonzero component perpendicular to the substrate surface will be excited by p-polarized infrared radiation. This provides a means of determining the average orientation of surface-confined molecules (32).

The Fresnel equation for a normal incidence reflectance measurement reduces to the simple expression

$$R = (n - 1)^2/(n + 1)^2 \qquad (7.17)$$

where n is the refractive index of the material. For absorbing samples such as polymers, the refractive index is complex, n^*, and is given by

$$n^* = n - ik \qquad (7.18)$$

where $i = \sqrt{-1}$ imaginary constant which takes into account refraction through n and absorption through k; k is the absorption index and is related to the absorption coefficient α by

$$\alpha = 2\pi k\nu \qquad (7.19)$$

where ν is the frequency in wavenumbers.

A further effect to be considered is that light undergoes a phase change upon reflection at an interface; the reflection coefficient, r^*, is related to the reflectance amplitude, r, and phase angle, θ, by

$$r^* = |r| \exp(-i\theta) \qquad (7.20)$$

and the measured reflectance, R is given by

$$R = |r^*|^2 \qquad (7.21)$$

Combining these equations, we obtain, for n and k at wavenumber ν

$$n = (1 - R)/(1 - 2R^{1/2} \cos \theta + R) \qquad (7.22)$$

$$k = 2R^{1/2} \sin \theta / (1 - 2R^{1/2} \cos \theta + R) \qquad (7.23)$$

Thus, if θ were determined at all frequencies, then n and k would be fully determined at each of the frequencies measured. This is possible by invoking the Kramers–Kronig relation, which enables θ to be determined from the measured reflectance, R:

$$\theta = (P/\theta) \int [\ln R/(v - v')] \, dv \qquad (7.24)$$

where P is the principal part of the integral, which means that the singularity at $v = v'$ is calculated as a Cauchy principal value. The optical constants can be determined for thin polymer films in this fashion (*26*).

Finished polymer products are routinely studied in a nondestructive fashion by specular reflectance. However, polymers show intrinsically rough surfaces, and therefore the specular reflection is superimposed with diffusely reflected light.

Infrared external reflection spectroscopy is used extensively in the study of thin films on metallic substrates. However, the external reflection spectrum of a thin free-standing film is usually severely distorted, so that interpretation of the spectrum is very difficult. But sometimes, external reflection spectroscopy is the only nondestructive method for measuring the spectra of thick, hard, and optically dark materials such as cured thermosetting resins and their composites, for which neither transmission nor internal reflection attenuated total reflectance techniques are suitable (*29*).

The Microspecular Reflection Method

FTIR microscopes provide reflectance measurements at near-normal angles of incidence. In the reflection mode, samples can be placed on the reference gold mirror and examined directly.

Microscopic reflection studies have been made of organic-rich shales (*27*), and microreflectance measurements allow the characterization of coal at the maceral scale and can be used to analyze coals of varying ranks (*28*).

Reflection–Absorption

Reflection–absorption (RA) occurs when a thin absorbing layer of a material is on the surface of a more reflecting substrate. The ideal case is a thin absorbing layer on a polished metallic substrate. The incident radiation passing through the absorbing layer is reflected at the metal's surface, and it then passes through the absorbing layer a second time before emerging as reflected radiation. Reflection–absorption spectra, often referred to as double-pass transmission spectra or *transflection,* are equivalent to absorption spectra. Radiation is transmitted by the sample until it is specularly reflected back through the sample. When normalized to account for path-length differences, transmission and RA spectra are the same.

The Reflection–Absorption Method

When the angle of incident radiation is greater than 60°, RA spectra are referred to as grazing angle reflection spectra. Grazing angle RA measurements are of interest for the analysis of very thin films.

We want to relate the RA of a thin film on a reflective substrate to its reflection properties. The reflectivity is given by

$$RA = -\log_{10}(R/R_0) \tag{7.25}$$

where R is the measured reflectance of the polymer-coated metal in air at a given frequency and R_0 the measured reflectance of the reflecting substrate. This relationship is analogous to a measured absorbance. Sometimes the relationship used is

$$RA = (R_0 - R)/R_0 \tag{7.26}$$

Oblique incidence has an advantage on rough sample surfaces, where a large component of diffuse scattering occurs from the surface. Illuminating the sample at near-grazing angles minimizes diffuse scattering from the rough surface, as this scattering is directed primarily in an upward direction. Meanwhile, scattering from the sample itself is directed primarily in the forward-scattering direction, which is the location of the detector.

Unfortunately for RA measurements, changes in reflection are not only associated with changes in k, but also with changes in n, resulting in loss of accuracy as the absorptivity decreases.

It has been pointed out that by using a polarizer, the p-polarized spectrum is enhanced by a factor of 2 (*33*).

Microreflection Absorbance Measurements

FTIR microscopes can be operated in the reflectance mode, and a grazing angle attachment is available (Figure 7.2) (*34*). This objective restricts the incident IR light to angles near grazing to the surface. The polar angles of the aperture are nominally 65–85°. The design of the microscope favors collection of the specular component relative to the diffuse component. The Spectra-Tech attachment has a numerical aperture (NA) =

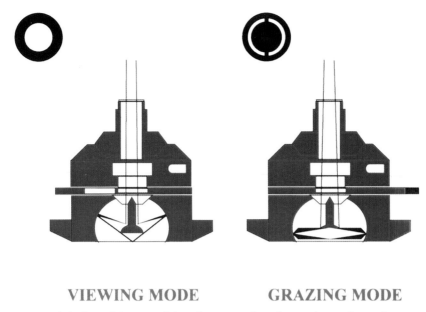

VIEWING MODE   GRAZING MODE

Figure 7.2. Optical diagram of the reflectance mode and a grazing angle attachment.
(Reproduced with permission from ref 34. Copyright 1991 International Society for
Optical Engineering.)

0.996, with a magnification of 30 at a working distance of 1 mm. The
grazing angle objective has a modified, overcenter, Schwarzschild reflecting lens design. Individual lens aperture masks are used to provide a near-
normal viewing mode and a grazing incidence spectroscopy mode. At
grazing incidence, the objective's throughput is 15–20%. Single-beam
spectra are ratioed to a background spectrum obtained from a gold-coated
mirror. The reflectance spectra obtained can be converted to absorbance
mode using the Kramers–Kronig transformation.

A recent study established that this grazing angle aperture shows no
angular dependence of incident intensity, and integrated intensity measurements can be quantitative, as verified by thickness measurement of a
dip-coated polyether on gold (*35*).

Mapping of chemically treated Al surfaces with and without adsorbed
epoxy deposits s has been done using the grazing angle method (*36, 37*).

Internal Reflection

Internal reflection was originally developed for IR examination of opaque
samples and has been widely used for identification of bulk polymer samples.

The Internal Reflection Method

Internal reflection spectroscopy (IRS), often referred to as attenuated total reflectance (ATR) is a well-established technique for obtaining absorbance spectra of opaque samples (*38, 39*). The mode of interaction is unique because the probing radiation is propagated in a high index-of-refraction internal reflection element (IRE). The radiation interacts with the sample of interest, which is in optical contact with the IRE, forming an interface across which a nonpropagating evanescent field penetrates the surface of the material of interest to a depth on the order of the wavelength of the radiation. The electric field at the interface penetrates the rarer medium in the form of an evanescent field whose amplitude decays exponentially with distance into the rarer medium (Figure 7.3) (*40*). When an IR absorbing material is in contact with the ATR prism, this evanescent wave interacts with the material, causing the attenuation of the propagating IR beam inside the ATR prism. Thus, an infrared spectrum can be obtained by detecting the absorbed radiation at the exit of the prism.

The major limitation of IRS is that its sensitivity is typically 3–4 orders of magnitude less than that of conventional transmission measurements performed using a cell with a 1-cm path length. This lower sensitivity is due to the low penetration depth of the evanescent wave into the absorbing medium, typically around 10 μm in the infrared.

An additional problem for ATR is that of reproducible optical contact. For soft samples and liquids, optical contact is not a problem, but for hard solids some difficulties can be encountered. For solids, pressure must be applied to obtain good contact between the sample and the crystal.

When an evanescent ray is totally internally reflected at the interface between two materials of different refractive index, the intensity of the

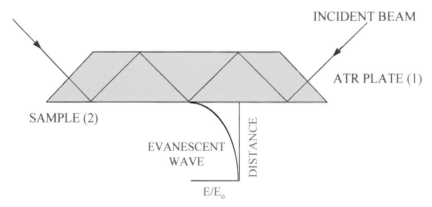

Figure 7.3. Optical path for an ATR experiment showing an evanescent wave. (Reproduced from ref 40. Copyright 1987 American Chemical Society.)

evanescent field extending into the medium of lower index decays exponentially with distance from the boundary, as shown by the relationship

$$I_{ev} = I_0 \exp(-\mathring{A}/d_p) \tag{7.27}$$

where $\mathring{A}$ is the distance normal to the optical interface, I_0 the intensity at $\mathring{A} = 0$, and d_p is the penetration depth, given by

$$d_p = (\lambda/4\pi n_2)[\sin^2\theta - (n_1/n_2)^2]^{-1/2} \tag{7.28}$$

where n_1 and n_2 are the refractive indices of the two media, respectively, and θ is the angle of internal reflection. The value of d_p is on the order of 0.5 to 5 μm.

Of more practical importance is the effective penetration depth, d_e, given by

$$d_e = (E_0^2 d_p n_2/n_1)/(2 \cos \theta) \tag{7.29}$$

The depth d_e is on the order of 0.3 to 10 μm.

For a two-layer system consisting of a thin surface layer on an infinitely thick substrate, the absorbance of a band in the ATR spectrum is related to the thickness of the surface layer. The thickness of the surface layer, t, can be calculated from the ATR spectrum from the absorbance of a band of the substrate, $A_s(t)$, according to

$$t = -(d_p/2)\ln A_s(t)/A_s(0) \tag{7.30}$$

where $A_s(0)$ is the absorbance of the substrate.

An alternate expression is

$$t = -(d_p/2)\ln\{[1 - A_1(t)]/A_1(\infty)\} \tag{7.31}$$

where $A_1(\infty)$ is the absorbance of the surface layer of infinite thickness (*41*).

To enhance the sensitivity of ATR spectra, one can increase the interaction of the evanescent wave with the absorbing medium by increasing the number of reflections per unit distance along the internal reflection element (IRE/sample) interface. A commercially available cell, called the Circle ATR cell (*42*), has a surface area that is approximately 7.5 times greater than that of a rectangular ATR crystal and has been used to obtain enhanced signal-to-noise ratio spectra of fibers and films (*43*). As shown in Figure 7.4, the fiber is wrapped around the ATR cell to enhance the path length (*44*).

In ATR, the spectral intensity varies as a function of the quality of

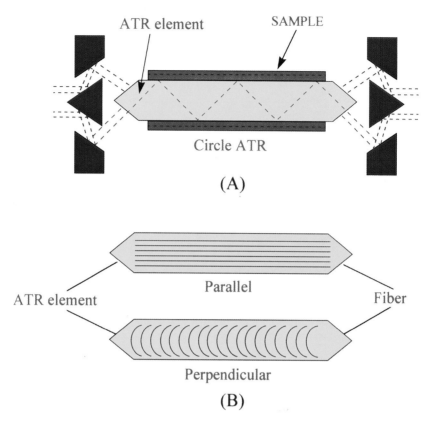

Figure 7.4. (A) Circular ATR configuration setup; and (B) parallel and perpendicular alignment of fibers with respect to a crystal axis. (Reproduced with permission from ref 44. Copyright 1993 John Wiley & Sons.)

optical contact. A quantitative measurement of the quality of the optical contact can be obtained for surfaces of thick or bulk films at an angle of incidence of 45° by noting that the reflectivity equations reduce to

$$\ln R_p/R_s = 2 \tag{7.32}$$

under ideal conditions. Here R_p and R_s are the p- and s-polarized components.

Variable-angle ATR can be used to determine concentration profiles as a function of depth without disrupting the sample. In general, as the angle of incidence approaches the critical angle, the field strength in the second medium is increased, thereby probing to an increasing depth (*45–48*). Thus, depth profiling of the sample can be carried out.

Micro-ATR

For microsamples, internal reflection is limited by the size of the illuminated portion of the internal reflection element. The active area of the crystal should not be significantly larger than the size of the sample. An ATR element has been designed which provides a six times linear reduction of the source image on the sampling surface and thus allows examination of smaller samples (*49*). Recent studies of micro-ATR have defined the best experimental conditions for establishing optical contact between the ATR crystal and the sample (*50*). This experimental approach has been applied to the analysis of the failure surfaces of adhesively bonded joints (*50*). ATR microscopic measurements have been used for direct measurement and identification of raw materials in textiles and of coated and impregnated substances on paper (*51*).

An ATR microscopic probe has been developed that allows one to examine the sample optically through the probe in the microscope, position the crystal in contact with the sample, and run the spectra (*34*) (Figure 7.5). The hemispheric ATR crystal is mounted at the focus of the Cassegrain objective, below the secondary mirror. In the survey mode, visible light at nearly normal incidence is selected to locate the area of measurement. In the contact mode, low-incident-angle visible light is used to detect contact of the sample with the ATR crystal surface. In the measurement mode, the ATR crystal is slid into position and the incident beam is optimized for total internal reflection. In the Spectra-Tech version, all of the available crystals, that is, ZnSe, diamond, silicon, and germanium can be used. However, Ge and Si are opaque and cannot be used in the survey or contact mode; an optical contact sensor can be used.

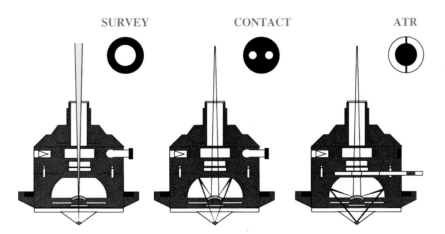

Figure 7.5. Diagram of a microscopic ATR cell.

The Spectra-Tech ATR objective is a coaxial, dual optic system, combining reflecting elements and a high-refractive ATR element for IR spectral measurements. The radiant energy is incident at 45° in the ATR probe. A high aperture is achieved by proper reflective elements and a hemispherical ATR crystal. Single-reflection ATR spectra are obtained.

A number of ATR crystals and objective configurations have become available.

Diffuse Reflectance

The Diffuse Reflectance Method

Diffuse reflectance results from the incident radiation that has traveled through some finite thickness of a material and is scattered or reflected from internal surfaces. Diffuse reflection spectra show a reduction in reflectance at frequencies where the absorption bands occur as a result of the radiation penetrating the sample to a distance comparable to its wavelength and being partially absorbed. On a macroscopic scale, the diffusely reflected radiation does not follow the simple laws of reflection: the angles of diffuse reflections are generally not equal to the angle of incident radiation. Diffuse reflection spectroscopy has been termed DRIFT (diffuse reflectance in Fourier transform). Although diffuse reflectance spectra closely resemble absorption spectra, there are fundamental differences. These differences, which include intensification of weaker bands and frequency shifts, are predicted by the Kubelka–Munk theory. It is necessary to apply a Kubelka–Munk correction to diffuse reflection data to create absorption-like spectra (*52, 53*). Good diffuse reflectors are samples that have rough surfaces and high scattering properties.

Microdiffuse Reflectance Measurements

Allen (*22*) has described a microscopic method for diffuse reflectance. The microscope is placed in the reflection mode, a small amount of potassium bromide is placed on a gold mirror, and a reference spectrum is run. The potassium bromide is removed and mixed with the sample and the process is repeated, resulting in the sample spectrum. The sample and background spectra are ratioed to remove instrument effects. Diffuse reflectance is particularly sensitive for measuring samples as small as 1 μg, and it is especially useful for analyzing small powdered polymer samples.

For nonfragmented samples, Allen (*22*) suggests a microscopic diffuse reflection measurement that utilizes silicon carbide paper (1200 grit); the background spectrum is obtained with the aperture fully open. The same carborundum paper is then rubbed with the sample to be analyzed, which transfers sufficient sample to be observed in diffuse reflection.

Microdiffuse Reflectance Detection for Thin-Layer Chromatography

FTIR microspectroscopy is an effective method to identify the eluting microcomponents obtained from thin-layer chromatography (TLC) (54). Microdiffuse reflection has been used for in situ detection as removal of the TLC spots is difficult (55). The primary limitation is the spectral interference of the TLC stationary phases, such as silica and alumina, with the spectra of the eluting compounds. These strong interfering absorbances result in eliminating some frequency regions from use for component identification. A recent report has outlined the advantages of using zirconia as a stationary phase for DRIFT–TLC (56). A new microchannel technique was also introduced which yields a smaller analyte spot size having a more concentrated amount of analyte for detection. Using the in situ microchannel DRIFT–TLC method, a detection limit of 1–10 ng was reported, with a possible limit of detection in the 100–1000-pg range (56). This represents an improvement of ~500 times in the minimum detectable quantity of solute compared to the use of microscope slides.

Infrared Instrumentation

Infrared spectroscopic instrumentation has a long and highly successful history. The types of instrumentation can be grouped into dispersion and multiplex methods. These two categories reflect the channels available when the spectra are acquired.

Dispersion Infrared Instrumentation

A scanning dispersion instrument utilizes a grating or prism as a geometric wavelength separation device to resolve the IR radiation into individual wavelength components (referred to as spectral resolution elements) (Figure 7.6). A means is provided in the instrument, such as an exit slit, to isolate a specific spectral resolution element for passage to the detector. The detector element is classically a broadband, single-channel element such as a thermocouple. The IR spectrum is obtained by moving (scanning) the grating over a given wavenumber region isolated by the exit slits after passing through the sample. The intensity of the spectral resolution element is determined by the intensity of the source, the band pass of the slits, and the sensitivity of the detector (57). These dispersion instruments were the workhorses of the infrared field from the 1940s until the advent of Fourier transform instrumentation.

Scanning IR spectrometers have provided valuable information, but they have a number of disadvantages arising primarily from their stepwise

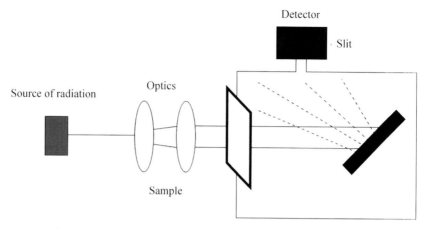

Figure 7.6. Optical diagram of a dispersive instrument. (Reproduced with permission from ref 44. Copyright 1993 John Wiley & Sons.)

or narrow band-pass nature in which the spectra are acquired. The nature of the optical dispersion process and the narrow exit slits limit the amount of energy falling on the detector to a small fraction of the available IR energy. In addition, moving and repositioning the grating or prism impose strict mechanical tolerances on the optical components. With scanning spectrometers it is difficult to increase the S/N by multiple scanning, as it is difficult to reproduce precisely the frequency positioning of the grating.

Multichannel Array Detectors

Multichannel techniques have developed by the use of dispersion monochromators and linear array detectors. The monochromator performs two functions: (1) a grating disperses the frequencies, and (2) a collimating lens focuses the geometrically dispersed light on a linear array detector. The theoretical spectral width, $\Delta\lambda$, of a linear array spectrometer is the product of the reciprocal dispersion and the length of the linear array. The reciprocal linear dispersion of the grating, $d\lambda/dx$, is given by

$$d\lambda/dx = b/mf \tag{7.33}$$

where b is the width of the grating ruling, m is the order of the grating, and f is the focal length of the focusing mirror in the monochromator. Typically, the expected spectral width is 0.5 μm in the first order and 0.25 μm in the second order. The instrument line shape is Gaussian. The band pass of the linear array spectrometer is the product of the reciprocal linear dispersion of the grating and the slit width. For a 2-mm slit, the band pass is 0.13 μm in first order and 0.06 μm (~27 cm^{-1}) in second order. An S/N value of 200 for the linear array spectrometer was measured with a root mean square (rms) noise of ~0.5%, so if 100 scans are averaged with a noise reduction having a square root dependence on the number of scans, the noise is reduced to ~0.05%, which will allow the detection of absorbance changes as small as 0.005 with a time resolution of 16 μs.

One type of array detector is based on an InSb system, with the size of each detector element on the order of 0.2 × 0.2 mm. The signal from each detector is converted to a voltage by a preamplifier and fed to a postamplifier and then to a data-acquisition board. The signals from the detector elements can be digitized in 16 μs and stored in the buffer memory.

Multiplex Infrared Instruments

Multiplex IR instruments overcome some of the difficulties of scanning spectrometers. *Multiplexing* refers to the simultaneous transmission of multiple resolution elements along the same channel. Multiplex instruments exist as two basic types: (1) nontransform multiplexing devices, such as array detectors (photodiode or charge-coupled device (CCD) arrays), where many wavelengths are measured simultaneously, and (2) transform multiplex devices, such as Fourier and Hadamard transform spectrometers. The transform multiplex instruments are frequency-division multiplexing instruments. The instruments use a single detection channel (that is, a single detector) to record the optical information that has been encoded. Each individual measurement in a set of encoded data contains information from all points in the spectrum. A Fourier transform (FT) spectrometer encodes the information in an interferogram (Figure 7.7) (*58*).

This encoding is accomplished by moving a mirror in a Michelson interferometer. In one arm of the interferometer there is a fixed mirror, and in the other, a movable mirror. These arms are separated by a beam splitter, which splits the incoming beam into the two arms. The overall path length of the two light beams determines the degree of constructive or destructive interference upon reflection to the detector. After the data are detected and stored, it is necessary to perform a mathematical transform operation (a Fourier transform) on the data set to convert it into a conventional spectrum.

The primary advantage of multiplexing the spectral resolution elements in the IR is the improvement in S/N. The improvement arises because the detector efficiency for each measurement is increased (*59*). Because the detector noise is constant and independent of the source, the S/N is improved because the multiplexing distributes the detector

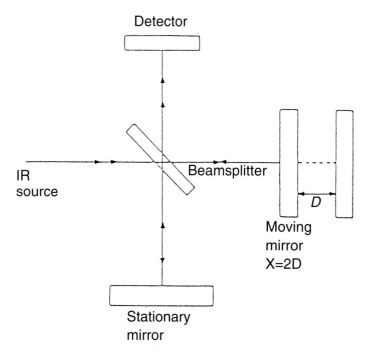

Figure 7.7. Optical diagram of an interferometer. (Reproduced from ref 10. Copyright 1992 American Chemical Society.)

noise over the intensities of many spectral resolution elements. This distribution process lowers the noise in each spectral resolution element more than if the spectral resolution element is measured individually by step scanning.

For infrared microscopy, FTIR instrumentation has a number of significant advantages over dispersive instrumentation (*58*):

- extremely high frequency–wavelength repeatability and accuracy (1:2000),

- greater optical throughput (80×),

- high scan rate (500×),

- absence of stray light (no need to measure 0% transmittance),

- reproducible resolution (mirror drive distance), and

- absence of grating and slit changes.

During the normal operation of an FTIR, the observed interferogram consists of a narrow region of strong information, called the *centerburst*, with wide wings having weak signals close to 0 V. To accurately sample both the stronger and weaker information with the same analog to digital (A/D) converter, a technique called *gain ranging* has been developed. In this technique, the centerburst region is sampled using no preamplifier gain, but outside of this region gain is applied to amplify the weak signals. The distance away from the centerburst where the gain is applied is called the gain ranging radius (GRR).

The recorded amplitude of an interferogram must be linearly proportional to the light intensity at the detector during acquisition if the transformed spectrum is to have the proper spectral intensity. The nonlinear responses of photoconductive detectors, such as mercury–cadmium–telluride (MCT) detectors can result from too much photon flux, improper current biasing, or preamplifier saturation. Nonlinearity produces spectrophotometric accuracies irrespective of the cause. Other sources of noise in FTIR include fluctuations in source output, interferometer instabilities, and fluctuations in detector preamplifier and filter electronics.

Changes in the quality of the purge of the sample compartment can lead to the appearance of infrared bands corresponding to changes in carbon dioxide and atmospheric water concentrations. There are a variety of optional experimental configurations available to enhance the instrumental capability for microspectroscopy. For example, a 100-μm-diameter MCT detector can be substituted for the more typical 250-μm model to obtain a theoretical S/N benefit of 2.5, based on a reduction of nonilluminated detector area.

Step-Scan Interferometers

The advantage of step-scan FTIR (*60*) over rapid scan is that the spectral multiplexing is removed from the time domain. In the rapid-scan operation, the moving mirror scans continuously, causing each IR wavelength to be modulated with a different Fourier frequency. Specifically, rapid-scan operation modulates each IR wavelength at its characteristic Fourier frequency, which is determined by the scan rate of the mirrors. In step-scan operation, the moving mirror is stopped at fixed positions during data collections. In this manner, spectral multiplexing is decoupled from the time domain. By moving the mirror stepwise, one obtains the time evolution of the changes in the interferogram at each sampling point, and from the data set, the spectral changes can be obtained by the Fourier transform at each time of interest. Thus, dynamic spectroscopy may be carried out using step-scan interferometry, without the need to deconvolute the time dependence of the sample response from any time dependence of the data collection process itself.

Step-scan interferometry has made a substantial impact on the use of infrared photoacoustic spectroscopy, which has resulted in improved depth profiling of polymer samples (*61, 62*).

In step-scan FTIR spectrometers, a frequency correlation can be applied to enhance the dynamic spectral analysis of the response of a liquid crystal film at the ac potential (*63*). This analysis produces two-dimensional FTIR correlation maps defined by two independent wavenumbers. Peaks appearing in the two-dimensional spectrum provide information about the degree of interaction among the functional groups associated with specific transition dipoles. The dynamic two-dimensional infrared technique has been particularly useful for looking at interactions in polymer blends (*64*).

Time-resolved spectroscopic measurements can be made using step-scan interferometers. Step-scan interferometry has also been useful in making IR emission measurements (*65*).

Infrared Microscopes

The historical development of IR microscopes has been documented (*66, 67*). In 1949, a modification of a reflecting microscope used to collect absorption spectra in the UV and visible regions, employing a Cassegrainian objective, was used to obtain the infrared spectrum of a microsample (*68*). In 1953, Perkin Elmer introduced the first commercial microscope (*69*), which had reflecting lenses of the Schwarzschild type. In 1980, V. Coates produced an IR microscope with a dedicated computer for storage of the single-beam spectra, and in 1983, a number of IR microscopes were developed, primarily for viewing microsamples. The first complete IR imaging microscope with a spectrometer embedded in it was developed by Spectra-Tech and called the IRμs Molecular Microanalysis System. Recently, a new confocal scanning IR microscope has been developed that can be used in transmission, reflection, or double-pass modes with or without differential phase contrast imaging (*70*).

The advantages of using an infrared microscope for microsamples rather than a beam condenser include the following:

- The user can view the sample.
- The sample can be positioned at a point that the IR beam is known to traverse.
- The microscope can give effective spatial selectivity.
- Complementary information can be obtained from the optical observation.
- The computer-controlled precision stage allows functional images to be obtained by spatial scanning.

- The optical requirements for an IR microscope include:
- exact positioning of the sample,
- spatial isolation of the sample from a larger matrix in the IR beam,
- capability to function in both the visible and the infrared spectral regions.

The practical problem is that no optical material with the corresponding properties of glass in the visible exists for the IR. This necessitates the use of reflecting optics employed on-axis. The Schwarzschild Cassegrain system is the choice and results in the visible-light path being coaxial with the IR spectrometer's light path (Figure 7.8) (*71*).

The photometric accuracy of infrared microscopes is limited by light diffraction effects, stray light, uniformity of the viewed sample, and specimen-induced refocusing of light by the sample. Remote, variable apertures are placed at conjugate image planes to define the sample area and to reduce stray light for spectroscopic measurements.

Modern Infrared Microscopes

Most commercial FTIR microscopes (including the Spectra-Tech IRμs Molecular Microanalysis System) use 15× and 10× condenser lenses of the noninfrared absorbing Cassegrainian type. An optical diagram of one of the current commercial instruments is shown in Figure 7.9. They are constructed from front surface mirrors and are mounted on-axis, so that the path for the visible light is perifocal and collinear with the infrared light. The microscope stage is computer controlled, with 1-μm positioning steps possible. The sample is not physically apertured because it is spatially isolated from the microscope apertures. For infrared microscopy, two apertures take up positions such that their images coincide in the sample plane (redundant aperturing [Spectra-Tech]). The apertures are a set of crossed knife blades in the form of a rectangle of adjustable size. The all-reflecting optical system is achromatic. The microscope also contains compensators, which correct for spherical aberration in the presence of samples, and IR transmitting windows that have a different refractive index from air.

The presence of the central obscuration in the Cassegrain system reduces the light throughput. The principal problem in IR microscopy that limits the spatial resolution is the presence of scattered light due to diffraction, which increases the noise. Diffraction is significant when the aperture dimensions approach the wavelength of the IR radiation. The main effect of diffraction is that at small aperture sizes, light spreads outside the specified area into the surrounding region. As higher spatial resolution is sought, the problem increases as the apertures are smaller, ultimately leading to loss of spectral quality and photometric accuracy. A

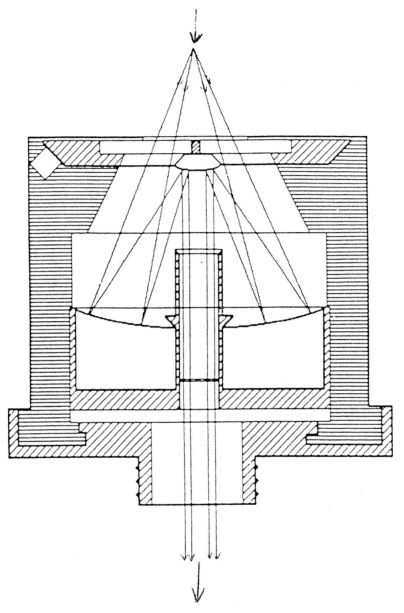

Figure 7.8. Optical diagram of the Cassegrain lens system. (Reproduced with permission from ref 71. Copyright 1989 Society of Spectroscopy.)

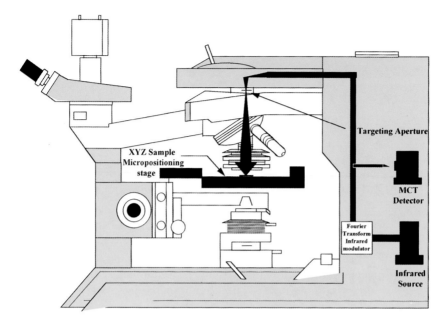

Figure 7.9. Optical diagram of an infrared microscope.

magnification of 50× can be obtained using commercial systems allowing a spatial resolution of 10–20 μm in the midinfrared.

With the Spectra-Tech IRμs, the field of view from the 15× Cassegrainian objective is captured by a video camera for display on the monitor, and an optical arrangement allows the image of the remote rectangular knife-edged aperture to be superimposed onto the overall video image. With this arrangement, a video image allows the operator to define the exact location where the spectrum will be recorded.

A new IR microscope design has a collector for focusing the IR beam and a device for imaging the sample on the input surface of the system. An operating mechanism moves the devices relative to each other and vertical to the centerline, where ATR can be performed (*72*).

Synchrotron Infrared Microscopes

A *synchrotron* IR light source improves the performance of an IR microscope about 100-fold. Synchrotron radiation provides enhanced *S/N* performance, improved spatial resolution, and extended spectral range. A synchrotron source uses accelerated electrons to produce high-intensity radiation in the form of a coherent light 1000 times brighter than ordinary sources. The brightness, spectral distribution, and low noise of synchrotron radiation are of particular value for infrared microspectroscopy.

Experiments have been performed on an infrared microscope with the synchrotron radiation source at the 2.5-GeV National Synchrotron Light Source at Brookhaven National Laboratory in Upton, NY. The high brightness of synchrotron light makes it possible to analyze a sample as thin as 5 μm in seconds, with the same precision possible on samples 50 times larger (*73*).

Spatial Resolution Enhancement Using Infrared Microscopes

Infrared microanalysis is a technique for the analysis of specimen microdomains of 5–250 μm. All-reflective optics permit both visual and infrared observation using either transmitted or reflected illumination. The spectral response for a selected region of a microdomain may be resolved from the surrounding sample matrix with an adjustable field stop, or aperture.

The infrared spectrum depends on the ratio of the sampled area to the area of the microdomain. As the apertured area is reduced in size, the area of the infrared absorption peaks resulting from the microdomain will increase (the microdomain region is rich in one component and deficient in another) as it begins to dominate the field of view. This is illustrated in Figure 7.10, which shows how the infrared spectrum obtained on an infrared microscope depends on the aperture size (*74*).

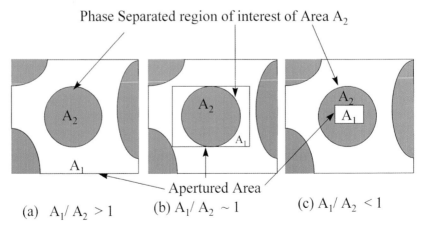

Phase Separated region of interest of Area A_2

(a) $A_1/A_2 > 1$ (b) $A_1/A_2 \sim 1$ (c) $A_1/A_2 < 1$

Figure 7.10. Schematic representation showing how the infrared spectrum obtained with an infrared microscope depends on aperture size. Three cases are shown: (a) $A_1/A_2 > 1$; (b) $A_1/A_2 \sim 1$; and (c) $A_1/A_2 < 1$. (Reproduced from ref 74. Copyright 1994 American Chemical Society.)

Near-Infrared Imaging

The near infrared (NIR) extends from 14,000 to 4000 cm^{-1} (0.7–2.5 μm), and the observed bands consist of overtones and combinations of fundamental mid-IR (MIR) bands. The NIR modes have several orders of magnitude less intensity than the fundamental modes in the MIR. Consequently, NIR can be used to examine samples with greater thicknesses, which would be totally absorbing in the MIR. This makes sample preparation considerably easier in the NIR than in the MIR. NIR spectra enhance the ability to identify polymers and to make quantitative analysis. Limitations in the NIR arise from extensive band overlap and the difficulty of making structural assignments to the observed frequencies. NIR microscopes can be constructed from visible microscopes and provide nearly comparable spatial resolution (*75*).

A recent development in NIR imaging is the use of focal-plane array (FPA) imaging detectors for the NIR region. A NIR imaging system using a silicon CCD detector and a tunable acoustooptic spectrometer has been described (*77*). Later, a system with several orders of magnitude higher sensitivity was described using an InSb FPA detector (*78, 79*). These systems can achieve a spatial resolution of 3–5 μm. Figure 7.11 shows the NIR image of stacks of poly(methyl methacrylate) (PMMA) films and illustrates the sensitivity of the NIR imaging system.

A NIR fiber-optic probe has been constructed for chemical imaging

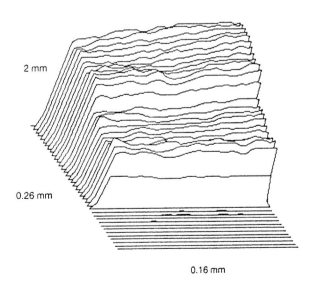

Figure 7.11. NIR image of stacks of PMMA films, illustrating the sensitivity of the NIR system. (Reproduced with permission from ref 78. Copyright 1994 Society of Spectroscopy.)

using point-by-point spatial measurement of arterial lesions in living tissue. The images were collected on a parallel vector supercomputer and displayed on a workstation. The results for images of aorta showing substantial low-density lipoprotein (LDL) accumulation are shown in Figure 7.12 (*80*). A near-IR Hadamard transform spectrometer has been described as the first step toward near-IR imaging using this technique (*81*). The spectrometer uses an acoustooptic, tunable filter for dispersion. These new microscopic instruments suggest that further utilization will be made in the near-IR in the immediate future.

Near-Field Near-Infrared Imaging

A near-field near-IR imaging microscope has been constructed (*76*). An optical fiber is pulled to a tip diameter of less than 200 nm, which is much smaller than the wavelength of near-IR photons. The fiber is also coated with aluminum or gold to dissipate the evanescent field. The distance between the probe and the sample is controlled by measuring the shear forces between the sample and the probe, and a feedback mechanism is used to position the probe. An InSb FPA camera is used to collect the spectrometric images in the near IR.

Midinfrared Microspectroscopic Imaging

Recently, it has been reported that it is possible, using infrared FPA detectors coupled to step-scan interferometers, to record data sets of 128 × 128 pixels at 16 cm^{-1} with total data acquisition times of 12 s (*84, 85*), which corresponds to 16,384 spatially resolved FTIR spectra.

> *The FPA detectors allow spectra from all pixels to be recorded simultaneously, while modulation of infrared radiation by the interferometer allows all the spectral frequencies across the wavelength range to be measured concurrently. This instrumental configuration furnishes for the first time chemically distinct midinfrared images and spectra with unprecedented speed and quality (85).*

A number of reports of this new type of instrument have been made (*86–93*). The coupling of an FPA detector to an IR microscope and a step-scan FTIR interferometer results in a *multiplex–multichannel* advantage (*89*). The multichannel advantage arises from the presence of multiple detector elements from which all the spectral frequencies are measured simultaneously. The step-scan approach yields spectral images and spectra with high speed and fidelity (*90*). A representation of a real-time microscopic instrument is shown in Figure 7.13. The frequency range is

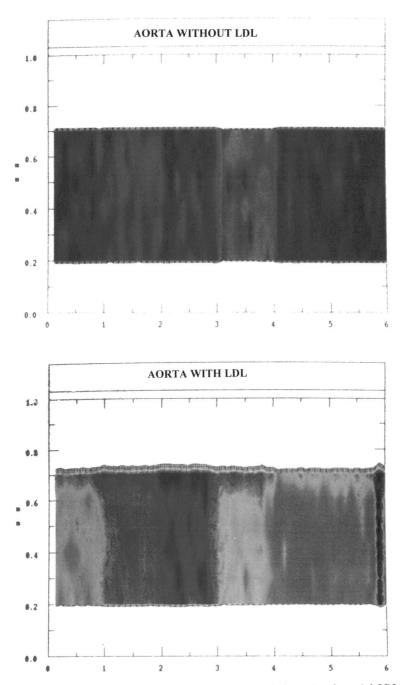

Figure 7.12. Images of an aorta showing (top) no and (bottom) substantial LDL accumulation. (Reproduced from ref 80. Copyright 1993 American Chemical Society.)

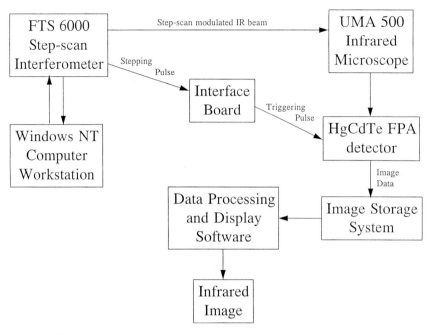

Figure 7.13. Schematic diagram of a real-time mid-IR microscopic imaging instrument.

determined by the FPA, and currently InSb detectors are being used that have a frequency range of 1.0 to 5.5 μm (10,000–1850 cm^{-1}) with 128 × 28 elements. However, an instrument was recently described using a 320 × 256 element InSb FPA, which achieves a theoretical spatial resolution of 1.3 μm/pixel with a 36× objective (*94*), although an effective spatial resolution of 8 μm/pixel was actually achieved.

The ideal FTIR FPA microscope would have a detector that would allow one to go to 25 μm into the infrared fingerprint region of the spectrum (*93*). It has been suggested that Si:As can be used for such a detector, as its range is 1–25 μm (10,000–400 cm^{-1}) (*93*). The step-scan interferometer has an interface board which generates a stepping pulse arising from the mirror position, that triggers a pulse to the detector for activating the data collection. The image data are transferred to a data storage system for the purpose of performing the fast Fourier transform and the image analysis (*93*).

Images of polymer dispersed liquid crystals (E-7 dispersed in epoxy resin) were obtained (*95*) in a 2-min measurement time, using an FTS 6000 (Bio-Rad) interfaced to a UMA 500 microscope accessory (Bio-Rad), with an ImagIR InSb FPA (FPA, Santa Barbara Focalplane) and a CaF$_2$ lens in place of the Cassegrainian condenser. Figure 7.14 shows three

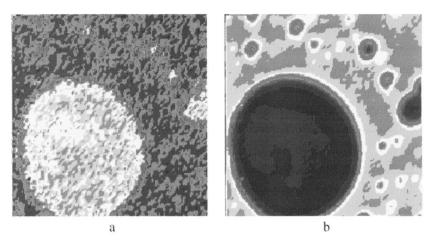

a   b

Figure 7.14. Infrared images of polymer dispersed liquid crystals (E–7 dispersed in epoxy resin), obtained in a measurement time of 2 min. These images were generated by plotting the intensity values of the liquid crystal nitrile (2251 cm⁻¹), the C–H stretching mode (2845 cm⁻¹), and the epoxy O–H (3470 cm⁻¹) peaks. The image at 3470 cm⁻¹ displays a higher concentration of polymer between the liquid crystal droplets than inside the droplets, which shows that the polymer is strongly segregated away from the liquid crystal droplets.

infrared images obtained at different frequencies from the same measurement. High-resolution images at any frequency in the range of 3700–1975 cm^{-1} can be obtained from this single experiment. These images were generated by plotting the intensity values of the liquid crystal nitrile (2251 cm^{-1}), the C–H stretching mode (2845 cm^{-1}), and epoxy O–H (3470 cm^{-1}) peaks. The image at 3470 cm^{-1} shows a higher concentration of polymer between the liquid crystal droplets than inside the droplets, which shows that the polymer is strongly segregated away from the liquid crystal droplets. The spatial resolution is approximately 8 μm. Figure 7.15, on the other hand, shows infrared traces taken from three different spatial positions in the image. Spatial variations of the component concentration can be monitored by extracting spectra from different regions and observing changes in spectral features. In this case, the polymer concentration in a polymer dispersed liquid crystal sample can be monitored by observing changes in the peak around 3470 cm^{-1}, attributed to the O–H stretch of the epoxy polymer. The spectral resolution is 16 cm^{-1}.

Infrared Microscopic Experiments

A large new class of IR microscopic experiments can be performed (*74*) including

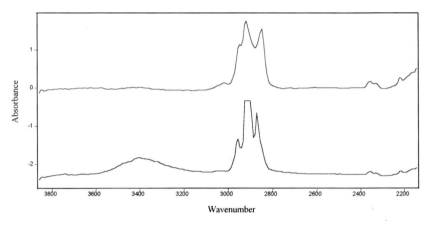

Figure 7.15. *Infrared traces taken from three different spatial positions in the image of E–7 dispersed in epoxy resin. Spatial variations of the component concentration are monitored by observing changes in spectral features. In this case, the polymer concentration in a polymer dispersed liquid crystal sample is monitored by observing changes in the peak around 3470⁻¹, attributed to the O–H stretch of the epoxy polymer.*

1. Determination of the spatial map of composition of a sample by obtaining spectra rastered through the pixels in the sample and measuring the intensities of the unique lines of each component of the mixture. The FTIR microscopy approach allows the detection of spatial differences in composition even for those micropolymer systems that cannot be observed optically because of small differences in the refractive indices (*82,96*).

2. Monitoring of the temporal changes in composition and structure by following the spectral changes for a given spatial position in the sample (*83*).

3. Monitoring of spectral changes from separate portions of the same sample that have been exposed to different chemical, physical, or environmental conditions (*36*).

4. Monitoring of the size and shape of the morphological subunits by taking the infrared spectrum over a range of apertured areas at the same position in the sample. The IR spectrum depends on the ratio of the sample area to the size of the compositional entity (*74*).

These new classes of microspectroscopic spatial experiments have proven to be of great value in the characterization of polymer samples, as the next chapter demonstrates.

References

1. *Practical Guide to Infrared Microspectroscopy;* Humecki, H. J., Ed.; Marcel Dekker: New York, 1995.
2. *Infrared Microscopy Theory and Applications;* Messerschmidt, R. G.; Harthcock, M. A., Eds.; Marcel Dekker: New York, 1988.
3. *The Design, Sample Handling, and Applications of Infrared Microscopes;* Rousch, P. B., Ed.; American Society for Testing and Materials: Philadelphia, PA, 1987.
4. *Microscopic and Spectroscopic Imaging of the Chemical State;* Morris, M., Ed.; Practical Spectroscopy Series; Marcel Dekker: New York, 1993; Vol. 10.
5. Lang, P. J.; Sommer, A. J.; Katon, J. E. In *Particles on Surfaces 2;* Mittal, K. L., Ed.; Plenum: New York, 1989; p 207.
6. Lang, P. In *Handbook of Surface Imaging and Visualization;* Hubbard, A. T., Ed.; CRC Press: Boca Raton, FL, 1994.
7. Katon, J. E. *Vib. Spectrosc.* **1994**, *7*, 201.
8. Holst, C. G. *Proc. SPIE Int. Soc. Opt. Eng.* **1993**, 1969.
9. Diem, M. *Introduction to Modern Vibrational Spectroscopy;* Wiley: New York, 1993.
10. Koenig, J. L. *Spectroscopy of Polymers;* American Chemical Society: Washington, DC, 1992.
11. Colthup, N. B.; Daly, L. H.; Wiberley, S. E. *Introduction to Infrared and Raman Spectroscopy,* 2nd ed.; Academic: New York, 1975.
12. Mielczarski, J. A.; Milosevic, M.; Berets, S. L. *Appl. Spectrosc.* **1992**, *46*, 1040.
13. Wilson, E. B., Jr.; Decius, J. C.; Cross, P. C. *Molecular Vibrations;* McGraw Hill: New York, 1955.
14. Painter, P.; Coleman, M.; Koenig, J. L. *The Theory of Vibrational Spectroscopy and Its Application to Polymeric Materials;* Wiley: New York, 1982.
15. Bellamy, L. J. *The Infrared Spectra of Complex Molecules,* 3rd ed.; Wiley: New York, 1975.
16. Bellamy, L. J. *Advances in Infrared Group Frequencies;* Barnes and Noble: New York, 1968.
17. Schrader, B. *Raman/IR Atlas of Organic Compounds,* 2nd ed.; VCH Publishers: New York, 1989.
18. Lin-Vien, D.; Colthup, N.; Fateley, W. G.; Grasselli, J. G. *The Handbook of Infrared and Raman Characteristic Frequencies of Organic Molecules;* Academic: San Diego, CA, 1991.
19. Abbot, T. P.; Felker, F. C.; Kleiman, R. *Appl. Spectrosc.* **1993**, *47*, 180.
20. Sommer, A. J.; Katon, J. E. *Microbeam Anal. (San Francisco)* **1988**, *23*, 207.
21. Krishnan, K. In *Proceedings of the SPIE-International Society for Optical Engineering;* Cielo, G. A., Ed.; International Society for Optical Engineering: Bellingham, WA, 1986; p 252.
22. Allen, T. J. *Vib. Spectrosc.* **1992**, *3*, 217.
23. Curry, C. J.; Whitehouse, M. J.; Chalmers, J. M. *Appl. Spectrosc.* **1985**, *39*, 174.
24. Lang, P. L.; Katon, J. E.; Schiering, D. W.; O'Keefe, J. F. *Polym. Mater. Sci. Eng.* **1986**, *54*, 381.
25. Tungol, M. W.; Bartick, E. G.; Montaser, A. *Appl. Spectrosc.* **1993**, *47*, 1655.
26. Graf, R. T.; Koenig, J. L.; Ishida, H. *Appl. Spectrosc.* **1985**, *39*, 405.
27. Lin, R.; Ritz, G. P. *Appl. Spectrosc.* **1993**, *47*, 265.
28. Mastalerz, M.; Bustin, R. *Int. J. Coal Geol.* **1993**, *24*, 333–345.
29. Ishino, Y.; Ishida, H. *Appl. Spectrosc.* **1992**, *46*, 504.
30. Harrick, J. *Am. Lab.* **1986**, *18*, 78.
31. Porter, M. D. *Anal. Chem.* **1988**, *60*, 1143.
32. Dubois, L. H.; Nuzzo, R. G. *Ann. Rev. Phys. Chem.* **1992**, *43*, 437.

33. Song, Y. P.; Petty, M. C.; Yarwood, J. *Vib. Spectrosc.* **1991**, *1*, 305.
34. Reffner, J.; Alexay, C. C.; Hornlein, R. W. *SPIE* **1991**, *1575*, 301.
35. Pepper, S. V. *Appl. Spectrosc.* **1995**, *49*, 354.
36. Fondeur, F.; Koenig, J. L. *J. Adhes.* **1993**, *40*, 189–205.
37. Fondeur, F.; Koenig, J. L. *Appl. Spectrosc.* **1993**, *47*, 1.
38. Harrick, N. J. *Internal Reflection Spectroscopy;* Harrick Scientific: New York, 1979.
39. Mirabella, F. M. *Spectroscopy* **1991**, *5*, 20.
40. Ishida, H. *Rubber Chem. Technol.* **1987**, *60*, 498.
41. Ohta, K.; Iwamoto, R. *Appl. Spectrosc.* **1985**, *39*, 418.
42. Bartick, E. G.; Messerschmidt, R. G. *Am. Lab.* **1984**, *15*, 92.
43. Tiefenthaler, A. M.; Urban, M. W. *Appl. Spectrosc.* **1988**, *42*, 163.
44. Urban, M. W. *Vibrational Spectroscopy of Molecules and Macromolecules on Surfaces;* John Wiley & Sons: New York, 1993; p 86.
45. Blackwell, C. S.; Degen, P. J.; Osterholtz, F. D. *Appl. Spectrosc.* **1978**, *32*, 480.
46. Fina, L. J.; Chen, G. *Vib. Spectrosc.* **1991**, *1*, 353.
47. Shick, R. A.; Koenig, J. L.; Ishida, H. *Appl. Spectrosc.* **1993**, *47*, 1237.
48. Huang, J.; Urban, M. W. *Appl. Spectrosc.* **1993**, *47*, 973.
49. Harrick, N. J.; Milosevic, M.; Berets, S. L. *Appl. Spectrosc.* **1991**, *45*, 944.
50. Buffeteau, T.; Desbat, T. B.; Eyquem, D. *Vib. Spectrosc.* **1996**, *11*, 29.
51. Satou, S.; Ikehara, Y.; Arime, M. *Kanzeri Chuo Bunsekishoho* **1994**, *33*, 93.
52. Kortum, G. *Reflectance Spectroscopy;* Springer-Verlag: New York, 1969.
53. Fuller, M. P.; Griffiths, P. R. *Anal. Chem.* **1978**, *50*, 69.
54. Brown, P. R.; Beauchemin, B. T. *J. Liq. Chromatogr.* **1988**, *11*, 1001.
55. Chalmers, J. M.; Mackenzies, M. W.; Sharp, J. L.; Ibert, R. N. *Anal. Chem.* **1987**, *59*, 415.
56. Bouffard, S. P.; Katon, J. E.; Sommer, A. J.; Danielson, N. D. *Anal. Chem.* **1994**, *66*, 1937.
57. Smith, A. L. *Applied Infrared Spectroscopy, Fundamentals, Techniques, and Analytical Problem-Solving;* Wiley-Interscience: New York,1979.
58. Griffiths, P. R.; deHaseth, J. A. *Fourier Transform Infrared Spectroscopy;* John Wiley & Sons, New York, 1986.
59. Compton, D.; Drab, J.; Barr, H. S. *Appl. Opt.* **1990**, *29*, 2908.
60. Palmer, R. A. *Spectroscopy* **1993**, *8*, 26.
61. Palmer, R. A.; Manning, C. J.; Rzepiela, J. A.; Widder, J. M.; Chao, J. L. *Appl. Spectrosc.* **1989**, *43*, 1193.
62. Smith, M. J.; Manning, C. J.; Palmer, R. A.; Chao, J. L. *Appl. Spectrosc.* **1988**, *42*, 546.
63. Gregoriou, V. G.; Chao, J. L.; Toriumi, H.; Palmer, R. A. *Chem. Phys. Lett.* **1991**, *179*, 491.
64. Marcott, C.; Dowry, A. E.; Noda, I. *Anal. Chem.* **1994**, *66*, 1065.
65. Tochigi, K.; Momose, H.; Misawa, Y.; Suzuki, T. *Appl. Spectrosc.* **1992**, *46*, 156.
66. Reffner, J. A.; Coates, J.; Messerschmidt, R. G. *Am. Lab.* **1987**, *19*(4), 86.
67. Reffner, A. *Microscopy Today* **1993**, *3*, 6.
68. Barer, R.; Cole, A. R. H.; Thompson, H. W. *Nature (London)* **1949**, *163*, 198.
69. Coates, V. J.; Offner, A.; Siegler, E. H., Jr. *J. Opt. Soc.* **1953**, *43*, 984.
70. Torok, P.; Booker, G. R.; Laczik, Z.; Falster, R. *Inst. Phys. Conf. Ser.* **1993**, *134*, 771.
71. Katon, J. E.; Sommer, A. J.; Lang, P. L. *Appl. Spectrosc. Rev.* **1989**, *25*, 173.
72. Messerschmidt, R. G. U.S. Patent 5 225 678, 1993.
73. Reffner, J. A.; Martoglio, P. A.; Williams, G. P. *Rev. Sci. Instrum.* **1995**, *66*, 1298.
74. Durraani, C. M.; Donald, A. M. *Macromolecules* **1994**, *27*, 110.
75. Treado, P. J.; Morris, M. D. *Appl. Spectrosc. Rev.* **1994**, *29*, 1.
76. Symons, W. C.; Lodder, R. A. *Abstracts of Papers*, PITTCON '95, New Orleans, LA; Pittsburgh Conference: Pittsburgh, PA, 1995; paper 646.

77. Treado, P. J.; Levin, I. W.; Lewis, E. N. *Appl. Spectrosc.* **1992,** *46,* 553.
78. Treado, P. J.; Levin, I. W.; Lewis, E. N. *Appl. Spectrosc.* **1994,** *48,* 607.
79. Lewis, E. N.; Gorbach, A. M.; Marcott, C.; Levin, I. W. *Appl. Spectrosc.* **1996,** *50,* 263.
80. Cassis, L. A.; Lodder, R. A. *Anal. Chem.* **1993,** *65,* 1247.
81. Turner, J. F., II; Treado, P. J. *Appl. Spectrosc.* **1996,** *50,* 277.
82. Kaito, A.; Kyotani, M.; Nakayama, K. *Macromolecules* **1991,** *24,* 3244.
83. Hsu, S. C.; Lin-Vien, D.; French, R. N. *Appl. Spectrosc.* **1992,** *46,* 225.
84. Lewis, E. N.; Levin, I. W.; Treado, P. J. U.S. Patent 5 377 003, 1994.
85. Lewis, E. N. In *Proceedings of the 29th Annual Meeting of MAS;* Editz E. G., Ed.; VCH Publishers: Breckenridge, CO, 1995; p 103.
86. Treado, P. J.; Morris, M. D. *Appl. Spectrosc. Rev.* **1994,** *29,* 1.
87. Lewis, E. N.; Treado, P. J.; Reeder, R. C.; Story, G. M.; Dowrey, A. E.; Marcott, C.; Levin, I. W. *Anal. Chem.* **1995,** *67,* 3377.
88. Lewis, E. N.; Levin, I. W. *Appl. Spectrosc.* **1995,** *49,* 672.
89. Marcott, C.; Story, G. M.; Dowry, A. E.; Reeder, R. C.; Noda, I. *Mikrochim. Acta* (in press).
90. Treado, P. J.; Levin, I. W.; Lewis, E. N. *Appl. Spectrosc.* **1994,** *48,* 607.
91. Lewis, E. N.; Levin, I. W.; Crocombe, R. A. Presented at the 10th International Conference on Fourier Transform Spectroscopy, Budapest, Hungary, 1995; paper B3.9.
92. Snively, C.; Koenig, J. L.; (Case Western University): Crocombe, R. A.; Wright, N.; McCarthy, W. J. (Bio-Rad): unpublished results.
93. Lewis, E. N.; Kidder, L. H.; Lester, D. S.; Levin, I. W. Presented at the 2nd International Symposium on Advanced Infrared Spectroscopy, Durham, NC, 1996; paper 3-2.
94. Marcott, C.; Reeder, R. C. Presented at the 2nd International Symposium on Advanced Infrared Spectroscopy, Durham, NC, 1996; paper A-7.
95. Snively, C. M.; Koenig, J. L.; Chen, P. Y.; Palmer, R. A. Presented at the 2nd International Symposium on Advanced Infrared Spectroscopy, Durham, NC, 1996; paper A-19.
96. Harthcock, M. A.; Atkin, S. C. *Appl. Spectrosc.* **1989,** *42,* 449.

8

Applications of FTIR Microspectroscopy

Introduction

FTIR microspectroscopy is an important source of structural and spatial information for polymer-based articles of commerce. A recent review gives references to the large number of studies in this area (*1*). Infrared mapping experiments have also been extremely valuable since their inception (*2*). The field has grown rapidly over the past 10 years, and it is difficult to separate it into segments, but several areas prominent in the study of polymers are discussed in this chapter.

Analysis of Contamination

Polymers are complex mixtures, and when the mixtures are homogeneous, the performance of the system is optimized. Contaminants usually inhibit performance. Inclusions typically phase-separate from the polymer matrix and "bloom" (diffuse) to the surface, giving the article an undesirable appearance. When contaminants are concentrated at the surface, the inclusions visually appear as fish eyes on the surface of the plastic article and are therefore appropriately termed "fish eyes."

Fluid inclusions can be trapped in the intracrystalline cavities. In the fabricating process of plastics, contamination can have a variety of sources, including residues from the molding equipment such as bearings, bushings, and seals. Other components in the processing chain for the particular engineering component can also contribute, such as impurities in additives, processing aids, etc.

However, the size and concentration of microcontaminants make them.difficult to analyze, so the chemical nature of contamination is often

difficult to determine. It is necessary to determine the nature of the undesirable impurity for its source to be identified and removed from the
manufacturing or fabrication process. Because of the wide range of pos-
sibilities and the small amount of material, there is no universal technique
for analysis of contaminants. For polymers, the problem is further complicated by the small amount of material in the presence of a large quantity of base polymer. Sometimes it is difficult to physically or chemically
separate the contaminant from the polymer matrix.

Microspectroscopic Identification of Contaminants

Infrared microsampling techniques have found wide application in this
area of identification of impurities. The samples to be examined are physically isolated and placed in microsampling attachments such as beam
condensers, masks, or microcells such as the diamond cell, and the spectra
are obtained.

The IR microscope allows one to view regions of the sample and record the spectrum of a particular spatial domain. Consequently, one can
obtain a spectrum from a spatially contaminated area and compare it with
an uncontaminated area of a sample, thus spatially isolating the contaminants. Spectral subtraction of the spatially resolved spectra can reduce
the spectral contributions of the major component relative to the contaminants. The overall reduced absorbance range allows one to use the computer to amplify the spectrum of the contaminant to the limit of the S/N.
Computer searches of the spectral database can generally result in identification if the spectrum of the contaminant is in the database. Problems
of spectral identification can occur if the inclusion is of a complex mixture
of a number of subcontaminants.

The approach to the identification of impurities is straightforward.
The limits of microinfrared analysis have increased from milligrams to
nanograms, which increases the potential for microscopic infrared spectroscopy to be successful in identification. Because of the fingerprint nature of the infrared spectra, once the spectrum has been spatially isolated
and spectrally purified by subtraction of the major component, identification is facilitated.

Infrared microsampling techniques are similar to those of optical microscopy and are generally easy to carry out. In some cases, it is possible
to physically isolate the imperfection by simply cutting it from the article
with a diamond knife or razor blade. Then the cut sample is pressed flat
between NaCl plates and mounted on the microscope stage. Using a suitable aperture to optimize the contributions of the impurity, the spectrum
is obtained. Gel inclusions in polyethylene can be identified by the IR
spectra using the approach just described and is illustrated in Figure 8.1
(3). The spatially isolated and chemically purified inclusion spectrum

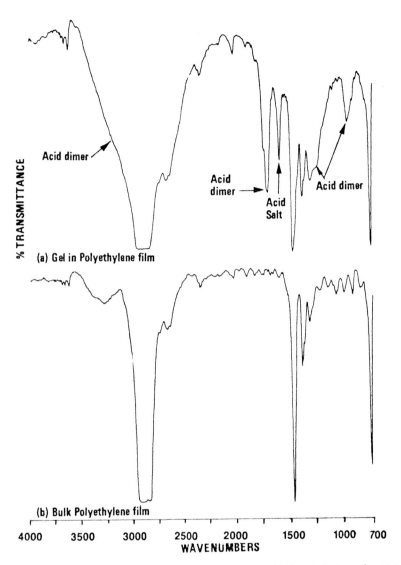

Figure 8.1. Infrared microtransmittance spectra between 4000 and 700 cm^{-1} of (a) a gel in polyethylene film and (b) bulk polyethylene film (Collection conditions for the two spectra: 8 cm^{-1} resolution, 64 scans, 150 × 150 μm aperture.) (Reproduced with permission from ref 3. Copyright 1987 American Society for Testing and Materials.)

shows bands due to acid dimers and acid salt structures originating from contamination by an ethylene–acid copolymer.

Infrared Mapping of Contamination

The IR experiments with computer-controlled microscope stages allow one to not only detect but also map the distribution of the impurities by spatially rastering the sample using the characteristic frequencies as sensors for the component. A large number of examples have been reported, but only a couple are discussed here to illustrate the method.

An infrared mapping experiment is useful in determining the distribution of the contaminants in the polymer sample. In Figure 8.2, functional group images of an imperfection in polyethylene are mapped. The image is based on absorptions at 1738 cm^{-1} and is represented by both axonometric and contour plots (4).

Infrared mapping for examination of nanostructures like semiconductor devices can be helpful in tracking the source of the contamination. The mapping of a contaminant on the surface of a semiconductor device has been reported (Figure 8.3) (5). A most interesting example is the detection of an acrylic fiber contaminant on a microcircuit die (6). The two-dimensional image clearly shows the spatial configuration of a fiber based on a plot of the absorbance of the 2968 cm^{-1} band due to the fiber.

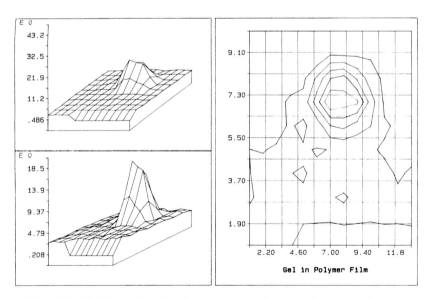

Figure 8.2. Mapping of functional group images of an imperfection in polyethylene. The image is based on absorptions at 1738 cm^{-1} and is represented by both axonometric and contour plots. (Reproduced with permission from ref 4. Copyright 1988 Marcel Dekker.)

Contaminate

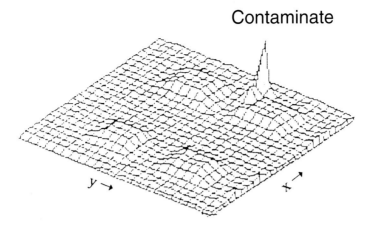

Figure 8.3. Compositional map of a contaminant on the surface of a semiconductor device. (Reproduced with permission from ref 5. Copyright 1991 International Society for Optical Engineering.)

Identification of Surface Imperfections

Microreflectance spectra can be used to study surface imperfections, whether on a painted surface or as part of a treated surface for adhesion promotion. A study has been made of a streak on a polyurethane surface (*4*). The reflectance–absorption map obtained of a streak in the film is shown in Figure 8.4. The spectra reveal that the mold release agent (a straight-chain hydrocarbon as characterized by the bands at 2920 and 2852 cm^{-1}) remained on the surface of the polyurethane, and the functional map shows the spatial distribution, which is consistent with the visual observations (*7*).

Infrared microreflectance spectra between 4000 and 700 cm^{-1} are shown in Figure 8.5 for (1) a streaked region in a blend of ethylene-acrylic acid (EAA) and ethylene methacrylic acid (EMA) copolymer on aluminum foil, and (2) a bulk film sample. The spectra show that the EMA copolymer is in higher concentration in the streaked region than in the bulk film (*3*).

Identification of Polymer Microsamples

An important industrial problem is the identification of polymer microsamples. In many cases, the quantity of available sample is limited, but IR microspectroscopy allows the analysis of samples in the nanogram range and has therefore become a most valuable tool for identification purposes (*7*). For qualitative identification, one can usually examine the

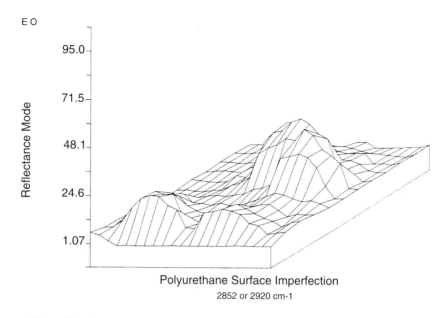

Figure 8.4. Functional group image of a surface imperfection in a polyurethane material. (Reproduced with permission from ref 4. Copyright 1988 Marcel Dekker.)

sample directly by the microscope in transmission or reflection or by one of the other microsampling techniques discussed earlier. Therefore, there is no matrix interference such as solvent, alkali halide, or mull, and the sample is recoverable for other tests without further contamination or destruction by the identification procedure.

The diamond anvil cell illustrated in Figure 8.6 functions as a sample-holding device as well as a device to press the sample into a thinner section (*8*). However, the diamond cell limits the energy throughput and contributes interference bands due to the presence of the diamond windows. On the other hand, the diamond cell allows study of samples subjected to different levels of pressure.

Identification of Fiber Samples

Infrared microscopy can be used to identify extremely small polymer samples such as fibers. Spectra of samples with diameters down to 10 μm can be obtained that are suitable for qualitative identification of the fibers (*9*). In forensic circles, this result has been a particularly important aspect of the utility of infrared microscopy. Traditionally, fiber identifications have relied on optical microscopic methods for visual comparisons of color, diameter, cross-sectional shape, birefringence, refractive index, and fluorescence. Modern fiber technology has developed a variety of new fibers,

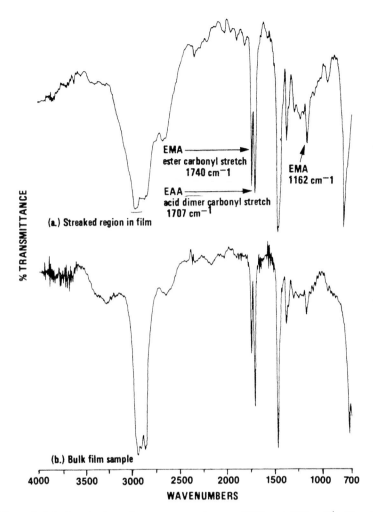

Figure 8.5. Infrared microreflectance spectra between 4000 and 700 cm^{-1} of(top) a streaked region in a blend of EAA and EMA copolymer on aluminum foil, and bottom a bulk film sample. The spectra show that the EMA copolymer is in higher concentration in the streaked region than in the bulk film. (Reproduced with permission from ref 3. Copyright 1987 American Society for Testing and Materials.)

including multicomponent or copolymer systems, which have similar textures not resolvable by traditional means. However, these modern multicomponent fibers do have internal compositional differences that can be detected by IR microscopy (*10*). A program is under way to develop a spectral database for identification of these modern fibers by IR microscopy (*11*).

In the simplest case of IR microscopy, a fiber of proper diameter can

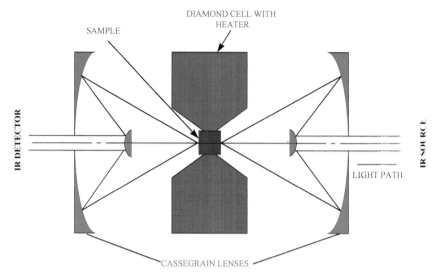

Figure 8.6. Schematic cross section of Cassegrain optics used with the diamond cell for infrared microscopic examination. (Reproduced with permission from ref 8. Copyright 1984 Society of Spectroscopy.)

simply be taped to a holder and the analysis performed in 1 min or so. Rectangular apertures are best used to mask out stray light, as they provide a much greater sampling area than circular apertures. Fibers should be positioned in the same optical orientation to prevent spectral polarization differences due to molecular orientation.

However, the commercial fibers most often have diameters too thick for microtransmission measurements using traditional sampling techniques. In this case one can use a rectangular aperture, which allows one to look at a "thinner" section, that is, the rounded portion of the fiber. Alternatively, one can "flatten" the fibers by various physical methods so that they have an appropriate optical thickness for microscopic examination (*12*). Because fibers are lightweight, a method of fixing the fibers on the specimen support has been suggested that involves using a commercial rubber cement which is diluted 1:1 with cyclohexane. The sample is fixed by the thin cement so it is attached for analysis (*13*).

IR microscopy can unequivocally identify the generic class and the subgeneric class of fibers. A particularly challenging type of fibers are the acrylic copolymers, which are not only the most common but a variety of them exist in the commercial market. To identify these acrylic copolymers, band ratios of the comonomers are measured and correlated with a collection of known fibers from different manufacturers (*9*).

IR microscopy has been used to identify different pigments loaded into single fibers of polypropylene. Using an internal band ratio method, it was also possible to make quantitative measurements of the loadings of the pigments at a level of 1% and above (w/w) (*14*).

Nonwoven fabrics have also been examined by IR microscopy (*15*). A "print bonded" nonwoven wiping cloth was examined, and the adhesive and fibers in the wiping cloth could be identified without physical separation of the components.

It has been observed that the spectra of single fibers are quantitatively different from the spectra of the same reference material. The ratios of the less intense bands to the stronger bands are considerably greater than in the reference spectra (*16*). This is a result of the variable path length arising from the circular cross section of the fiber.

A note of caution has been made concerning use of the spectra for *quantitative* purposes, such as measuring the dichroic ratio to determine the degree of orientation (*17*). An aperture suitable for the observation of the spectra of a single fiber has only a 3% transmission, which is a factor of 3 smaller than one would expect assuming homogeneous illumination. The additional loss in transmission is most likely associated with specimen-induced defocusing, which produces a higher level of stray light. In regions of low absorption, the fiber acts as a lens with a relatively high radius of curvature which defocuses the beam. Flattening of the fiber and the use of physical masking improve the potential for quantitative measurements by minimizing diffraction and stray light effects.

Identification of Particulates

Particulates also present a challenge because of their small diameter. IR microspectroscopy allows one to view specific regions of the specimen and obtain spectra. Particles with diameters as small as 10 μm can be easily examined and, with special attention, particles as small as 5 μm can be identified. If the diffraction limit dictated by the Rayleigh criterion is 10 μm at 1000 cm^{-1}, how is it possible to identify particles with a diameter of 5 μm? The diffraction limits the resolution of spectral features from adjacent materials that absorb infrared light, but diffraction does not prevent the identification of an isolated particle that has a diameter less than the diffraction limit. Isolated particles are frequently observed, and the lower limit to the detectable diameter is the *S/N* ratio of the FTIR microscope (*18*). A recent report indicates the utility of IR microscopy for this purpose (*19*).

Polymer dust particles have been analyzed using infrared microscopic techniques. It was found that cellulose is the dominant component and kaolin was a secondary component in airborne particles (*20*).

One of the particle contaminants that we have all experienced and wondered about is the common fly speck. Using micro-ATR (attenuated

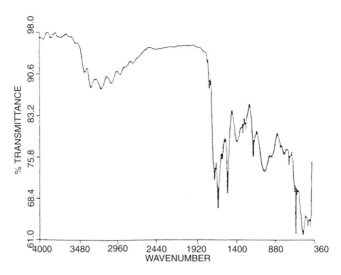

Figure 8.7. Internal reflection spectrum of a fly speck measured with a silicon hemisphere. (Reproduced with permission from ref 21. Copyright 1991 Society of Spectroscopy.)

total reflection) techniques, the spectrum of a fly speck was measured with a Si hemisphere and is shown in Figure 8.7 (*21*). Surprise, it is a proteinaceous substance!

Identification of Layers in Multilayered Structures

Polymer systems in packaging applications are often used as a combination of several polymers to form laminates. The different layers can have a variety of functions, including adhesion, prevention of diffusion of oxygen and moisture, mechanical improvement, etc. These laminates may have as few as two or up to 10 layers. Chemical separation or physically peeling the adhesively bonded layers apart is difficult and often impractical. However, if such multilayered structures are microtomed in cross section, IR microanalysis of the layers using the IR microscope can be made. An example of examination of a seven-layer laminate has been reported (*3*). IR microscopy can be used to identify layers as thin as 10 to 15 μm. Spectral identification of five layers in a laminate has also been reported (*22*). Similar results have been obtained by Gal and Toth (*23*), who demonstrated that each layer produced a discrete spectrum using computerized spectral manipulation techniques.

It is desirable to isolate unique spectra of multilayer systems where the layers are less than 10 μm thick (which is below the physical resolution of the microscope). The observed spectra of a multilayer system will therefore be a mixture spectra arising from contributions due to adjacent layers. A factor analysis method was evaluated to isolate the spectrum of a given layer by simulation. A structure was simulated that had two thick layers [polyethylene terephthalate (PET) and a vinylidene chloride containing copolymer (VCP)] with a 5-μm ethylene–vinyl acetate copolymer (EVA) layer. The results show the estimated spectra for the 5-μm layer of EVA with the standard spectra (24).

Plant cell walls are complex layered structures composed of several different polymers, principally cellulose, ligin, and noncellulosic polysaccharides. FTIR microspectroscopy has been used to study the differences in the chemistries of the inner and outer surfaces of nutshells. The outer surfaces were suberized, while the inner surfaces contained absorbances indicative of lignin or tannins or both (25). Structural differences in the cell walls of bamboo and potato tubers were also studied.

Identification of Polymer Additives

Plastics are everywhere—in automobile parts, in components for houses and buildings, and in packaging for everything from food to electronic parts. Plastics would not be able to perform such diverse functions without the utilization of a very broad range of plastic additives. Without them, some plastics would degrade during processing and, over time, the polymers would lose impact strength, discolor, and become statically charged, to list just a few problems. Additives not only overcome these and other limitations but also can impart improved performance to the final product.

There is a need for new and improved additive technologies to meet increasing performance demands from end-users. Alternative additives are needed as a result of proposed legislation coupled with public demands for environmentally friendly and recyclable plastic products. There are also needs to improve the quality and consistency of the additives.

Most commercial engineering plastics are complex multicomponent systems designed for specific applications. A physical separation is usually necessary if the identity of the individual components is required. In order to accomplish such separations, difficult wet chemical or chromatographic methods are often required.

The problem with identification of additives is that the concentration is low and cannot not normally be observed spectroscopically. To overcome this problem, extraction procedures are often used to concentrate a small amount of the additive. The sample can then be deposited on an IR plate for analysis (22). It is possible to deposit a 10-mL sample containing the solvent and still obtain a spectrum sufficient for identification.

A recent example that has been reported is the identification of a lubricant in a nylon sample (22).

For cross-linked samples like elastomers and rubbers, it is difficult to obtain a sample that is sufficiently thin for analysis. A recommended approach for samples of this kind is the pyrolysis of the specimen and the observation of the spectrum of the pyrolysis products. It has been reported that samples as small as 1 μg can be analyzed (26).

A recently developed method that should be valuable for additive analysis is an extension of the classical HPLC–FTIR technique, which is illustrated in Figure 8.8.

The current approach is the use of a new device termed the LC Transform. With this device, the LC eluants are deposited on a rotating sample collection disc while the solvent is simultaneously evaporated. A nozzle with a heated nebulizer gas eliminates the solvent while depositing the sample on the collection disc. The disc holding the deposited fractions is then placed on a rotating stage of the optics module, and spectra are obtained for each of the chromatogram components (27). The collection discs are of germanium, and the spectra are obtained by reflection. This approach allows the simultaneous identification of a number of additives in one step. The technique has also been used to map the distribution of monomers in copolymer samples (28).

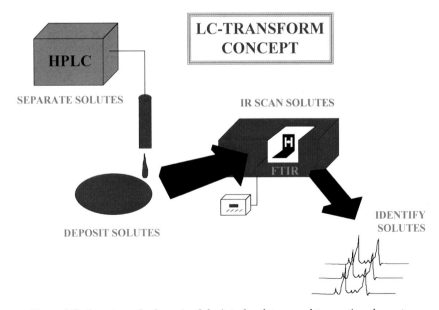

Figure 8.8. Concept and schematic of the interface between gel permeation chromatography and FTIR. (Reproduced with permission from ref 27. Copyright 1996 Society of Spectroscopy.)

A particularly general method of additive analysis utilizes high-resolution, capillary, supercritical fluid chromatography (SFC) in conjunction with an infrared microspectrometer to detect the deposited polymer additives as they elute from the column. In contrast to liquid chromatography, where a solvent is involved which produces spectral interferences if not removed, SFC with carbon dioxide has no such medium, as the carbon dioxide volatilizes on decompression. The eluant can be directly deposited onto a small area of KBr as the mobile phase evaporates away. The eluted additive sample (typically 200 μm) is then positioned in the IR microscope beam for analysis (*29*). This combination of high-resolution capillary SFC with the identification capabilities of FTIR for the qualitative analysis of nonvolatile chemical additives has been applied to a commercial polypropylene system (*29*).

Microanalysis of Multicomponent Phases in Polymers

Polymorphism

Polymorphism is the existence of a substance in different forms in the solid state. The substance has the same chemical composition in each form, but the solid-state structure is different in each form. Only a single form is thermodynamically stable, but different polymorphs can be kinetically stable and exist at the same temperature. Polymorphic forms often exhibit optical behavior that can be observed by polarized light microscopy. When one solid state form transforms into another, optical changes will occur for most samples. Although optical microscopy has been the simplest method for studying polymorphic forms, they can also be distinguished on the basis of their IR spectra. IR also has the advantage of differentiating between a polymorphic transition and the occurrence of a chemical reaction. In addition, in IR microscopy, the separate phases can be isolated and the IR spectra can be used to determine the molecular structure.

Crystalline–Amorphous Phases

Polymers often exhibit a variety of polymorphic crystalline forms as well as an "amorphous" phase. For most semicrystalline polymers, the crystals have a variety of shapes and orientations, which produce differences in birefringence. Under a light microscope, it is often possible to separate this type of sample into its individual components as a function of crystallinity or morphology. With IR analysis, it is possible to identify the chemical composition of the different morphological entities. For polyurethanes, the "soft" segments are chemically, and therefore spectroscopically, different from the "hard" segments. For polyethylene, the crystalline phase shows specific bands that are due to the ordered domains, whereas the amorphous chains show isolatable bands characteristic of the irregular structure.

Polymer Liquid Crystalline Mesophases

Polymer liquid crystals (PLCs) are considered to be unique materials. The simplest view is that PLCs can be thought of as assemblies of liquid crystal molecules strung together by polymer strings or chains. This is certainly an oversimplification. Because of the antagonistic conflict between polymer backbone entropy (driving for disorder) and mesogenic orientational ordering (striving for order), the resultant PLC structure depends on the resolution of this antagonism. The distinctive features of these systems are due to the coupling between the nematic order and the configurations of the PLCs. As a result, the structures are unique, and there are a number of mesomorphic properties which are unique to PLCs. Infrared microscopic examination allows one to identify these mesophases. Phase-separated samples can exhibit more than one mesophase, and occasionally these mesophases are spatially isolated, in which case infrared microscopic examination can be used to identify the contributing mesophases (*30*).

The sample shown in Figure 8.9 is a thermoset polymer dispersed liquid crystal, consisting of epoxy binder system EPON 828 and Capcure 3-800, and liquid crystal ZLI 1957/5. The ratio of the epoxy component to the liquid crystal component is 2 to 1. The droplet illustrated is approximately 143 μm in diameter. The sample thickness is contained between two salt plates, with Mylar spacers defining the 25-μm sample cavity. As the thickness of the cell is much smaller than the diameter of the droplet, the droplet is actually a flattened disk rather than spheroid, as the name droplet might imply.

The droplet viewed under normal light is defined reasonably well, but it is obvious that in general the sample is rather grainy in appearance, making computer-edge detection methods difficult. Viewed under polarized visible light, the sample clearly shows the typical liquid crystal droplet textures. The infrared maps of these two optical cases also show evidence of the droplet textures under polarized infrared radiation. The infrared maps were generated by using a 33-μm-diameter square aperture and incrementing spectral acquisition points by steps of 11 μm. This arrangement allowed for a sufficient S/N ratio, as well as high enough data resolution. Spectra were acquired at a resolution of 4 cm^{-1}, from 4000 to 650 cm^{-1}, through a 15× power Cassegrainian objective. Each spectrum was a result of 64 coadded scans.

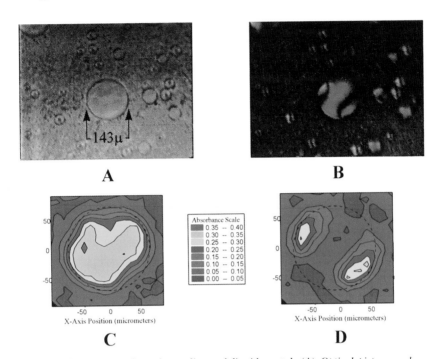

Figure 8.9. Images of a polymer dispersed liquid crystal. (A) Optical picture under normal light. (B) Optical picture under polarized light. (C) Infrared image under 0° polarized radiation. (D) Infrared image under 90° polarized radiation. The dashed contours in C and D represent the physical boundary of the droplet indicated in A.

Admixtures in Cements

A series of cements were examined to identify the ferrite, alite, and belite crystals in cement phases (*31*). The effect of bonding of organic additives on the various phases was then examined. An IR examination of various admixtures used in slurries for both structural and oil-well cements has been reported. Three lignosulfonates, a sulfonated naphthalene–HCHO condensate, a sugar–lignosulfonate mixture, and a calcium gluconate based material were examined. The IR spectroscopic results were found to be useful in characterizing these admixtures (*32*).

FTIR Microanalysis of Phase-Separated Polymer Systems

Introduction to Polymer Dispersed Liquid Crystals

Polymer dispersed liquid crystals (PDLCs) are an important area of liquid crystal technology (*33–39*). Nematic LC microdroplets are dispersed in a polymer matrix. These films are used like LCDs for applications such as displays, switchable windows, and light shutters. PDLC devices operate on the principle of electrical modulation of the LC refractive index to match or mismatch the refractive index of an optically transparent, isotropic solid(40–41). PDLC films are made between conductive, transparent substrates, and can be switched from being opaque to being transparent with the application of an electric field.

There are three basic methods of preparing PDLCs: polymerization-induced phase separation (PIPS), solvent-induced phase separation (SIPS), and thermally induced phase separation (TIPS). Also, other methods exist to make dispersions of LCs and polymers, such as nematic curvilinear aligned phase (NCAP) and gel dispersions. The NCAP procedure involves emulsifying an LC in an aqueous mixture containing the encapsulating medium. The NCAP films start with inhomogeneous solutions, whereas the phase-separated PDLCs start with homogeneous solutions from which the LCs separate. The gel dispersions use cholesteric LCs with nematic LCs dispersed in a UV-cured acrylate polymer. These displays can be colored, depending on cell thickness. The gels can be made clear or translucent when a voltage is applied across the films.

PIPS is a useful method when prepolymers are miscible with the low-molecular-weight LCs. The polymer matrix may be a thermoplastic, but it is usually a thermoset capable of cross-linking. The process begins with a homogeneous solution containing monomeric or oligomeric materials and LC. The polymerization occurs through a condensation reaction, free radical polymerization, or photoinitiated polymerization, depending on the starting materials. Epoxies are common polymer binders used to make PDLCs in the PIPS process. Often, other agents are added, such as a catalyst or a material to help match the polymer's refractive index to the LC's ordinary refractive index. The LC solubility decreases as polymerization occurs, until the LC finally phase-separates into droplets. The droplets are able to grow until droplet morphology is locked by polymer gelation. Droplet size and morphology are determined by events occurring between the time of droplet nucleation and polymer gelation. The rate of polymerization is one factor that determines droplet size. For example, changing the curing rate and curing temperature of an epoxy PDLC affects polymerization. In general, higher cure temperatures produce smaller droplets because the polymer cures faster, and there is less time for droplet growth.

SIPS is used for thermoplastics and liquid crystals dissolved in a common solvent. This procedure also is advantageous if the polymer melts above the decomposition temperature. The evaporation of the solvent causes phase separation, and subsequent droplet formation and growth, to create a PDLC film. Essentially the same time line diagram as described for PIPS is followed. The droplet size can be controlled by the solvent evaporation rate. In general, the faster the evaporation rate, the smaller the droplets. In addition, the polymer will be plasticized by the dissolved liquid crystal, which will decrease the glass-transition temperature and enhance polymer processability.

Thermally induced phase separation (TIPS) involves heating the polymer–LC solution to a homogeneous state. As the solution cools, the miscibility gap is passed so that droplets form and grow. The droplet size depends on the cooling rate and the concentrations of the components. Generally, increased LC concentration increases the size of the droplets. Also, larger droplets are formed with a slow cooling rate. If the cooling rate is fast, more of the LC remains dissolved in the polymer matrix. For some systems, droplets as large as 100 μm can be formed. Although droplets of such a large size are not useful in practical display applications, they are useful for theoretical study.

Infrared Microscopic Studies of PDLCs

One of the first considerations in studying any type of LC or PDLC by infrared spectroscopy is the type of window/substrate to be used. One of the most important features of LC technology is the ability to "switch" the materials with an applied field. For most display applications, a conductive and transparent coating, such as indium tin oxide (ITO), is used on a glass substrate. Vossen (*40*) presents a useful review of these types of coatings. However, glass is an IR-absorbing species, so other types of windows must be employed for IR. There are several options. Germanium is a semiconductive, IR-transparent substrate. The resistivity is 46×10^3 Ω cm. Although the resistivity is relatively high, germanium conducts well enough that an electric field can be established. The disadvantage of using Ge is that it is silver colored and not visibly transparent. Also, Ge is quite reflective, so much of the IR energy is lost.

A second window material is platinum or gold-plated sodium chloride. Salt plates can be purchased with either of these coatings. The biggest disadvantage of this substrate is poor adhesion between the coating and the salt plates. The platinum is easily rubbed off the salt plate by fingers and electrodes. Another disadvantage is that the coating must be very thin—in the range of hundreds of angstroms. The thin coating causes a high resistivity, necessitating a higher voltage to apply the electric field. Also, if the coating is too thick, then the infrared transmittance is seriously reduced.

Another type of coating is indium sesquioxide (In_2O_3). Chen et al. (*41*) developed and characterized these coatings for the IR wavelength range. These films do not have the high reflectivity and highly dispersive optical constants of gold, copper, and silver coatings. Those undesirable properties result from their high free-carrier concentrations and low elec-tron mobilities. Chen et al. found that the transparency and conductivity of In_2O_3 in the IR range of 1–12 μm (10,000–833 cm^{-1})were due to their high electron mobility, which permits an appropriate electron concentra-tion (41). They used a special reactive thermal evaporation process, which is simpler than other techniques such as sputtering or spraying. A diffu-sion-pumped vacuum with a resistance-heated source was used to evapo-rate a mixture of In_2O_3 and metallic In onto a zinc sulfide (ZnS) substrate. The resistivities were on the order of 20–50 Ω cm with 65–80% transmis-sion.

Yet another type of IR-transparent coating was developed by Bachte-mann and Schulz (42), who used thin carbon films produced by evapo-ration. Their technique is unique in that they evaporated the carbon films directly onto their polymer sample. They heated two carbon rods, with the sample 15 cm away from each rod. The polymer films were 9–13 μm thick, and care was taken not to heat the sample too much and not to let spark particles hit the sample. In the wavenumber range of 350–800 cm^{-1}, greater than 95% transmission was achieved. The resistance was about 100 kΩ measured over 10 mm. This technique could not be used for liquid crystals, but would be quite useful for free-standing PDLC films.

With regard to pure liquid crystals in particular, surface treatment of the substrates is another important factor to consider when taking IR spec-tra. IR is an excellent technique for studying orientation, with or without the presence of an applied field. A preferential orientation is required for dichroic studies, and a liquid crystal can be aligned by surface treatments. There are two types of alignment: homogeneous and homeotropic. Ho-mogeneous alignment occurs when the LC molecules are parallel to the surface, and homeotropic alignment occurs when the LC molecules are perpendicular to the surface. Also, the angle between the long axis of the LC and the surface is often referred to as the tilt angle. There are many chemical surface coatings processes as well as many "rubbing" techniques.

Knowing the initial alignment of the LC, studies about the electric-field-induced orientation transition can be performed. Gregoriou et al. (43) switched a nematic liquid crystal, 5 cyanobiphenyl (CB) and studied the transition from homogeneous to homeotropic alignment using step-scan, two-dimensional FTIR. They initially aligned the LC by rubbing pol-yvinyl alcohol with a cotton swab. Using polarized IR data, they showed that the molecules do align with the electric field and that the transition can be represented by a 90° rotation of the nematic director. From the two-dimensional FTIR data, they show that the rigid core of the 5 CB molecule reorients as a unit, and that the more flexible alkyl chain reo-rients faster than the core.

Hatta studied the reorientation of 5 CB in the boundary region of a twisted nematic cell by an electric field using attenuated total reflectance

(ATR) spectroscopy (44). He used a silicon ATR prism and a SiO-coated silicon plate for electrodes. He found that it is possible to characterize the anisotropic structure of the LC–substrate boundary layer at the electrode surface. He also confirmed that the molecular alignment at the surface in the absence of the applied field was uniaxial, as suspected.

In a later study, Hatta et al. used modulated ATR spectroscopy (45) to study the boundary region again. This time they used an untreated germanium ATR crystal and a rubbed ITO-coated glass plate so the 5 CB molecules were perpendicular to the plane of light incidence. They showed that the molecules at the germanium surface (20 μm from the rubbed ITO surface) had the same orientation as the molecules near the ITO surface. Therefore, the orientation at the top electrode was due to long-range ordering in the nematic phase from the counter electrode. Also, by changing the depth of penetration of the IR beam, they showed that the field-induced alignment was increasingly suppressed closer to the germanium surface.

The birefringence of a liquid crystal is another parameter that can be measured by IR spectroscopy. Wu et al. have measured the infrared bi-refringence of LCs throughout the 2–16 μm (5000–625 cm^{-1}) range (46, 47). The data were calculated from measurements of the phase differences that occur when monochromatic polarized light propagates through the LC, which has an anisotropic refractive index. As expected, the bire-fringence Δn, (where n is the refractive index), of the LC does not de-crease with increasing wavelength. There is significant resonant enhance-ment of Δn by specific molecular absorption bands, causing unusually large and nearly constant values of Δn (47). Large birefringence is desir-able for LC devices because the device can be made thinner to produce faster switching times.

Linear diattenuation also can be determined by IR. Chenault et al. (48) describe linear diattenuation as a measure of the tendency of a sam-ple to linearly polarize incident light that is unpolarized. Diattenuation is not desirable for an electrooptical modulator. The larger the linear diat-tenuation, the further from orthogonal are the initial and final polariza-tion states (48).

West and Ondris-Crawford (49) have studied PDLC shutters by IR spectroscopy. They measured the IR absorbances of the PDLC in the ON and OFF states to find the amount of liquid crystal dissolved in the poly-mer binder (Fig. 8.9). Any LC in the binder will not be affected by the applied field and will remain randomly oriented. Only the LC confined in the droplets will exhibit dichroic properties. They found that films having larger droplets have a higher percentage of LC dissolved in the binder, not accounting for light scattering. The larger droplets produce more scattering. They determined that approximately 20% of the LC is dissolved in the binder in a PDLC of E7–polyvinylpyrrolidone.

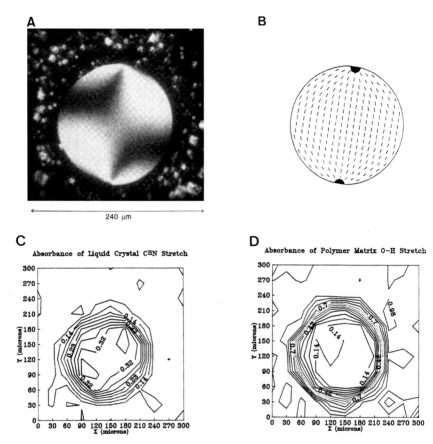

Figure 8.10. (A) Optical image of a polymer dispersed liquid crystal. (B) Schematic of the orientation of the liquid crystal in the epoxy matrix. Also shown are a map of the same region, obtained using an infrared microscope (C) for the liquid crystal and (D) the epoxy level. (Reproduced with permission from ref 51. Copyright 1993 Society of Spectroscopy.)

Infrared Mapping of PDLCs

The optical image of a PDLC is shown in Figure 8.10A, and a schematic of the orientation of the LC in the epoxy matrix is shown in Figure 8.10B. A map of the same region, made using an infrared microscope, is shown in Figure 8.10C for the LC, which is a eutectic mixture of phenylcyclohexanes, and in Figure 8.10D for the epoxy (EPOM 828 cured with Capcure 3-800) level. The inverse relationships between the absorbance confirm the existence of the LC in the droplet surrounded by the epoxy resin(50, 51) These IR textures change with IR polarization (Figure 8.11) and applied voltage (Figure 8.12).

0 POLARIZED IR (Absorbance C≡N Stretch)

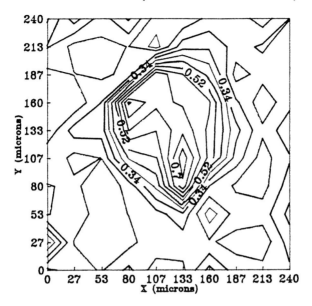

90 POLARIZED IR (Absorbance C≡N Stretch)

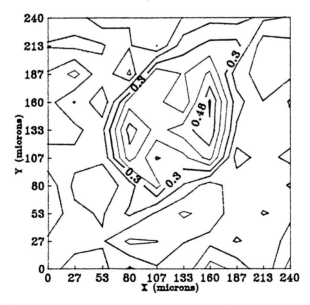

Figure 8.11. Two infrared images showing the effect of IR polarization based on the nitrile stretching mode of the liquid crystal. (Reproduced with permission from ref 51. Copyright 1993 Society of Spectroscopy.)

ELECTRIC FIELD OFF (Absorbance C≡N Stretch)

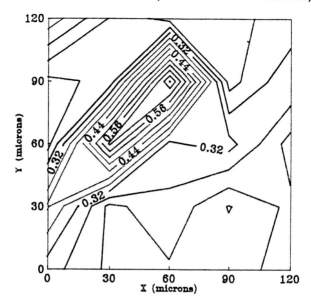

ELECTRIC FIELD ON (Absorbance C≡N Stretch)

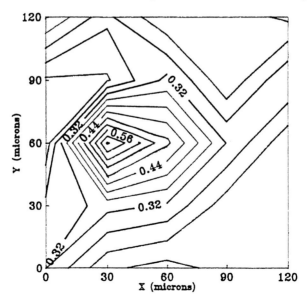

Figure 8.12. Two functional group images demonstrating that the effect of an applied electric field within the droplet changes with the field. (Reproduced with permission from ref 51. Copyright 1993 Society of Spectroscopy.)

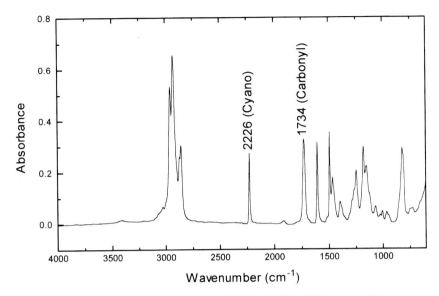

Figure 8.13. Infrared spectrum of the E7/PBMA 70/30 PDLC system. The cyano peak is characteristic of the liquid crystal, whereas the carbonyl peak is unique to the polymer. (Reproduced with permission from ref 52. Copyright 1994 Society of Spectroscopy.)

The smaller droplets require a higher voltage for switching because of the presence of greater surface anchoring. With greater anchoring, the voltage required to align the droplet is lower. Details of the switching process suggest that the nitrile groups and the pentyl chain groups switch more easily than the phenyl groups. The influence of the polymer matrix on the bipolar and radial droplet configurations within the PDLC films has also been examined by infrared microspectroscopy (*50*). With the change of polymer from poly(*n*-butyl methacrylate) to poly(isobutyl methacrylate) (PiBMA) with the E7 LC (a eutectic LC mixture consisting mostly of 4-pentyl-4'-cyanobiphenyl [5 CB]), the droplet configuration changes from bipolar to radial. With PiBMA, the droplet size grows with time, which is related to the low glass-transition temperature of the polymer (*52*). A detailed study of the growth of these PDLCs has been made (*51*). Recently, a detailed study of the phase-separation process has been made (*54*). Studies of the orientational changes induced by an applied electric field have also been reported (*54*).

The infrared spectrum of a 100-μm region of the E7–PiBMA 7:3 PDLC is shown in Figure 8.13 (*52*). The composite spectrum is simply an addition of the pure components with minor shifts in some of the peaks. The cyano peak at 2226 cm⁻¹ is unique to the liquid crystal, whereas the carbonyl band at 1728 cm⁻¹ is specific for the PiBMA polymer. The

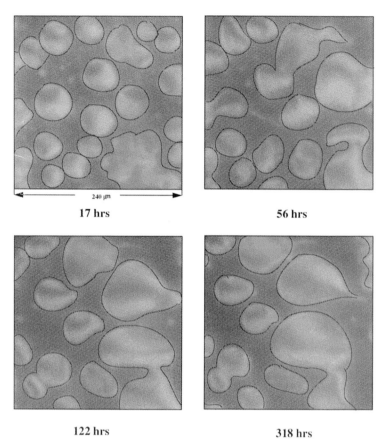

17 hrs   56 hrs

122 hrs   318 hrs

Figure 8.14. Optical images of a 240-μm region under cross polarizers at different times. (Reproduced with permission from ref 52. Copyright 1994 Society of Spectroscopy.)

changes in the peak intensities of these peaks with position give the spatial information about the composition of each component. By monitoring these alterations in the intensities within a particular region as a function of time, one can monitor the kinetics of the growth of droplets.

The optical images of the PDLC under crossed polarizers are shown in Figure 8.14 at different times. An outline of the droplets is drawn to facilitate observation of the changes that occur within the system. It is seen that the smaller droplets disappear over a period of time, and the number of droplets decreases as the average radius increases.

Infrared microscopic images obtained using a 24×24 μm aperture are shown in Figure 8.15. The images are based on the absorbances of the carbonyl peak. Different colors are associated with the different absorbances. Red in the color chart corresponds to the maximum absorb-

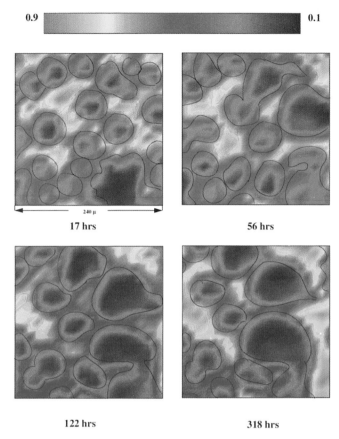

0.9 ▮ 0.1

17 hrs   56 hrs

122 hrs   318 hrs

Figure 8.15. IR spectroscopic images based on the absorbance of the carbonyl peak. The images reflect the concentration of the polymer within the system. (Reproduced with permission from ref 52. Copyright 1994 Society of Spectroscopy.)

ance of 0.9, while blue represents the minimum absorbance of 0.1. The blue regions indicate very low concentration of polymer and hence a high concentration of liquid crystal. The concentration fluctuations that occur over a period of time are clearly seen.

The driving force for the process is the purification of the LC-rich phase. With the progress of time, the droplets tend to get richer in LC, the thermodynamic force driving them to pure LC droplets. The growth of the LC droplets occurs through the combination of coalescence and Ostwald ripening. The average radius increases with time as the number of droplets decreases. This process follows temporal power laws.

The basic requirement for the control of the droplet size in PDLC is a knowledge of the composition phase diagram for the specific LC in the thermoplastic material. A new approach to making this type of measure-

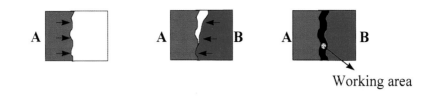

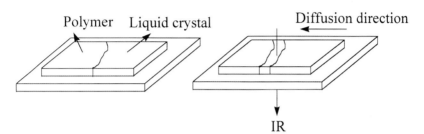

Figure 8.16. Diagram of the contact diffusion method, showing the region of measurement for FTIR microscopy. (Reproduced with permission from ref 53. Copyright 1994 Society of Spectroscopy.)

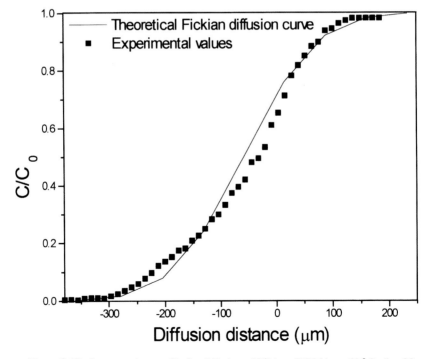

Figure 8.17. Concentration profile for diffusion of E7 into PiBMA at 60 °C after 20 min. (Reproduced with permission from ref 53. Copyright 1994 Society of Spectroscopy.)

Diffusion direction

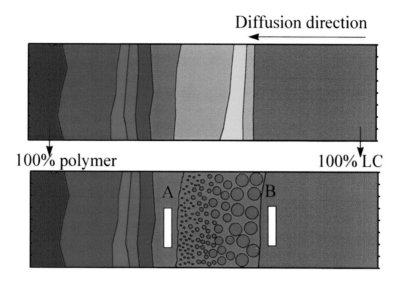

Amount of LC dissolved in the polymer matrix = A/B

Figure 8.18. Diagram of the relationship between concentration gradient and particle formation. (Reproduced with permission from ref 53. Copyright 1994 Society of Spectroscopy.)

ment is through the use of a contact diffusion method. In this method, the LC is placed in contact with the thermoplastic material, as shown in Figure 8.16.

The diffusion of the LC is allowed to take place at a predetermined temperature, resulting in a gradient of concentration arising from the Fickian diffusion. An example of this process is shown in Figure 8.17. Then the sample is quenched to room temperature and the phase separation occurs, with the dispersed particles attaining a size that reflects the concentration of the LC in that zone. A diagram of this process is shown in Figure 8.18. An optical image demonstrating this result is shown in Figure 8.19. An IR functional image is then obtained. The results are shown in Figure 8.20. In this fashion, it is possible in a relatively short time to obtain the necessary information concerning the composition dependence of the PDLCs.

FTIR Microspectroscopy of Adhesive Systems

The formation of strong, reproducible, and reliable adhesive bonds is an important technological problem. The phenomenon of adhesion is de-

Diffusion direction

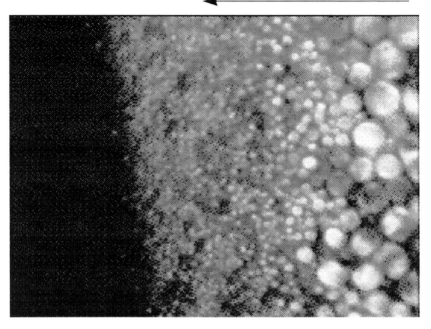

Figure 8.19. Optical image of the phase-separation process in the presence of a concentration gradient. (Reproduced with permission from ref 53. Copyright 1994 Society of Spectroscopy.)

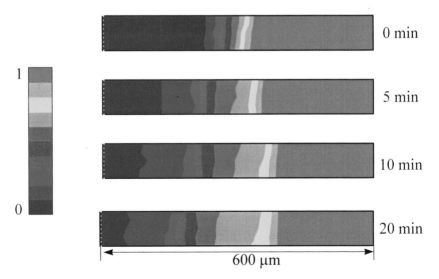

Figure 8.20. Infrared map of concentration profiles at different times. (Reproduced with permission from ref 53. Copyright 1994 Society of Spectroscopy.)

scribed in terms of four mechanisms: mechanical interlocking, diffusion, electronic interaction, and adsorption. The most dominant mechanisms are adsorption and mechanical interlocking. The difficulty limiting adhesion studies and their improvement has been the lack of information concerning the composition and structure of the adhesive bond and the nature of the distribution across the interface.

Adhesive performance and reliability depend on the adhesive–substrate interface behavior. In order to develop adhesive systems, it is important to relate the peeling strength of adhesive samples to the interactions occurring at their interfaces. The physical properties of the surface, including roughness and surface energy of the substrate material, as well as the chemical properties (e.g., chemical composition of the substrate material), influence the resulting adhesive bond. The presence of impurities such as moisture at the interface can have a serious effect on the tensile and shear strengths of the adhesive bond. In order to understand the molecular mechanics of the adhesive bond, it is important to probe the molecular structure of the adhesive in contact with the surface and relate this structure to the adhesive strength of the adhesive–substrate bond. The molecular forces and the conformation energies of polymer chains at surfaces differ from those of bulk phase since polymer molecules at surfaces have less motional freedom. The surface energy is therefore higher than the energy of the bulk phase. The excess free energy available at the solid surface provides the necessary driving force for the adhesive bond to form.

The concentration of the different adsorption sites on solids can be probed by IR microspectroscopy. The adsorption of the adhesive layer on the substrate surface can be followed and changes in chemical structure determined. Microscopic mapping experiments are useful for a determination of the uniformity of the chemical surface as well as the adhesive. Bond-line corrosion is a major degradation mechanism in bonded interfaces. The exposure of the adhesive to stress, temperature, and environmental probes like water can also be followed with IR microspectroscopy.

In this case, infrared spectroscopy has the advantage of being able to be used to study the samples under operating conditions rather than in a high vacuum system. Reflection measurements probe the surface and are highly sensitive, so that small changes in structure can be easily determined.

Polymer–Polymer Interfaces

Vibrational microspectroscopy can be used to study an adhesive coated interface (55). A polymer–polymer interface consisting of an ethylene–acrylic acid (EAA) copolymer (15% acrylic acid), which was sprayed with a urethane paint to a thickness of 100 μm, has been studied. Microtomed

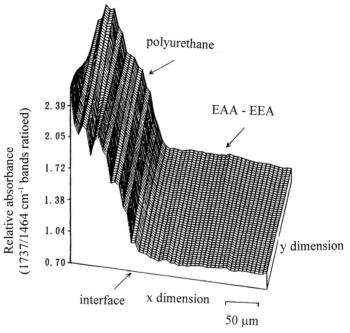

Figure 8.21. Microspectroscopic functional group image based on the 1737/1464 cm⁻¹ ratio of absorption frequencies, for the interface between a coated polyurethane and an ethylene–ethyl acrylate copolymer whose ethyl ester group has been partially hydrolyzed (EAA–EAA). (Reproduced with permission from ref 56. Copyright 1992 Society of Spectroscopy.)

samples (5 μm thick) were cut vertical to the coated surface and examined by microtransmission using an aperture set at 10 × 200 μm, and the mapping was accomplished by moving the sample in steps of 10 μm. It was observed that the absorbance of the carboxyl group is reduced near the polyurethane interface, reaching a depth of 60 μm of the EAA layer. This indicates that the polyurethane has penetrated to a depth of 60 μm from the interface of the sample. Figure 8.21 shows the microspectroscopic functional group image based on the 1737/1464 cm⁻¹ ratio of absorption frequencies (56).

It is also possible to study the formation of a polymer–polymer interface by interfacial diffusion (57). Figure 8.22 shows a schematic of the experimental technique used for observing the formation of a polymer–polymer interface by interdiffusion, as observed by transmission FTIR. A pure film of one polymer is placed in contact with a pure film of the second polymer, and then the polymers will slowly diffuse and ultimately achieve phase equilibrium across the polymer–polymer interface, yielding a single homogeneous phase. The mutual diffusion of poly(ethylene-*co*-

Pure PVME film

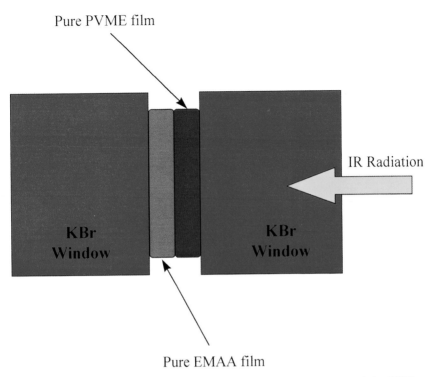

IR Radiation

KBr Window   **KBr Window**

Pure EMAA film

Figure 8.22. Schematic of the experimental technique used in which transmission FTIR is used to study the formation of a polymer–polymer interface by interdiffusion. Reproduced from ref 57. Copyright 1992 American Chemical Society.)

methacrylic acid) (EMAA) and poly(vinyl methyl ether) (PVME) has been studied, and typical spectra as a function of time are shown in Figure 8.23 (57). The dominant band at 1700 cm^{-1} corresponds to the pure EMAA copolymer, and there is no significant contribution detected at 1730 cm^{-1} because of the acid–ether hydrogen bonding interactions. With increasing time, during which the EMAA and PVME diffuse together, there is an increase in the 1730-cm^{-1} band at the expense of the 1700-cm^{-1} band. Using a mathematical model of the diffusion process and the spectroscopic measurements, one can measure a diffusion constant for the EMAA of 6.5×10^{-1} cm^2/s. Concentration profiles of the polymer–polymer interface as a function of time can be calculated and are shown in Figure 8.24 (57).

Polymer–Steel Interfaces

Zinc-coated steels are used extensively (as much as 30%) in the automobile industry. Some elements, when added to hot-dip galvanized steel,

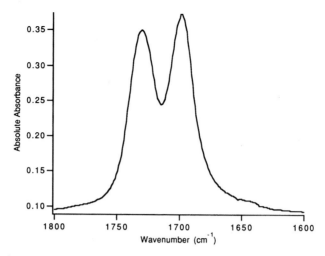

Figure 8.23. FTIR spectra of the EMAA-PVME diffusion couple as a function of time at 110 °C. (Reproduced from ref 57. Copyright 1992 American Chemical Society.)

improve the adhesion properties. FTIR microscopy can be used to determine the nature of the chemical reaction occurring at the interface between the substrate and the bulk adhesive (58). A three-point flexure test was used to induce failure of the interface of the galvanized steel and the epoxy adhesive. The failed delamination zone was analyzed using grazing angle incidence to determine (1) whether a thin layer of adhesive remains on the substrates, and (2) whether this layer originates from an interphase such that the chemical composition differs from that of the bulk adhesive.

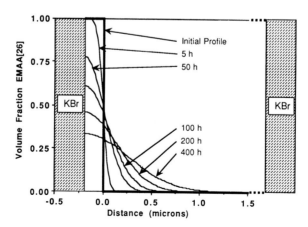

Figure 8.24. Theoretical concentration profiles of EMAA during the formation of a polymer–polymer interface as a result of mutual diffusion. (Reproduced from ref 57. Copyright 1992 American Chemical Society.)

The IR analysis was performed with an IR aperture of 100 μm and clearly indicates that a thin layer of polymer remains on the metal after fracture. This interfacial polymer is compositionally different from the bulk of the adhesive. The modified zone of adhesive near the coated steel surface is greatly enriched in nitrile groups (*58*).

Dicy reacts readily when heated with zinc substrates and is used for corrosion protection. The reduction of dicy by metallic zinc is likely to occur. The spectral changes observed after heating the dicy on zinc show the formation of a definite compound. The major changes are relative to the nitrile band. Upon end-on coordination to a metal through nitrogen lone-pair orbitals, this band is known to shift toward high frequencies (*59*). A shift toward low frequencies suggests that the dicy acts as a π-acceptor and therefore undergoes a side-on coordination. In this context, such a reaction should be governed by acid–base (electron acceptor–donor) properties of the substrate surface (*59*).

Polymer–Aluminum Interfaces

Environmental failure of the adhesive-bonded aluminum joint can be caused by displacement of the adhesive (or coating–primer) from the adherent by ingress of moisture causing weakening or rupture of secondary bonds. This weaking can result from the transformation of the Al oxide into a weaker oxide, or hydrolysis of the adhesive (or coating–primer) layer near the adherent surface. A decrease in the fatigue life of the aluminum adherents due to the formation of a fatigue-sensitive oxide layer can also occur.

Surface treatments can create a substrate structure that is more suitable for mechanical interlocking and increase the long-term durability of the joints by the creation of stable oxides (*60, 61*). The surface pretreatments and primers for Al can be divided into the following types:

- cleaning methods (e.g., methanol and bath pretreatments such as alkaline baths),
- etching methods (e.g., flourinated phosphoric acid etch or a sodium hydroxide etch),
- mechanical methods (e.g., brushing, grinding, and grit blasting),
- anodizing (e.g., in phosphoric acid, chromic acid, or sulfuric acid),
- adhesion promoters (e.g. ,silanes),
- water-based primers, and
- primers using inhibitors other than chromates.

FTIR microspectroscopy has been used to study the chemical differences in the anodized layers on the surface of aluminum (62). Studies have been made for several different anodization treatments of an Al alloy by grazing-angle IR microspectroscopy (62). The results for the chromic acid anodization (CAA) process are shown in Figure 8.25.

A *chemical sectioning procedure* was developed in which a boiling sulfuric acid solution was used to dissolve the anodic oxides from the surface of the sample, so chemical differences as a function of distance from the original surface could be obtained as a result of measuring spectra of films of different thicknesses. The spectra show chemical differences as a function of the thickness of the anodized layers. The broadband at 1150 cm^{-1} decreases, while the broadband at 950 cm^{-1} is relatively unaffected as the thicknesses of the film decrease. The broad band at 1150 cm^{-1} is associated with the longitudinal optical phonons of the hydroxides. The band at 950 cm^{-1} is associated with the Al–O–Al species. Thus, the hydroxides and water are preferentially concentrated near the surface of the treated samples, while the alumina (of the kind seen in the untreated sample) is concentrated near the metal surface. The bands associated with the contaminating species are concentrated near the surface, as they are not observed after a short period of dissolution. However, bands associated with carbonates remain. The carbonates either are incorporated in the oxide network or they are within the pores that penetrate through the barrier layer.

An infrared microprobe study of the epoxy adhesive on the chemically treated Al (63, 64) demonstrated that the chemical nature of the surface influences the polymerization of the epoxy system. A *spatially resolved method* was developed to compare the effects of the different treatments on the same film. In order to enhance the differences, a single film of epoxy was spin-cast on the Al sample, and the spectra were obtained from three different spatially resolved regions of the same sample preparation (Figure 8.26) (63).

As shown in Figure 8.27, the intensities of the 2210- and 2160-cm^{-1} bands due to the dicyandiamide curative for the uncured epoxy system are dominant for the (A) untreated Al surface, and decrease in going from the (B) sulphuric acid anodized (SAA)–untreated, and (C) treated portions for a film 0.1 μm thick. Thus the amount of curative in any portion of the film depends on the nature of the Al surface, and the curative segregates preferentially on the untreated Al surface. In contrast, the uncured epoxy is preferred on the anodized Al samples. The cured epoxy has a cross-linked structure that depends on the adsorption process on the substrate. The cross-linked structure influences the nature of the interfacial bonds.

The loss of interfacial strength of polymer–Al interfaces due to exposure to humidity is well-established, but the mechanism by which water

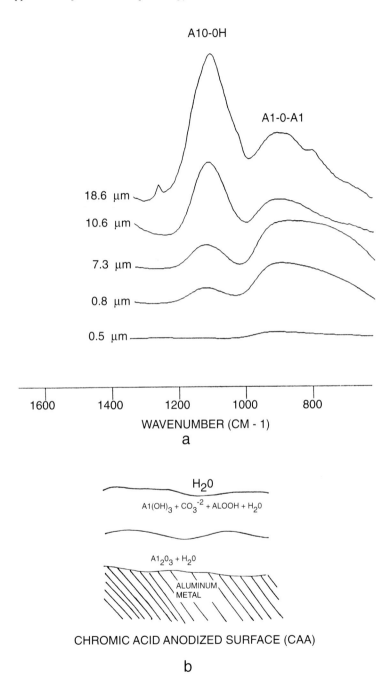

Figure 8.25. (a) Microreflection FTIR measurements on chromic acid anodized Al alloy. (b) Schematic cross section of the sample. (Reproduced with permission from ref 62. Copyright 1993 Gordon & Breach.)

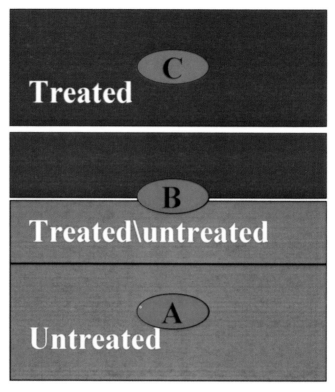

Figure 8.26. Regions of and adhesive-bonded aluminum joint sampled by spatial IR spectroscopy. (A) Untreated. (B) Treated–untreated. (C) Treated. (Reproduced with permission from ref 63 . Copyright 1993 Society of Spectroscopy.)

interacts with the adhesive joints is not known. Initially, oxide dissolution by water was thought to be the primary reaction in environmental exposure. The transformation of anodized surfaces to boehmite and then to bayerite was responsible for the formation of a weak boundary layer. Corrosion has also been suggested as a mode of environmental decay for adhesive bonds involving Al. FTIR microscopy has been used to re-examine this problem, as it can discern the nature of the moisture-induced chemical reactions in the epoxide and oxide layers (65). The spatially resolved technique just discussed was used to compare the ingress of water in the joints. The reflectance spectra obtained were a composite of the absorbances of the epoxy and the aluminum oxides. Functional group maps were generated for both the anodized surfaces and the untreated, cleaned Al surface. Figure 8.28 is a three-dimensional plot of the Al–O–Al vibrational band at 1104 cm^{-1} as a function of the spatial coordinates on the sampled area of the CAA–untreated sample after exposure to humidity for three months (65).

Figure 8.28 shows a rough surface image with new absorbance peaks.

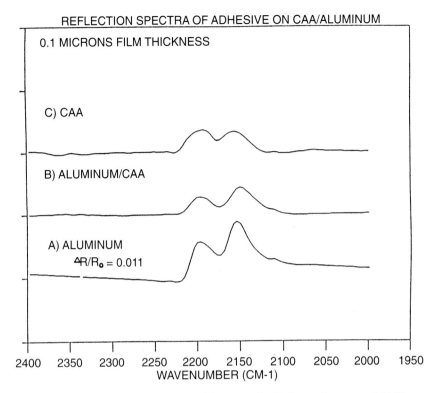

Figure 8.27. Microreflectance spectra of the uncured adhesive on the untreated Al (A), Al–CAA interface (B), and CAA surface (C). (Reproduced with permission from ref 63. Copyright 1993 Society of Spectroscopy.)

These new absorbance peaks are not associated with surface roughness in the original sample before exposure. Thus, they are the consequences of water interaction and are associated with the formation of new boehmite. The effect of water on the anodic aluminum–epoxy joints is to transform the oxides of the anodic film into boehmite. In the untreated Al portion of the sample, there appears to be both oxide and oxyhydroxide formation occurring simultaneously. This same effect, to a lesser extent, is observed for the CAA–untreated Al samples (65). The only organic degradation of the epoxy observed was the formation of carboxylate species, which were found at the outer surface of the epoxy film (the air–epoxy interface).

FTIR Micromapping of Interfaces in Fiber-Reinforced Composites

The use of modern spectroscopic techniques has led to remarkable improvement in the molecular understanding of the interfacial structure.

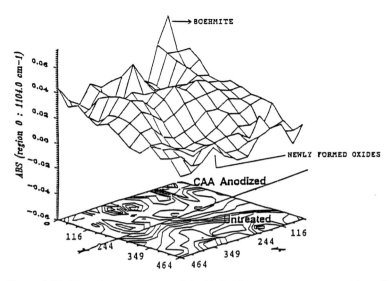

Figure 8.28. Three-dimensional plot of the Al–O–Al vibrational band at 1104 cm⁻¹, as a function of the spatial coordinates on the sampled area of the CAA–untreated sample, after exposure to humidity for three months. (Reproduced with permission from ref 62. Copyright 1993 Gordon & Breach.)

Although FTIR spectroscopy has been proven to be a powerful analytical method that can yield important structural information about fiber-reinforced composites, detailed spectroscopic information concerning the interfacial structure cannot be obtained conventionally(*67, 68*).

With the development of FTIR microspectroscopy, detailed analysis of the spatial distribution of chemical species at the fiber–matrix interphase can be performed. The coupling of an infrared spectrometer to a microscope allows the in situ analysis of the interfacial region and the detection of fiber–matrix interactions. Localized infrared microspectroscopy has been used to characterize the interphase of fiber–matrix composites. Moreover, infrared mapping was performed to analyze the interactions between the matrix resin and different surface treatments and types of fibers.

Kevlar Fiber Composites

Poly(*p*-phenyleneterephthalamide) (PPTA) fibers, known commercially as Kevlar, reinforced with epoxy have become one of the primary composites used in tensile loading applications. However, the transverse properties of these composites are relatively low because of the poor adhesion between the fibers and the matrix resin. The low adhesion between Kevlar and matrix resin is believed to be a result of the fiber's surface morphology and chemical structure. The smooth surface of Kevlar precludes adhesion via mechanical interlocking, and its chemical inertness limits adhesion through chemical bonding. A number of methods have been used to improve the bonding of the epoxy to the surface of the fiber, including wet methods. Infrared mapping has been used to study the effects of chemical reactions on the surface of Kevlar (*66*) and the spatial distribution of chemical species at the fiber–matrix interphase. Kevlar fibers with surface metalation were compared with clean and dried Kevlar fibers. The localized microsampling technique is shown in Figure 8.29.

The treated and untreated fibers are aligned parallel to each other on a KBr plate. Thin films of epoxy with an aliphatic polyamine curing agent, DEH 26 (tetraethylenepentamine), were deposited on the fibers and cured for 30 min at 30 °C in a vacuum oven. The samples were mounted on the motorized IR micropositioning stage, where the mapping experiments were performed on a 90-μm^2 area with the use of 24 × 24 μm square apertures. The map generated from this experiment contains 11 rows consisting of 11 spectra with 9-μm step increments in both the x and y directions. A functional image of the N–H stretching vibration at 3326 cm^{-1} due to the Kevlar fiber is shown in Figure 8.30. The IR intensities are represented by colors, with red indicating the areas of highest intensities and blue the lowest. In Figure 8.30, the image displays a lower level of intensity along the surface of the reacted fiber as compared with

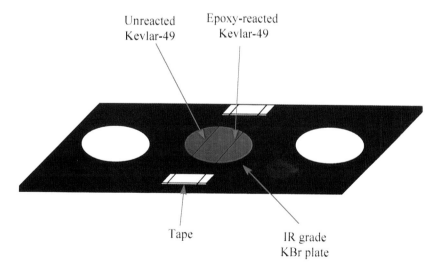

Figure 8.29. Schematic representation of the alignment of single filaments of unreacted and epoxy-reacted Kevlar-49 on an IR-grade KBr plate before the casting of thin films of epoxy and amine curing agent. (Reproduced with permission from ref 66. Copyright 1995 Society of Spectroscopy.)

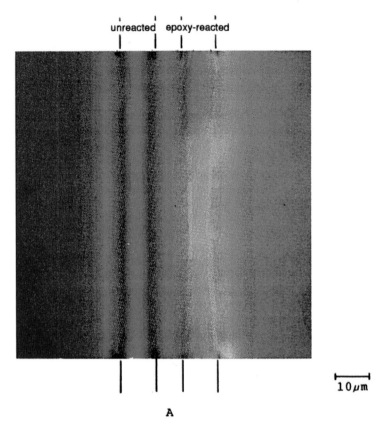

Figure 8.30. IR functional group images of reacted and unreacted fibers of the N–H stretching vibration at 3326 cm⁻¹ for the fiber and the C–H deformation mode of the epoxide ring at 916 cm⁻¹ (Reproduced with permission from ref 66. Copyright 1995 Society of Spectroscopy.)

the unreacted fiber. This result signifies a reduced population of N–H functional groups. This decrease in intensity levels provides evidence that most of the amide group protons were abstracted and replaced with epoxy functionalities. In Figure 8.30, the 916-cm⁻¹ band due to the epoxide ring is shown. Lower intensities appear along the surface of both the treated and untreated fibers in comparison with the bulk matrix. However, much lower intensities appear along the surface of the epoxy-reacted fiber, displaying a much higher level of cure. This advanced level of cure is attributed to the increased chemical reactivity between the fiber and the matrix. The decrease in intensity levels along the surface of the untreated fiber confirms that the amine group in Kevlar has accelerated the curing process.

Glass Fiber Systems

Fiberglass-reinforced plastics have been widely used in several applications since their industrial appearance many years ago. The high-performance properties of these composite materials are not the sum of the properties of their components. The interfacial region (the region between the fiber and the matrix) is also considered of great importance. It is at the interfacial region where stress concentrations develop because of (1) differences between the reinforcement and the matrix-phase thermal expansion coefficients, due to loads applied to the structure, and (2) cure shrinkage (in thermosetting matrices) and crystallization (in some thermoplastic matrices). The interphase can also serve as a nucleation site, a preferential adsorption site, and a locus of chemical reactions. For these reasons the interphase is considered a major factor affecting mechanical and various physical properties of fiber-reinforced composites.

The important effect of the interphase on the properties of glass-fiber-reinforced composites has led to a considerable effort to understand it, control it, and even specifically modify it. One of the methods that has been proposed to improve adhesion between the glass fiber and matrix is the use of organofunctional silane coupling agents. The most widely accepted theory for the interpretation of the mechanism of reinforcement suggests that the coupling agent forms covalent bonds to both the glass surface and the resin matrix.

One of the main limitations for the composite samples that can be used for transmission FTIR analysis is that of the thickness. The sample must be thin enough to transmit light without saturation of the infrared peaks. Glass fiber–epoxy composites were specifically designed for FTIR microspectroscopic analysis. Glass fiber monofilaments (10 μm thick) were extracted from segments of yarn and then aligned on an aluminum disk. Epoxy resin was then added on top of the glass monofilaments. With the use of a hydraulic press, the appropriate pressure was applied, giving a thin composite sample. The thickness of the samples prepared was 12–14 μm. The resin was cured for 7 days at 25 °C, followed by postcure for 90 min at 70 °C. The cured composite sample was mounted onto the motorized IR microscope stage, and localized transmission IR spectroscopy was then performed (70). Localized transmission spectroscopy was performed on the composite samples using a 10 × 140 μm upper aperture and a lower aperture of the same dimensions. Mapping experiments were also performed in an area 80 × 80 μm in size using a 24 × 24 μm upper and lower variable aperture. All spectra were obtained with a resolution of 2 cm^{-1} and 250 scans.

Three different types of glass fiber were used in order to generate different interactions in the interfacial region: (1) as-received glass fibers, (2) glass fibers heat-treated at 600 °C for 24 h, and (3) E-glass fibers

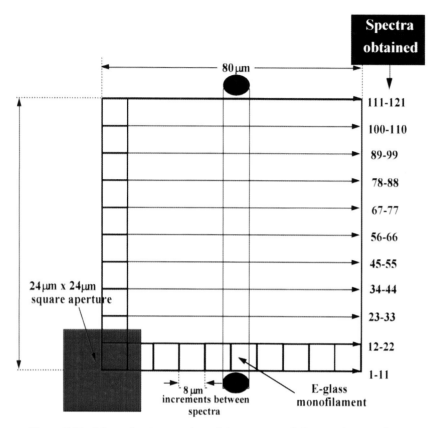

Figure 8.31. Schematic representation of the geometry of the mapping experiments. (Reproduced with permission from ref 70. Copyright 1996 Society of Spectroscopy.)

treated with an aminosilane coupling agent, γ-aminopropyltriethoxysilane (γ-APS) from Petrarch Systems Inc.

To characterize the interactions between the glass fibers and the epoxy matrix, spatially resolved transmission spectra were obtained from (1) the fiber embedded within the matrix, (2) the bulk matrix 40 μm away from the fiber center, and (3) a glass monofilament in the air. For the extraction of the two pure components (E-glass and bulk matrix) from the fiber–epoxy spectrum, an automatic spectral subtraction program was used. The program utilizes an iterative least squares procedure to optimize the scaling factor and generates the difference spectrum characteristic of the interphase (70).

Infrared mapping experiments were also performed on every sample in an 80 × 80 μm area, with the glass monofilament being in the center. By using a 24 × 24 μm aperture, with 8-μm increments between each spectrum, an infrared map consisting of 121 spectra was produced

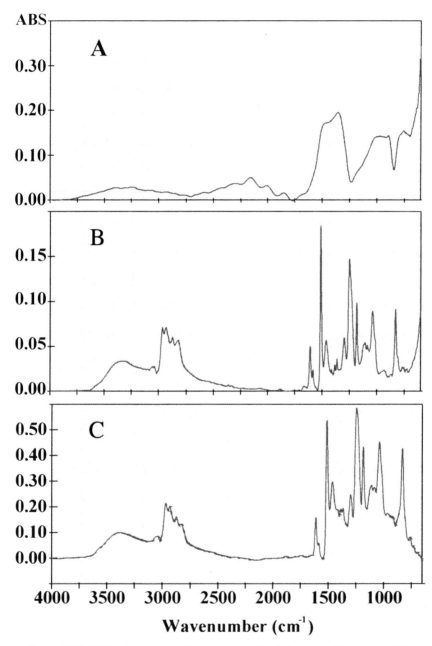

Figure 8.32. FTIR microspectra of (top) an as-received glass monofilament, (middle) the bulk epoxy matrix, and (bottom) the fiber embedded within the matrix. (Reproduced with permission from ref 70. Copyright 1996 Society of Spectroscopy.)

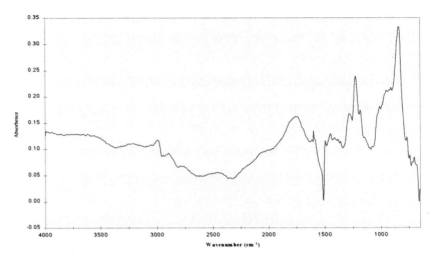

Figure 8.33. The difference spectrum characteristic of the interphase for samples prepared with as-received glass fibers. (Reproduced with permission from ref 70. Copyright 1996 Society of Spectroscopy.)

(Figure 8.31). All 121 spectra obtained from the mapping experiment were normalized to a constant thickness (using a scaling factor that was optimized by an iterative least squares procedure), giving as a final result, a "thickness-corrected" infrared map.

The FTIR microspectra of an as-received E-glass monofilament, the bulk matrix (40 μm away from the fiber center), and the fiber embedded within the matrix are shown in Figure 8.32 (70). The difference spectrum characteristic of the interphase is shown in Figure 8.33.

The results for the mapping experiments made in samples prepared with as-received glass fibers are shown in Figure 8.34. Contour plots for the 1510- and 1229-cm^{-1} peaks are shown in the figure. The different colors in these contour plots illustrate different levels of IR intensities. The difference spectrum is characteristic of the interphase for samples prepared with as-received glass fibers.

The difference spectra characteristic of the interfacial region, along with the contour plots of the epoxy/glass-fiber-reinforced composites, suggest that specific interactions between the fiber and the matrix are taking place. These interactions depend on the type of glass fibers used, since distinct spectral differences in the results are observed by comparing as-received, heat-cleaned, and silane-treated glass fibers. Thus, the E-glass surface exhibits different properties when heat-treated or silane-treated, resulting in differences in the structure of the interfacial region of the epoxy–E-glass composites.

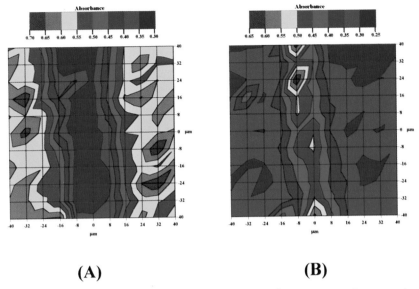

(A)   **(B)**

Figure 8.34. Contour plots of the 1510- and 1229-cm⁻¹ peaks for samples prepared with as-received glass fibers. (Reproduced with permission from ref 70. Copyright 1996 Society of Spectroscopy.)

When as-received glass fibers are used, the difference spectrum that characterizes the interfacial region shows that the most distinct changes occur at the frequency region of 1700–650 cm⁻¹. The spectrum is dominated by strong bands near 1569, 1510, 1229, and 1184 cm⁻¹. These bands involve different molecular motions of the aromatic rings found in the epoxy structure. The peaks for the aromatic C = C stretching modes (at 1569 and 1510 cm⁻¹) appear negative, whereas peaks that involve the in-phase (Ar–O) stretching modes in C_6H_5OC (at 1229 cm⁻¹) and the in-plane aromatic CH deformation (at 1184 cm⁻¹) appear positive. These results are confirmed by the infrared contour plots. Along the fiber surface, lower levels of IR intensities appear in the contour plot of the aromatic C = C stretching band (1510 cm⁻¹). On the other hand, the contour map of the 1229 cm⁻¹ peak (in-phase (Ar–O) stretching mode in C_6H_5OC shows a clear increase in the IR intensities along the fiber surface.

At this point it must be mentioned that the difference spectrum shows peaks at 1384 and 1360 cm⁻¹, which are characteristic of the gem-dimethyl structure of the epoxy resin, that is, the $C–(CH_3)_2$ group that is surrounded by the two aromatic rings. The contour map of the 1384-cm⁻¹ peak indicates at the fiber surface is affecting this specific band, resulting in increased intensities in the interfacial region.

The results just mentioned indicate that a portion of the epoxy structure is preferentially attracted to the fiber surface. This portion of the structure involves the aromatic rings and the gem-dimethyl group:

$$R_2NH + CH_2 \overset{O}{\triangle} CH\text{~~~~~} + HOH \longrightarrow$$

$$\left[R_2\overset{+}{N} \underset{H^+}{\text{------}} CH_2 \text{------} CH\text{~~~~~} \overset{\overset{H^-OH}{\diagdown}}{\underset{O}{\diagup}} \right] \longrightarrow$$

$$\left[\underset{R_2N}{\overset{H^+}{}} \text{---} CH_2 \text{---} \underset{\overset{|}{OH^-}}{\overset{OH}{CH}}\text{~~~~~} \right] \longrightarrow R_2N \text{---} CH_2 \text{---} \underset{\overset{|}{OH}}{CH}\text{~~~~~} + HOH$$

Because some bands associated with certain molecular structures exhibit higher intensities and some lower along the fiber surface, it is possible that a preferred alignment of the aromatic and gem-dimethyl groups exists at the interfacial region. In other words, the presence of the glass fiber is causing the restriction of certain molecular motions of the epoxy resin, resulting in structural changes at the interfacial region. These structural changes are translated as a preferred conformation of the epoxy resin at the fiber surface.

Another interesting feature in the difference spectrum that characterizes the interphase of composites prepared with as-received glass fibers, is a strong peak located at 841 cm^{-1}. This peak corresponds to the NH wagging mode of the curing agent and is one of the most characteristic vibrational modes for liquid aliphatic amines. Higher intensities along the fiber surface indicate a nonuniform distribution of the functional groups in the mapping area. The fiber surface is acting as an attractive site for the curing agent, revealing that specific interactions are taking place between the functional groups of the curing agent and the as-received glass fiber.

In the case of composites that were prepared with heat-cleaned glass fibers, spectral subtraction resulted in the difference spectrum characteristic of the interphase. A careful examination of the 1700–650-cm^{-1} frequency region reveals certain similarities with the results for the composites prepared with as-received glass fibers. Certain bands (like the aromatic C = C stretching at 1510 cm^{-1}) have a lower intensity at the interphase, whereas others (like the in- and out-of-phase (=C–O) stretching in C_6H_5OC at 1254 and 1230 cm^{-1}, the in-plane aromatic CH deformation at 1185 cm^{-1}, and the gem-dimethyl motions at 1381 and 1360 cm^{-1}) have a higher one.

These results indicate that the same preferred conformation of the epoxy resin, as the one found in samples with as-received glass fibers, exists in the interfacial region of epoxy-heat-cleaned glass fiber composites. This same conformation, where the aromatic and the gem-dimethyl portions of the epoxy resin are attracted at the fiber surface, suggests that there are similarities in the surface characteristics of the as-received and heat-cleaned glass fibers.

The peak near 842 cm^{-1} (NH wagging mode) found in the difference spectrum of epoxy-heat-cleaned glass fiber composites also suggests the existence of some similarities between as-received and heat-treated glass fibers. The peak appears positive in the difference spectrum, and its contour plot (Figure 8.33) reveals higher intensities along the fiber surface. In other words, the heat-cleaned glass fiber surface attracts the amine curing agent, a feature that also appears in the case of as-received glass fibers.

The difference spectrum that characterizes the interfacial region of composites prepared with heat-cleaned glass fibers also has some other interesting features. A positive peak appears at 866 cm^{-1}, which corresponds to the epoxide deformation. This peak is used to characterize the extent of cure of the epoxy resins, and its intensity value decreases with increasing extent of cure. Higher intensities are observed along the fiber surface, indicating lower extent of cure in this particular region. This means that the curing of the epoxy resin is inhibited in the interfacial region of composites prepared with heat-cleaned glass fibers.

The mechanism of curing of epoxy resins involves the reaction of the amine with the epoxy group through the active amine hydrogen as shown in reaction 8.1. Next, the secondary amine reacts further with the epoxy groups (reaction 8.2). An important side reaction of the curing process is the homopolymerization of epoxy groups through etherification of pendant hydroxyl groups formed by reactions 8.1 and 8.2. This reaction is shown in reaction 8.3. Reaction 8.3 is base catalyzed. The basicity of the surface acts as a catalyst for reaction 8.3, leading to it being favored at the interfacial region. As a result, the etherification side reaction 8.3 predominates over reactions 8.1 and 8.2. This predomination means that the main reactions of the curing process are inhibited at the interfacial region, giving a possible explanation for the inhibition of cure of the epoxy resin that was observed by examining the epoxide deformation band (866 cm^{-1}):

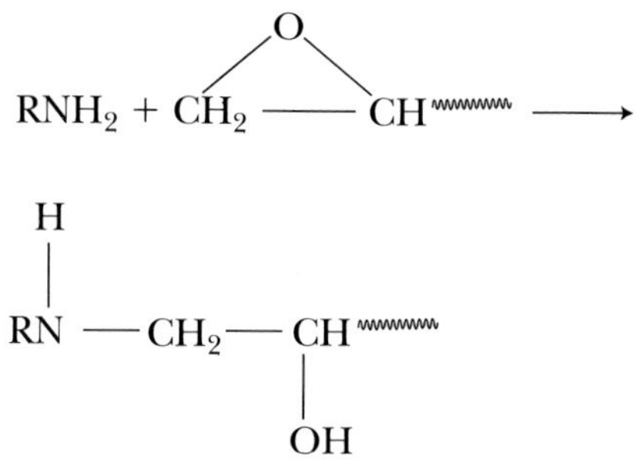

$$RNH_2 + CH_2 \overset{O}{\triangle} CH \text{\textasciitilde} \longrightarrow$$

$$\underset{RN}{\overset{H}{|}} - CH_2 - CH \text{\textasciitilde} \atop \underset{OH}{|}$$

(1)

$$\underset{RN}{\overset{H}{|}} - CH_2 - CH \text{\textasciitilde} \atop \underset{OH}{|} + CH_2 \overset{O}{\triangle} CH \text{\textasciitilde} \longrightarrow$$

(2)

$$RN \overset{CH_2 - CH \text{\textasciitilde} \atop |\, OH}{\underset{CH_2 - CH}{}} \atop \underset{OH \text{\textasciitilde}}{|}$$

$$\text{\textasciitilde}CH\text{\textasciitilde} + CH_2 \overset{O}{\triangle} CH \text{\textasciitilde} \longrightarrow \atop \underset{OH}{|}$$

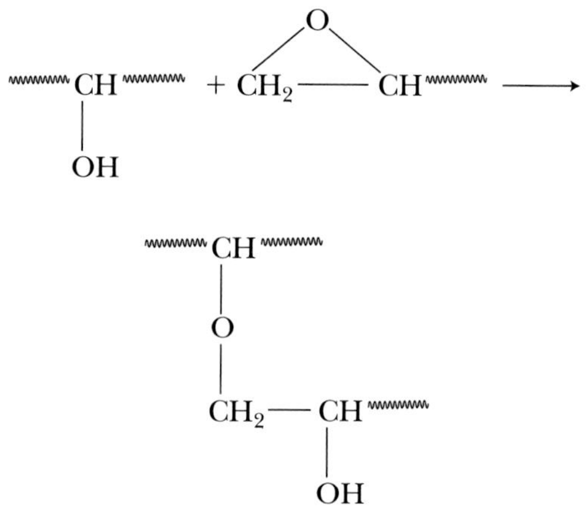

(3)

$$\text{\textasciitilde}CH\text{\textasciitilde} \atop \underset{O}{|} \atop CH_2 - CH \text{\textasciitilde} \atop \underset{OH}{|}$$

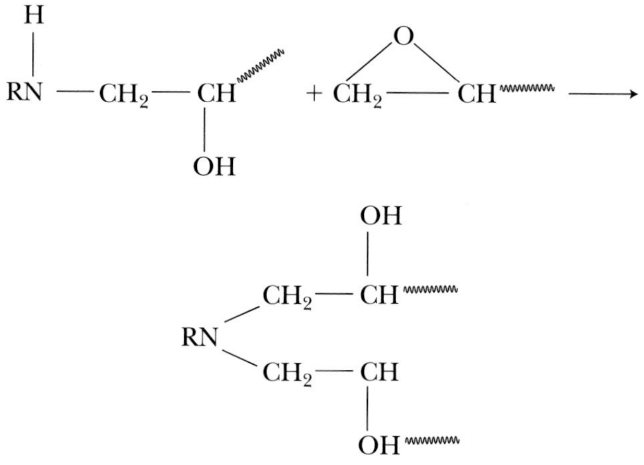

The aminosilane coupling agent was one of the first organofunctional silanes to find wide acceptance as a coupling agent in reinforced plastics. The high reactivity and versatility of the amino functionality have led to an increasing number of diverse applications. Numerous studies have indicated that hydrolyzed γ-aminopropyltriethoxysilane exists in two different forms: one a chelate ring, and the other a nonring extended structure. The γ-APS coupling agent exists as a multilayer on the E-glass fiber surfaces and forms a polyaminosiloxane.

The use of the γ-APS coupling agent usually improves the adhesion between the matrix and the reinforcing agent. Silanes can anchor themselves to the glass surface via Si–O–Si bonds. In addition to the hydrolyzable groups, the silane contains an organic nonhydrolyzable group, which can potentially react with the matrix resin. In the case of an epoxy resin, this functional group may be an amine or epoxy group. In other words, the establishment of covalent bonds between the silane and both the glass surface and the organic matrix results in an important contribution to adhesion between the fiber and the matrix.

For samples prepared with silane-treated glass fibers, the difference spectrum that characterizes the interfacial region was obtained. The aromatic $C = C$ stretching bands at 1609 and 1510 cm^{-1} appear negative in the difference spectrum, suggesting that the specific molecular conformation is affected by the presence of the silane-treated glass monofilament. The infrared map for the 1510-cm^{-1} band indicates a decrease in the absorbance values of this band along the fiber surface. The difference spectrum characteristic of the interphase also reveals other bands that are present in the interfacial region, such as those related to the 1472, 1455, 1383, 1295, 1227, and 1183 cm^{-1} bands, which appear positive. The infrared map of the C–$(CH_3)_2$ methyl deformation peak at 1384 cm^{-1} is indicative of increasing intensity values for this band at the interfacial region.

A very interesting feature for the difference spectrum appears at 1252 cm^{-1}. This band, which characterizes the out-of-phase (Ar–O) stretching in C_6H_5OC, has a negative value in the difference spectrum. If one also considers the contour plot of this band, then a decrease in its peak intensity along the fiber surface can be presumed. Because the above feature does not appear in the case of as-received glass fibers and appears substantially different in the case of heat-cleaned glass fibers, differences in the glass surface when silane treatment is applied can be deduced. These differences result in a different preferred conformation of the epoxy resin at the interphase as compared to the one observed in samples prepared with as-received and heat-cleaned glass fibers.

The fact that silane treatment alters the surface characteristics of glass is confirmed by the presence of a negative band located at 842 cm^{-1} (amine NH wagging mode) in the difference spectrum. The contour plot of this peak reveals lower intensity values in the interfacial region. This indicates a preferential repulsion of the curing agent from the glass fiber surface. If one also considers the previously mentioned results, where the curing agent was attracted to the interphase when as-received and heat-treated glass fibers were used, then it is clear that silane treatment has a significant effect on the surface properties of the glass fibers.

Another interesting phenomenon in the difference spectrum characteristic of the interfacial region of composites prepared with silane-treated glass fibers, is a series of negative bands (2965, 2932, 2873, 2834 cm^{-1}) in the frequency region 3600–2600 cm^{-1}. These peaks mainly involve the motions of methyls and methylenes found in the epoxy resin. Contour maps of the 2965- and 2873-cm^{-1} bands clearly indicate decreasing absorbance values for the aforementioned bands along the fiber surface. Again, this kind of behavior may be attributed to a preferred conformation of the epoxy resin in the interfacial region.

Effects of Moisture on Composite Interfaces

Sensitivity to moisture is one of the principal limitations for industrial usage of fiberglass-reinforced composites. Their mechanical properties are irreversibly degraded by exposure to a moist environment, a phenomenon that is often accompanied by the debonding of the resin from the filler. Moreover, composite degradation is enhanced by tensile, shear, and compressive stress, especially when these are applied cyclically. The poor reliability of glass-fiber-reinforced composites necessitates the examination of these materials under moist conditions.

The fiber-reinforced composite samples were exposed to a 75 °C liquid water environment, giving a 90% relative humidity. After appropriate exposure times (30 and 60 days), the samples were removed from the humidity chamber for IR mapping studies (71).

To directly observe the spectral differences in the sample after exposure to moisture, mapping subtraction has been utilized. Mapping subtraction is based on the concept of spectral subtraction that has been extensively used in the past few years. A difference infrared map is generated by subtracting two maps, one of which has been multiplied with a scaling factor in order to correct for nonuniform sample thickness and instrumental differences.

Because IR mapping involves spectral examination of sample areas, special care must be taken when using mapping subtraction to ensure that the same area is being examined every time. Differences in the sampling area geometry need to be eliminated. As a result, the scaling factor used in mapping subtractions would compensate only for instrumental differences. These differences are often minimal, resulting in a scaling factor very close to unity.

The use of the microscope motorized stage is an important factor to ensure that the same sample area is being examined. The motorized stage enables one to move the sample in increments as small as 1μm and to measure the distance of specific points from a reference point. To assure image registration, a coordinate system was set up on the sample, as illustrated in Figure 8.35. Six individual points, each having a known distance from a reference point (origin), are marked in a geometrical pattern. After exposure to moisture the sample was remounted on the motorized stage, and by using the known distances of the six points from the origin, the reference point can be determined.

The difference maps obtained from samples prepared with as-received glass fibers are shown in Figure 8.36. The term "difference map" refers to the result of the subtraction between the initial map (before exposure to moist conditions) and the final map (after exposure to moist conditions for 30 and 60 days). Figure 8.36 shows the difference maps for the OH stretching vibration (3450 cm^{-1}) after exposure to moist conditions for 30 and 60 days. The different colors in these maps illustrate different levels of subtracted IR intensities.

The general conclusion from these maps is that more water molecules are found along the fiber surface than in the bulk matrix. The above

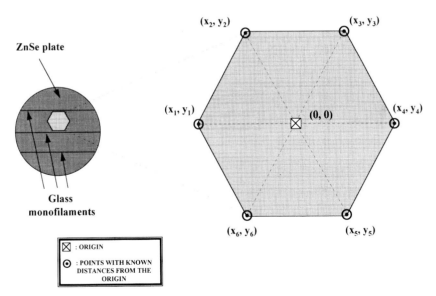

Figure 8.35. Schematic representation of the coordinate system used in the mapping experiments. (Reproduced with permission from ref 71. Copyright 1996 Society of Spectroscopy.)

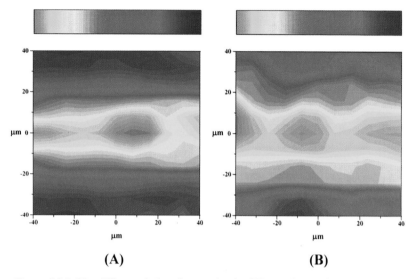

(A)   **(B)**

Figure 8.36. The difference infrared maps for the OH stretching vibration after (A) 30 and (B) 60 days of exposure to moisture, for samples prepared with as-received glass fibers. (Reproduced with permission from ref 71. Copyright 1996 Society of Spectroscopy.)

phenomenon is attributed to the presence of microvoids at the interfacial region. These microvoids, which result from the incomplete wetting of the fiber from the resin, enable water molecules to penetrate the interfacial region more easily. Moreover, because of their hydrophilic character, the glass fibers are acting as an attraction site for water molecules. The overall result of these phenomena is the preferential absorption of water molecules at the interfacial region rather than in the bulk matrix.

When samples prepared with as-received glass fibers are examined, the difference infrared maps suggest that more water molecules are found in the interfacial region than the bulk matrix. This trend is observed for exposure to moisture after both 30 and 60 days. After 60 days, however, the difference map suggests that water molecules are starting to affect a larger sampling area, indicating migration of these molecules to the bulk epoxy matrix.

Heat treatment of the glass fibers removes impurities and molecular water, and most silanol groups condense to form siloxane bonds. Samples that were prepared with these fibers show an interesting behavior when exposed to moist conditions. After 30 days of exposure to moisture, water molecules are concentrated on the sides of the glass monofilament; the glass fiber surface is essentially free of water molecules. After an additional 30 days of exposure to moisture, however, the water molecules seem to migrate to the glass surface, resulting in a higher water concentration at the interfacial region.

Difference infrared maps were also obtained after 30 and 60 days of exposure to moisture for samples prepared with silane-treated glass fibers. These maps indicate preferential attraction of water molecules to the interphase rather than the bulk matrix. This trend is evident in both maps (after 30 and 60 days of exposure to moisture), although there is no significant change in water concentration after the first 30 days of exposure to moist conditions.

The maximum absorbance values of the OH stretching vibrational mode for all samples examined are shown in Figure 8.37. These values correspond to the relative concentration of water molecules at the glass fiber surface. The plot indicates that samples prepared with as-received glass fibers exhibit the highest water uptake when exposed to moist conditions. This trend can be attributed to the presence of impurities at the glass fiber surface. These impurities facilitate the creation of microvoids at the interfacial region, and thus the penetration of water molecules. Composites that were prepared with silane-treated glass fibers exhibit the lowest water uptake at the interfacial region. This is not surprising if one considers that silane treatment is often applied in order to reduce the moisture-induced degradation of composites. Silane treatment is considered to protect against attack by moisture in the interfacial region in composite materials. This is attributed to the hydrophobicity of the siloxane bonds formed by the coupling agent. Moreover, coupler molecules block

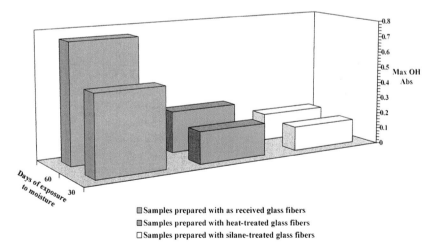

Figure 8.37. Diagram showing the maximum absorbance values for the OH stretching vibration. (Reproduced with permission from ref 71. Copyright 1996 Society of Spectroscopy.)

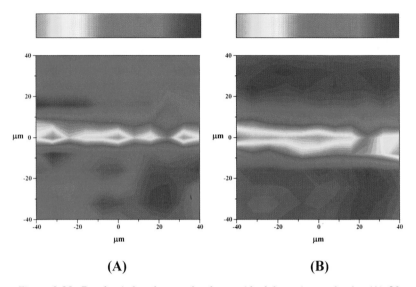

Figure 8.38. Regular infrared maps for the epoxide deformation peak after (A) 30 and (B) 60 days of exposure to moisture, for samples prepared with silane-treated glass fibers. (Reproduced with permission from ref 71. Copyright 1996 Society of Spectroscopy.)

glass surface silanols by selective adsorption or eradicate them by condensation reaction, diminishing the adsorptive potential of the substrate for water.

Regular infrared maps for the epoxide deformation peak (916 cm^{-1}), obtained under the same moisture exposure conditions for samples prepared with silane-treated glass fibers, are shown in Figure 8.38. Higher intensity values for the epoxide deformation peak are observed in the interfacial region, suggesting that in the proximity of the glass fiber the extent of cure of the epoxy resin is smaller than in the bulk matrix. Therefore, the interfacial region contains unpolymerized (or partially polymerized) epoxy resin.

The existence of unpolymerized or partially polymerized resin at the interfacial region of composites prepared with heat and silane-treated glass fibers can be explained as follows: epoxy resins are nonhomogeneous materials and consist of micelles and granules of high-density polymer separated by narrow boundary regions of lower molecular weight material. Polymerization initiates from random points and proceeds radially, and as these regions approach one another, they are unable to coalesce into a homogeneous network. Instead the polymerization is terminated, leaving unreacted or partially reacted material. The presence of a filler facilitates the occurrence of these phenomena, resulting in the existence of unreacted epoxy groups at the interfacial region.

A higher degree of curing is present at the interfacial region for composites prepared with as-received glass fibers after being exposed to moisture. This is attributed to the presence of water molecules at the interfacial region during the exposure to moisture. Water molecules catalyze the curing of the unreacted epoxy resin by opening the epoxy rings. This is caused by hydrogen bonding effects and is represented by Scheme 8.1.

These reactions are believed to occur at the interfacial region of composites prepared with as-received glass fibers but not of the ones prepared with heat- and silane-treated glass fibers. This is most probably due to the differences in water concentration at the interphase. There is a critical water concentration necessary for Scheme 8.1 to be initiated, and this concentration exists only in the interfacial region of composites that were prepared with as-received glass fibers.

Moreover, water molecules catalyze a polymerization reaction, resulting in a highest extent of cure at the interfacial region of composites prepared with as-received glass fibers. Because a critical water concentration is necessary for the polymerization reaction to be initiated, no further curing was observed at the interphase of composites prepared with heat- and silane-treated glass fibers when these were exposed to moisture.

FTIR Microspectroscopy of Diffusion Processes in Polymers

Diffusion in polymers is an important process which determines, in many instances, the utility of the polymer for applications in certain environmental conditions. Diffusion also plays a role in special applications such as drug delivery (72).

The utility of additives in polymers is determined to some degree by the rate of loss through diffusion to the surface and the relative mobility of the additive within the polymer. Diffusion of additives in polymers can be characterized by sorption curves and concentration profiles. Sorption curves are determined by the uptake of additive into a sample with time but are limited because of the low solubility of typical additives in polymers. FTIR microspectroscopy can be used to follow the diffusion of substances in polymers as the concentration profile of the additive can be measured over small distances with time. Because the FTIR technique monitors many frequencies at the same time, the concentration profiles of several different additives can be measured simultaneously. FTIR microspectroscopy can be used to determine the diffusion coefficients of the additives in the polymers.

One of the first examples of such studies was that of the diffusion of an antioxidant in cross-linked polyethylene using a microbeam FTIR accessory (73). The "diffusion-in" experiments allow the additive to diffuse into the polymer from an external reservoir. A model for this experiment assumes that diffusion occurs only in one direction, that is, it can be approximated as diffusion in a semi-infinite medium. The concentration, C, of the additive is given by

$$(C_0 - C)/C_0 = \mathrm{erf}\{x/[2(Dt)^{1/2}]\}$$

where C_0 is the concentration of the additive on the polymer surface, x is the distance from the surface, D is the diffusion coefficient, and t is the diffusion time. In the IR experiment, the concentration is related to the IR intensity, A, as

$$A/A_0 = \mathrm{erfc}\{x/[2(Dt)^{1/2}]\}$$

The IR measurements are made as a function of position, and the diffusion coefficient is determined by a nonlinear least squares fitting.

The diffusion of the UV stabilizer Cyasorb UV531 (4-alkoxy-2-hydroxybenzophenone with $R = C_8H_{17}$) in polypropylene was characterized by the "diffusion-in" concentration profiles using an infrared microprobe to make measurements along the diffusion axis at 26-μm increments (74). The intensities of the bands at 699 and 628 cm^{-1} associated with the Cyasorb were measured. Figure 8.39 shows a three-dimensional plot of the IR spectra of Cyasorb UV531 in polypropylene at different distances from the surface of the polypropylene after a diffusion experiment at 80 °C (74).

A similar approach was used to study the diffusion of bovine serum albumin (BSA) in amylopectin gels (75). A tube of gel is prepared, and the exposed end is suspended in a solution of BSA in NaCl in D_2O. After exposure to the BSA, the tube is sealed and mounted on the microscope and scanned every 80 μm to obtain a concentration (and hence the diffusion) profile. The amide peak at 1648 cm^{-1} is plotted versus distance. In this fashion, it is possible to measure the rate at which an externally applied entity (such as a pathogen) penetrates a gel matrix.

Diffusion of probe particles through an entangled polymer gel is characterized by a mesh size, ξ. The particle can move through the mesh of the gel with little hindrance if the particle is significantly smaller than ξ, but with increasing difficulty as the particle size (radius R) becomes comparable to ξ. The following equation has been proposed:

$$D/D_0 = \exp(-bc^k)$$

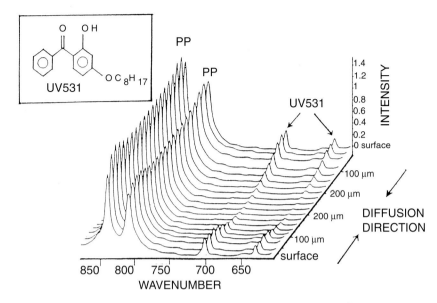

Figure 8.39. Three-dimensional plot of the IR spectra of Cyasorb UV531 in polypropylene at different distances from the surface of the polypropylene, after a diffusion experiment at 80°C. (Reproduced with permission from ref 74. Copyright 1992 Society of Spectroscopy.)

where D_0 is the diffusion coefficient in a gel of zero concentration, b is a constant that depends on the radius of the probe molecules and the molecular weight of the polymer, c is the concentration of the gel–solution, and k is a constant, expected to be in the range 0.5–1.0 and whose exact value depends largely on the polymer species. Gels of rodlike polymers should have a value of k of 0.5 (*76*). A value of k for the BSA–amylopectin system is 0.93, which is in the predicted range of 0.5–1.0 (*75*). However, the diffusion is influenced by the heterogeneity of the amylopectin gels.

An infrared microscopic investigation has been made of the interfacial reaction and interdiffusion of poly(styrene-*co*-maleic anhydride) (SMA) with a bis(amine)-terminated poly(tetrahydrofuran) (PTHF) (*78*). Two-layer specimens were prepared and microtomed perpendicular to the surface, with the FTIR measurements performed at different distances from the interface. For FTIR microscopy, a spot size of approximately 25 μm and a step of 5 μm were used. The imidization reaction was followed using the peak at 1702 cm^{-1}. The band at 700 cm^{-1} was used as an internal thickness band for the kinetic studies. The diffusion of the SMA–PTHF system was studied using the 1780-cm^{-1} band of the anhydride. It was found that the PTHF diffused into the SMA phase prior to the imide formation, indicating that the diffusion across the interface is faster than the reaction between the amine and anhydride (*78*).

FTIR Microspectroscopy of Polymer Orientation

A high degree of alignment is the basis of improved mechanical, optical, and electrical properties in almost all polymers. Consequently, it is necessary to understand the correlations among molecular order, material properties, and fabrication procedures. Experimental characterization of the degree and direction of alignment is required to understand how molecular design and processing strategies lead to the ultimate state of alignment.

Oriented polymers are generally anisotropic, the properties along a particular direction being affected according to the conditions of stretching along that direction. Uniaxial materials are produced by stretching in one direction, whereas biaxial materials are produced by stretching in two mutually perpendicular directions.

In polymer structures, orientation is not complete and distributions of directions of planes or axes will be observed. It is also possible for several orientation types to coexist, as is frequently found in metallurgical specimens.

The orientation process can involve three competing but interrelated deformations:

1. Elastic or Hookean deformation resulting from valence-angle deformation.

2. Viscous flow deformation resulting from slippage of molecules. This deformation is nonrecoverable.

3. Deformation resulting from uncoiling of polymer chains.

The ideal orientation process leads to maximum uncoiling and chain alignment and to minimum viscous flow or chain slippage.

Orientation is a highly time-dependent process because polymers are viscoelastic materials, and their molecular relaxation is also a highly rate and temperature dependent process. When the temperature is too low, that is, when the polymer is in the glassy state, stretching requires excessively high stresses and can lead to rupture. At too high temperatures, that is, in the purely viscous state, deformations can easily be applied, but they may not result in orientation because rapid relaxation can exceed the rate of deformation.

Polymer drawing involves the extension of an entangled molecular network. The macromolecules form a transient network, with entanglements acting as nonlocalized junctions. The tensile behavior of the polymer can be affected by the extension and the alignment of its chain segments. For low degrees of alignment, deformation at the molecular level is mostly related to torsional rotations of chain segments around the chain backbone. There is an uncoiling of the chain as extended conformers are formed as well as an alignment of the segments with the direction of the mechanical load.

Injection-molded articles of amorphous polymers exhibit an anisotropy in the optical and mechanical properties due to frozen-in orientation of the polymer chains. To a large extent, this orientation is created during the nonisothermal mold-filling stage; polymer molecules tend to orient while flowing under the action of the prevailing stress field. If the molten polymer is cooled rapidly to a temperature below its glass-transition temperature, the polymer molecules will not have sufficient time to relax their orientation and return to a random configuration.

There is also a distribution of orientation in the injection-molded article, because polymer molecules at different locations of the final molded part are generally exposed to different thermal conditions due to the ever-present thermal gradients in the molds.

The measurement of molecular anisotropy by polarized infrared spectroscopy is a well-established technique. The intensity of a vibrational band depends quantitatively upon the mutual orientation of the molecule and the electric field vector of the radiation. The advantage of using infrared rather than X-ray diffraction is that one can study the conformers found in either the crystalline or amorphous regions simultaneously, whereas X-ray diffraction resolves only the crystalline phase.

It is the usual practice to measure the dichroic ratio of orientation-sensitive bands. The dichroic ratio is expressed as the ratio of radiation polarized parallel to the principal direction relative to that of radiation polarized perpendicular to the principal direction. Such an approach has been used to study the orientation of polyethylene tubing by infrared microspectroscopy (*77*). The bands at 2016 cm^{-1} and 1894 cm^{-1} have been assigned to all-trans conformers in polyethylene, the former existing in both the amorphous and crystalline phases and the latter only in the crystalline regions. Substantial differences were observed in the polyethylene tube with a wall thickness of 0.6 mm, indicating that substantial orientation existed in the tube. Similar results were obtained on a polyethylene-sheathed copper cable, where the polyethylene coating was 150 μm, using reflection–absorbance sampling.

FTIR Microspectroscopy of Lubricants on Metal Surfaces

FTIR grazing-angle microscopy has been used to examine a perfluoro-polyether lubricant of the Krytox 16256 type on a gold surface (*69*). The approach used quantitative reflection microspectroscopy. The film was applied to a gold-coated silicon substrate by dip-coating from a solution. The film thickness was determined by ellipsometry in the visible to be 47.4 nm, and the calculated thickness from the infrared measurements were 46, 57.5, and 59.5 nm. The difference in thickness reflects the nonuniform nature of the film and the fact that the ellipsometry measures require a large surface region, whereas the IR method probed local regions are much smaller.

FTIR Microspectroscopy of Crack Growth in Polymers

It is possible to use FTIR microscopy to characterize the crack tip region of unfilled rubber and compare the molecular structure and orientation of the material in the crack tip region with those of the bulk (79). Cracks were formed in the natural rubber system in two different ways: stressing the sample and cutting with a razor blade. A crack was formed in the rubber surface when the sample was strained to 629% elongation. A cross section of this sample was then microtomed at − 85 °C into slices approx-

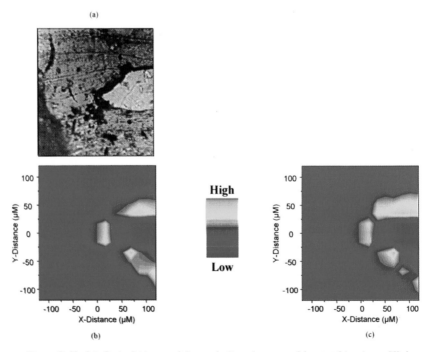

Figure 8.40. (a) Optical picture of the crack tip region created by stretching in unfilled natural rubber. (b) Map of the 837-cm⁻¹ peak. (c) Map of the 1665-cm⁻¹ peak.

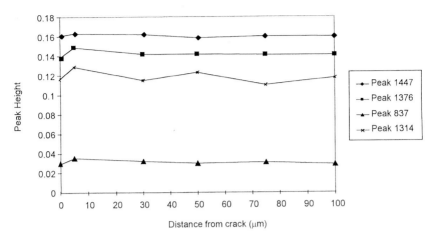

Figure 8.41. Infrared map of the crack tip region created by a razor cut in unfilled natural rubber. (a) Map of the 837-cm^{-1} peak. (b) Map of the 1665-cm^{-1} peak.

imately 0.5 μm thick. A 40 × 40 μm aperture was used to scan specific portions of the crack formation. Figure 8.40 compares the optical photograph with the IR maps obtained using the 837^{-1} and 1665-cm^{-1} bands due to natural rubber. The crack itself has no material present and is therefore represented by low absorbances. Surrounding the crack tip are areas of significantly higher absorbances. The absorbances are high in areas adjacent to the crack but then gradually decrease to an average value in the bulk.

A similar analysis was carried out for a sample in which the crack tip was created by cutting the rubber with a razor blade. In this case, the crack was very narrow, with a maximum width of about 25 to 30 μm. The infrared map of the razor crack is shown in Figure 8.41.

Because of the aperture size (60 × 60 μm), there are no regions of extremely low absorbances indicating a crack through the material. These results show no significant differences in the absorbances in the area of the crack tip region and in the bulk of the material.

The contour plot of the stress-induced crack tip most likely can be attributed to concentrations of material due to stress. Crack growth implies that there are regions in the vicinity of the crack tip where retraction of material occurs. However, this is not the only phenomenon responsible for increasing the absorbances. The molecular structure in the damage zone is different from the structure in the bulk of the material. Orientation analysis of the crack tip region indicates that there is a higher degree of orientation present of higher stress around the crack tip region, and that significant degrees of orientation occur only in the stressed samples. The area of highest stress is at the crack tip.

References

1. Katon, J. E. *Vib. Spectrosc.* **1994,** *7,* 201.
2. Harthcock, M. A.; Atkin, S. C. *Appl. Spectrosc.* **1989,** *42,* 449.
3. Harthcock, M. A. *In The Design, Sample Handling, and Applications of Infrared Microscopes;* Rousch, P. B., Ed.; American Society for Testing and Materials: Philadelphia, PA, 1987; p 85.
4. Harthcock, M.; Atkin, S. C. *In Infrared Microscopy Theory and Applications;* Messerschmidt, R. G.; Harthcock, M. A., Eds.; Marcel Dekker: New York, 1988.
5. Weesner, F. J.; Carl, R. T.; Boyle, R. M. *SPIE* (Bellingham Wash.), **1991,** *31,* 1575, 486,.
6. Ward, K. J. *Proc. SPIE Int. Soc. Opt. Eng.* **1989,** *1145,* 212.
7. Harthcock, M. A.; Lentz, L. A.; Davis, B. L.; Krishnan, K. *Appl. Spectrosc.* **1986,** *40,* 210.
8. Miller, P. J.; Piermarini, G. J.; Block, S. *Appl. Spectrosc.* **1984,** *39,* 680.
9. Krishnan, K. *Polym. Reprints,* Polymer Div., American Chemical Society, Washington, DC, **1984,** *25,* 1182.
10. Tungol, M. W.; Bartick, E. G.; Montaser, A. *Appl. Spectrosc.* **1986,** *40,* 210.
11. Tungol, M. W.; Bartick, E. G.; Montaser, A. *Appl. Spectrosc.* **1990,** *44,* 543.
12. Kirkbride, K. P. *Forensic Exam. Fibres* **1992,** 181–218.
13. Katon, J. E.; Lang, P. L.; Schiering, D. W.; O'Keefe, J. F. *In The Design, Sample Handling, and Applications of Infrared Microscopes;* Rousch, P. B., Ed.; American Society for Testing and Materials: Philadelphia, PA, 1987; p 49.
14. Bouffard, S. P.; Sommer, A. J.; Katon, J. E.; Godber, S. *Appl. Spectrosc.* **1994,** *48,* 1387.
15. Reffner, J. A.; Coates, J. P.; Messerschmidt, R. G. *Am. Lab.* **1987,** April, 86.
16. Bartick, E. G. *In The Design, Sample Handling, and Applications of Infrared Microscopes;* Rousch, P. B., Ed.; American Society for Testing and Materials: Philadelphia, PA, 1987; p 64.
17. Chase, D. B. *In The Design, Sample Handling, and Applications of Infrared Microscopes;* Rousch, P. B., Ed.; American Society for Testing and Materials: Philadelphia, PA, 1987; p 4.
18. Ward, K. J. Presented at PITTCON, Chicago, IL, March, 1993; paper 433.
19. Lang, P. J.; Sommer, A. J.; Katon, J. E. *Particles on Surfaces 2;* Mitral, K. L., Ed.; Plenum: New York, 1989; p 207.
20. Lang, P. L.; Katon, J. E.; Bonanno, A. S.; Pacey, G. E. *Infrared Microscopy Theory and Applications;* Messerschmidt, R. G.; Harthcock, M. A., Eds.; Marcel Dekker: New York, 1988; p 41.
21. Harrick, N. J.; Milosevic, M.; Berets, S. L. *Appl. Spectrosc.* **1991,** *45,* 225.
22. Misco, E. V.; Guilmette, I. W. In *The Design, Sample Handling, and Applications of Infrared Microscopes;* Rousch, P. B., Ed.; American Society for Testing and Materials: Philadelphia, PA, 1987; p 97.
23. Gal, T.; Toth, P. *Can. J. Appl. Spectrosc.* **1992,** *37,* 55.
24. Pell, R. J.; McKelvy, M. L.; Harthcock, M. A. *Appl. Spectrosc.* **1993,** *47,* 634.
25. Steward, D. *Appl. Spectrosc.* **1996,** *50,* 357.
26. Humecki, H. J. In *The Design, Sample Handling, and Applications of Infrared Microscopes; .* Rousch, P. B., Ed.; American Society for Testing and Materials: Philadelphia, PA, 1987; p 97.
27. Provder, T.; Kuo, C-Y.; Whited M.; Huddleson, D. *Polym. Reprints,* Polymer Div., American Chemical Society, Washington, DC, **1995,** 340. 28 Liu, M. X.; Dwyer, J. L. *Appl. Spectrosc.* **1995,** *50,* 349.
28. Liu, M. X., Dwyer, J. L. Appl. Spectrosc. **1995,** *50,* 349.

29. Raynor, M. W.; Bartle, K. D.; Davies, H. L.; Williams, A.; Clifford, A. A.; Chalmers, J. M.; Cook, B. W. *Anal. Chem.* **1988**, *60*, 427.
30. Wall, B.; Koenig, J. L. submitted for publication in *Appl. Spectrosc.*
31. Price, R.; Caveny, B. *Proc. Int. Conf. Cem. Microsc.* **1992**, *14*, 114–133.
32. Bensted, J. *Cemento* **1993**, *90*, 151.
33. West, J. L. *Liquid-Crystalline Polymers;* Weiss, R. A.; Ober, C. K., Eds.; ACS Symposium Series 435; American Chemical Society: Washington, D.C., 1990; p 475.
34. Doane, J. W.; Golemme, A.; West, J. L.; Whitehead, J. B., Jr.; Wu, B.-G. *Mol. Cryst. Liq. Cryst.* **1988**, *165*, 511.
35. Doane, J. W. *Liquid Crystals: Applications and Uses;* Bahadur, B., Ed.; World Scientific: 1990; Teaneck, N.J., Ch. 14.
36. Doane, J. W.; Yaniv, Z. *Proc. SPIE Int. Soc. Opt. Eng.* **1989**, *1080.*
37. Doane, J. W.; Yaniv, Z. *Proc. SPIE Int. Soc. Opt. Eng.* **1990**, *1257.*
38. Drzaic, P. S.; Efron, U. *Proc. SPIE Int. Soc. Opt. Eng.* **1991**, *1455.*
39. Vaz, N. A. *Proc. SPIE Int. Soc. Opt. Eng.* **1989**, *1080*, 2.
40. Vossen, J. L. *Phys. Thin Films* **1977**, *9*, 1.
41. Chen, T.-C.; Ma, T.-P.; Barker, R. C. *Appl. Phys. Lett.* **1983**, *43*, 901.
42. Bachtemann, A.; Shulz, E. *Thin Solid Films* **1987**, *152*, 135.
43. Gregoriou, V. G.; Chao, J. L.; Toriumi, H.; Palmer, R. A. *Chem. Phys. Lett.* **1991**, *179*, 491.
44. Hatta, A. *Mol. Cryst. Liq. Cryst.* **1981**, *74*, 195.
45. Hatta, A.; Amano, H.; Suktaka, W. *Vib. Spectrosc.* **1991**, *22*, 371.
46. Wu, S.-T.; Efron, U.; Hess, L. D. *Appl. Phys. Lett.* **1984**, *44*, 1033.
47. Wu, S.-T. *Opt. Eng.(Bellingham, Wash.)* **1987**, *26*, 120.
48. Chenault, D. B.; Chipman, R. A.; Johnson, K. M.; Doroski, D. *Opt. Lett.* **1992**, *17*, 447.
49. West, J. L.; Ondris-Crawford, R. *J. Appl. Phys.* **1991**, *70*, 3785.
50. McFarland, C. A.; Koenig, J. L.; West, J. L. *Appl. Spectrosc.* **1993**, *47*, 598.
51. McFarland, C. A.; Koenig, J. L.; West, J. L. *Appl. Spectrosc.* **1993**, *47*, 321–329.
52. Challa, S. R.; Wang, S. Q.; Koenig, J. L. *Appl. Spectrosc.* **1994**, *49*, 267.
53. Challa, S. R.; Wang, S. Q.; Koenig, J. L. submitted for publication in *J. Appl. Spectrosc. (Engl. Transl.).*
54. Wall, B.; Koenig, J. L. *MacroAkron* **1994**.
55. Urban, M. W. *Vibrational Spectroscopy of Molecules and Macromolecules on Surfaces;* Wiley Interscience: New York, N.Y.; 1993.
56. Nishioka, T.; Nakano, T.; Teramae, N. *Appl. Spectrosc.* **1992**, *46*, 1904.
57. High, M. S.; Painter, P. C.; Coleman, M. *Macromolecules* **1992**, *25*, 797.
58. Gaillard, F.; Romand, M. J. *Proceedings of the Adhesion Society;* 1993; p 108.
59. Panda, P. K.; Mishra S. B.; Mohapatra, B. K. *J. Inorg. Chem.* **1980**, *42*, 443.
60. Kinloch, A. J. *Adhesion and Adhesives;* Chapman & Hall, London, 1987.
61. Thrall, E. W.; Shannon, R. W. *Adhesive Bonding of Aluminum Alloys;* Marcel Dekker, New York, N.Y., 1985.
62. Fondeur, F.; Koenig, J. L. *J. Adhes.* **1993**, *40*, 189–205.
63. Fondeur, F.; Koenig, J. L. *Appl. Spectrosc.* **1993**, *47*, 1.
64. Fondeur, F.; Koenig, J. L. *J. Adhes.* **1993**, *43*, 289.
65. Fondeur, F.; Koenig, J. L. *J. Adhes.* **1993**, *43*, 263.
66. Mavich, A. M.; Ishidia, H.; Koenig, J. L. *Appl. Spectrosc.* **1995**, *49*, 149.
67. Ishida, H.; Koenig, J. L. *Polym. Eng. Sci.* **1978**, *18*, 128.
68. Ishida, H.; Koenig, J. L. *J. Polym. Sci. Polym. Phys. Ed.* **1979**, *17*, 615. 69 Pepper, S. V. *Appl. Spectrosc.* **1995**, *49*, 354.
70. Arvanitopoulos, C.; Koenig, J. L. *Appl. Spectrosc.* **1996**, *50*, 1–10.
71. Arvanitopoulos, C.; Koenig, J. L. *Appl. Spectrosc.* **1996**, *50*, 11–18.
72. Nerella, N.; Drennen, J. K. *Appl. Spectrosc.* **1996**, *50*, 285.
73. Sheu, K.; Huang, S. J.; Johnson, J. F. *Polym. Eng. Sci.* **1989**, *29*, 77.

74. Hsu, S. C.; Lin-Vien, D.; French, R. N. *Appl. Spectrosc.* **1992,** *46,* 225.
75. Cameron, R. E.; Jalil, M. A.; Donald, A. M. *Macromolecules* **1994,** *27,* 2713.
76. Cukier, R. *Macromolecules* **1984,** *17,* 252.
77. Chalmers, J. M.; Everall, N. J.; Local, A. *Proceedings of the 29th Annual Meeting of MAS;* VCH Publishers: Breckenridge, CO, 1995; p 123.
78. Schafer, R.; Kressler, J.; Neuber, R.; Mulhaupt, R. *Macromolecules* **1995,** *28,* 5037.
79. Neumesiter, L.; Koenig, J. L. submitted for publication in *Appl. Spectrosc.*

9

Raman Microscopy and Imaging

Introduction To Raman Spectroscopy

The *Raman effect* occurs when a sample is irradiated by monochromatic light, causing a small fraction of the scattered radiation to exhibit shifted frequencies that correspond to the sample's vibrational transitions (*1–5*). Lines shifted to energies lower than the laser frequency are produced by molecules in the ground state, while the slightly weaker lines at higher frequency arise from molecules in the excited vibrational states. These new lines, the result of the inelastic scattering of light by the sample, are called *Stokes* and *anti-Stokes* lines, respectively (Figure 9.1). Elastic collisions resulting in no change in frequency yield *Rayleigh scattering* and appear as the much more intense, unshifted component of the scattered light.

In normal Raman scattering, a molecule is excited to a *virtual state,* which is a quantum level relating to the electron cloud distortion created by the electric field of the incident light. A virtual state does not correspond to a real eigenstate (vibrational or electronic energy level) of the molecule but rather is a sum over all eigenstates of the molecule.

Raman scattering is a process of reradiation of scattered light by dipoles μ induced in the molecules by the incident light and modulated by the vibrations of the molecules. In normal Raman scattering by molecules in isotropic media, the dipoles are simply those that result from the action of the electric field component $\mathbf{E}$ of the incident light on the molecules

$$\mu = \alpha\mathbf{E} \tag{9.1}$$

where α is the molecular (dipole) polarizability. When a beam of light is incident upon a molecule, it can be either absorbed or scattered. Scatter-

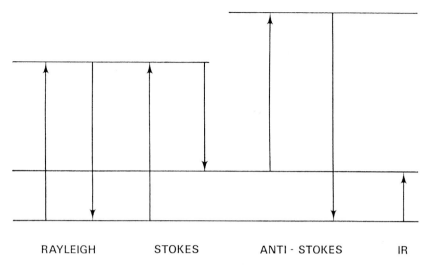

RAYLEIGH   STOKES   ANTI - STOKES   IR

Figure 9.1. Diagram of Stokes and anti-Stokes lines with elastic scattering. IR, infrared.

ing can be either elastic or inelastic. The electric field of the incident light induces a dipole moment, **P,** in the molecule, given by

$$\mathbf{P} = \alpha\mathbf{E} \tag{9.2}$$

where **E** is the electric field and α is the polarizability of the molecule. Because the electric field oscillates as it passes through the molecule, the induced dipole moment in the molecule also oscillates. This oscillating dipole moment radiates light at the frequency of oscillation in all directions except along the line of action of the dipole. The electric field is an oscillating function that depends upon the frequency of the light, ν_o, according to

$$\mathbf{E} = \mathbf{E}_0 \cos 2\pi\nu_0 t \tag{9.3}$$

where $\mathbf{E}_o$ is the impinging electric field and t is time.
    Substitution in the previous equation gives

$$\mathbf{P} = \alpha\mathbf{E}_0 \cos 2\pi\nu_0 t \tag{9.4}$$

The polarizability, α, depends on the motion of the nuclei in the molecule. The motion of the nuclei of a diatomic molecule can be expressed in terms of the normal coordinate of the vibration, x_1, and the dependence of α on x (the change in internuclear separation with vibration) can be approximated by a series expansion

$$\alpha = \alpha_0 + (\delta\alpha/\delta x) + \ldots \tag{9.5}$$

The normal mode is a time-dependent vibration with a frequency, ν_1. This dependence can be expressed as

$$x_1 = x^0{}_1 \cos 2\pi\nu_1 t \tag{9.6}$$

where $x^0{}_1$ is the equilibrium position. Substitution gives

$$\mathbf{P} = (\mathbf{E}_0 \cos 2\pi\nu_0 t)[\alpha_0 + (\delta\alpha/\delta x)x](x^0{}_1 \cos 2\pi\nu_1 t) \tag{9.7}$$

From basic trigonometry

$$\cos \theta \cos \phi = [\cos (\theta + \phi) + \cos (\theta - \phi)]/2 \tag{9.8}$$

Substitution gives

$$\mathbf{P} = \alpha_0\mathbf{E}_0 \cos 2\pi\nu_0 t + (1/2)\mathbf{E}_0 x^0{}_1 \, \delta\alpha/\delta x[\cos 2\pi(\nu_0 + \nu_1)t + \cos 2\pi$$
$$(\nu_0 - \nu_1)t + \cos 2\pi(\nu_0 - \nu_1)t] \tag{9.9}$$

This complex equation demonstrates that three lines are predicted in the light scattered by a diatomic molecule. The α_0 term represents the light that is not shifted in frequency (Rayleigh scattering). If $\delta\alpha/\delta x$ does not equal 0, Raman lines are shifted higher and lower in frequency than ν_o by ν, the frequency of vibration of the molecule. If $\delta\alpha/\delta x$ equals 0, the second term is 0, and no Raman lines are observed. This change in polarization with the vibrational motion of the nuclei, $\delta\alpha/\delta x$, is the basis for the selection rule governing the Raman activity of a vibrational mode.

The molecule will scatter light at the incident frequency. However, the molecule vibrates with its own unique frequencies. If these molecular motions produce changes in the polarizability, α, the molecule will further interact with the light by superimposing its vibrational frequencies on the scattered light at either higher or lower frequencies.

When a beam of photons strikes a molecule, most of the photons are scattered elastically, and this elastic scattering is termed Rayleigh scattering. Rayleigh scattering is responsible for the sky appearing blue because scattering is more efficient at shorter wavelengths. A few photons (1 in 108) undergo inelastic scattering or Raman scattering. These Raman or inelastically scattered photons have different frequencies and produce a spectrum of frequencies in the scattered beam that constitute the Raman spectrum of a molecule. The photons that lose energy appear on the lower

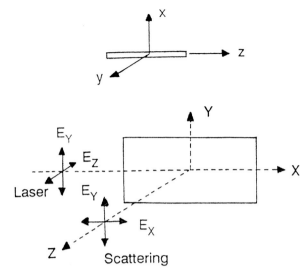

Figure 9.2. Coordinate system for the Raman scattering experiment. The laser beam path is along X, with vertical polarization along Y or horizontal polarization along Z. The Raman-scattered radiation is collected along Z and analyzed either horizontally along X or vertically along Y. The rod models a local chain segment and shows the attached molecular coordinate system. E is the electric vector of light. (Reproduced with permission from ref. 10. Copyright 1993 Society of Spectroscopy.)

frequency side of the exciting line and are called Stokes lines, and the photons that gain energy appear at higher frequencies and are called anti-Stokes lines. The infrared (IR) frequency is the difference between the two levels, so Raman spectroscopy arises from the same energy levels as IR spectroscopy, but by a quite different route.

In the usual Raman experiment, the observations are made perpendicular to the direction of the incident beam, which is plane-polarized (Figure 9.2).

Resonance Raman Spectroscopy

Resonance Raman takes place when the frequency of the exciting line is similar to the absorption frequency of a chromophore in the sample (6). The Raman signal from the chromophore can be enhanced by 7 or 8 orders of magnitude. Resonance Raman occurs when the energy of the exciting beam is close to an electronic energy level of the molecule and the excitation makes one term in the sum of eigenstates dominate over all others. When this occurs, the electron cloud surrounding the molecule is more readily distorted by the electric field of the incident light. The transition has a higher probability, and there is an enhancement of the Raman process. The vibrational modes enhanced are generally totally symmetric and distort the molecule along directions of electron density changes between the ground and the resonant electronic excited states.

Surface-Enhanced Raman Spectroscopy (SERS)

It has been observed that the Raman effect for molecules in contact with noble metal surfaces is resonantly enhanced by several orders of magnitude (7). In order to achieve large surface Raman signals, it is necessary that the metal surfaces be specially prepared in one of several ways, which renders them rough or finely divided on a length scale comparable to optical wavelengths. Under favorable conditions, SERS-active metals enhance the surface Raman intensity by up to ca. 10^5-fold, giving strong Raman signals. When the incident radiation interacts with the surface, it causes the free electrons to oscillate with the incident electric field and polarizes the noble metal particles (8). This creates a strong local electric field at the particle surface known as a *surface plasmon*. When a molecule is in close proximity to a noble metal particle, the molecule is polarized by the electric field of the noble metal particle. This leads to an enhancement of the Raman signal because the Raman scattering is proportional to the square of the local electric field. The enhancement originates, in part, from the interaction of the optical frequencies with the electric field of the noble metal particles.

However, the Raman resonance enhancement due to the surface decreases very quickly as a function of distance, and little enhancement is obtained for molecules a few monolayers away from the surface. As a result, SERS is surface-selective. The scattering from a polymer film on a SERS-active substrate arises almost entirely from the first few molecular layers adjacent to the substrate as long as the film is less than 1000 in thickness. Thus SERS can be used for in situ nondestructive characterization of polymer interfaces (9).

Advantages of Raman Spectroscopy Compared to IR Spectroscopy

Raman spectroscopy has a rich history of applications in the characterization of materials including polymers. Raman spectroscopy represents the richer half of the vibrational spectral information and has several advantages relative to IR spectroscopy (10):

- Little sample preparation is necessary. Scattering from the sample is the basic Raman technique (i.e., optically thin light-transmitting samples are not required).

- The spectral information is richer; the Raman selection rules are less restrictive than IR, so theoretically more modes are observable.

- A single instrument has the ability to reach the low-frequency region below 400 cm^{-1} where lattice frequencies occur for solids and polymers.

- Raman can be used to study aqueous solutions because water is a weak Raman scatterer and therefore introduces little spectral interference or overlap.

Raman spectroscopy makes microscopic and mapping analysis simpler than with IR spectroscopy for the following reasons:

- Microsamples are easily studied because the Raman laser beam has an inherently small diameter (100 μm) and so no beam-condensing techniques are required. The flux of the Raman radiation is inversely proportional to the diameter of the focus of the laser beam at the sample; therefore the optimum sample for Raman spectroscopy is a microsample.

- The spatial resolution of Raman is on the order of 1 μm × 1 μm, whereas IR is limited to around 20 μm × 20 μm.

- Samples can be supported on glass slides; glass is a weak Raman scatterer but a strong IR absorber so traditional optical measurements can be made simultaneously.

- Raman scattering is inherently a scattering process, and back scattering at 180° is particularly useful for mapping purposes because positioning of the beam is simple.

As a consequence of these advantages, Raman microscopy has been widely used for the study of materials, and recent instrumental advances promise to continue the development of this interest in the use of micro-Raman spectroscopy. The inherent limitation of Raman microscopy is the presence of interfering fluorescence for many industrial samples. Another potential limitation is the necessity of concentrating the laser energy in a very small area on the sample; such high photon densities can cause thermal or photoinduced degradation effects in the samples.

The scattering volume of a sample in a 90° scattering geometry is $\pi D^2 l/4$, where D is the diameter of the beam and l is the length of the beam inside the sample. The Raman-scattered signal from a given sample is proportional to

- the number of molecules per unit volume,

- the total volume of sample illuminated,

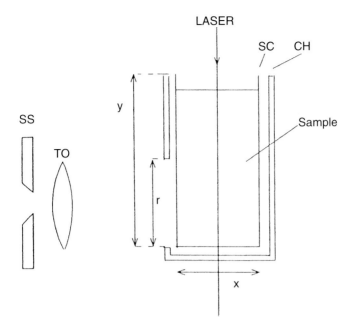

Figure 9.3. Front view of the sample cell holder: r, effective excitation path length collected by the transfer optics (TO) in the spectrometer; y, excitation path length; x, collection path length; SC, sample cuvette; CH, cuvette holder; SS, spectrometer entrance slit. (Reproduced with permission from ref 8. Copyright 1993 Society of Spectroscopy.)

- the excitation-laser irradiance per unit area, and
- the Raman-scattering cross-section of the molecules.

Consider the sample geometry shown in Figure 9.3 (*11*). The total Raman intensity (photons $sr^{-1} s^{-1}$) from a sample illuminated by a collimated laser beam of power I_0(photons s^{-1}) is given by

$$I = I_0 \, \sigma N r \tag{9.10}$$

where σ is the Raman cross-section ($cm^2 \, sr^{-1}$molecule^{-1}), N is the number density of scatterers (molecules cm^{-3}), and r is the excitation path length (cm) of the laser beam in the sample observed by the spectrometer (*11*). In this expression, it is assumed that both the excitation and collection path lengths are constant. The transmittance decreases exponentially with collection path length x so the intensity variation produced as a function of the lateral movement of the laser beam will be

$$I_x = I_0 \, \sigma N r e^{-\gamma x} \tag{9.11}$$

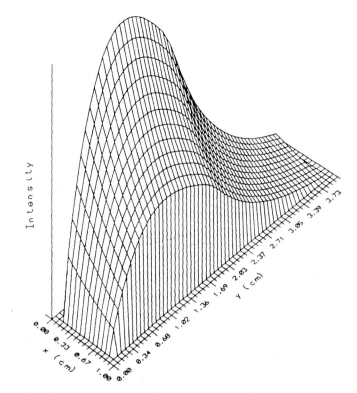

Figure 9.4. Raman intensity profile generated by a collimated laser beam passing through a sample cuvette (1 cm × 4 cm) using the geometry of Figure 9.3. (Reproduced with permission from ref 15. Copyright 1993 Society of Spectroscopy.)

where γ is the sample coefficient of absorption (cm^{-1}) at the Raman frequency of interest. Figure 9.4 shows the Raman intensity profile generated by a collimated laser beam passing through a sample cuvette (1 × 4 cm) using the geometry of Figure 9.3 (8). Figure 9.4 shows that the intensity is a maximum close to the wall facing the transfer optics. Clearly, the maximum Raman signal depends on proper selection of sample volume and illumination.

Back scattering (180°) geometry is also often used and has several advantages relative to the 90° scattering experiment. First, it is much easier to position the sample in the spectrometer, especially in the case of thick samples. Second, it allows the determination of the distribution of orientation in nonhomogeneous samples since the diameter of the focused laser beam on the sample is only approximately 200 μm. Finally, birefringence effects are less important since the scattered radiation comes from a small volume near the surface of the material (8).

In general, the induced polarizability is not necessarily in the direction of the incident beam. The relationship between the induced polarization and the electric field of the source is

$$P_x = \alpha_{xx}E_x + \alpha_{xy}E_y + \alpha_{xz}E_z \qquad (9.12)$$

$$P_y = \alpha_{yx}E_x + \alpha_{yy}E_y + \alpha_{yz}E_z \qquad (9.13)$$

$$P_z = \alpha_{zx}E_x + \alpha_{zy}E_y + \alpha_{zz}E_z \qquad (9.14)$$

where α_{xx}, α_{yy}, and α_{zz} are the components of the principal axes of the polarizability ellipsoid, and α_{xy}, α_{yz}, and α_{xz} are the other components(*12–14*). Consequently, the *Raman-scattered light emanating from a random sample is polarized to a greater or lesser extent.* For randomly oriented systems, the polarization properties are determined by the two tensor invariants of the polarization tensor (i.e., the trace and the anisotropy).

The *depolarization ratio* is defined as the intensity ratio of the two polarized components of the scattered light that are parallel and perpendicular to the direction of the propagation of the (polarized) incident light (*11*). The polarization of the incident beam is perpendicular to the plane of propagation and observation. For this geometry (Figure 9.5) (*15*) the depolarization ratio is defined as the intensity ratio

$$\rho = VH/VV \qquad (9.15)$$

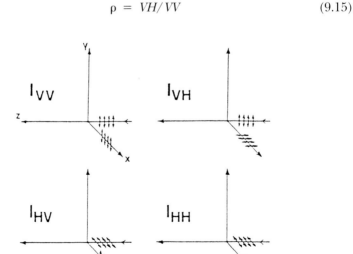

Figure 9.5. Configurations for polarization measurement using the 90° scattering geometry. (Reproduced with permission from ref 15. Copyright 1993 Society of Spectroscopy.)

where, for the right angle scattering experiment, V is perpendicular to the scattering plane and H is in the scattering plane. An alternate notation expressed in terms of the laboratory coordinate system is

$$A(BC)D \tag{9.16}$$

where A is the direction of travel of the incident beam, B and C are the polarization of the incident and scattered light, respectively, and D is the direction in which the Raman-scattered light is observed. Generally, the incoming beam is along the X-axis, the scattered beam is along the Z-axis, and the Y-axis is perpendicular to the scattering plane.

The Raman depolarization ratio is always less than or equal to $3/4$. For a specific scattering geometry, this polarization ratio is dependent upon the symmetry of the molecular vibration giving rise to the line. The various instrumental factors that can influence the determination of Raman polarizations and band profiles have been considered (15–17). Errors can be introduced as a result of the misorientation of the optical axis, the retardation tolerance of the polarization scrambler, missetting of the incident beam polarization, or misorientation of the optical axis of the polarization analyzer. In the past, it was necessary to consider the nonideal behavior of the polarization analyzer, but with modern polarization analyzers the extinction ratio is better than 10^{-4} so this error is negligible.

In solids, the problem of polarization is even more complicated but the results are more rewarding. In solids, the molecular species are oriented with respect to each other. Therefore, the molecular polarizability ellipsoids are also oriented along definite directions in the solid. Because the electric vector of the incident laser beam is polarized, the directionality in the crystal can be utilized to excite and obtain Raman data from each element of the polarizability ellipsoid. With the laser polarization along Z and collection along Z, a spectrum from the α_{zz} component of the tensor is obtained. By rotating the analyzer 90°, thereby collecting X-polarized light while still exciting along Z, α_{zx} is obtained.

To study the orientation of the sample, these experiments must be related to the molecular polarizabilities. The appropriate relationships have been worked out (10, 12). For an isotropic system, there are orientationally invariant terms δ and γ, which are defined as

$$\delta = 1/3 \sum \alpha_{ii} \tag{9.17}$$

and

$$\gamma^2 = 1/3 \sum_{i,j} [(\alpha_{ii} - \alpha_{jj})^2 - 6\alpha_{ij}^2] \tag{9.18}$$

Table 9.1. Contribution to Spectrum for Three-Dimensional Isotropic Systems

Experiment	Geometry	Symmetric	Anisotropic
VV	X(YY)Z	δ^2	$4\gamma^2/45$
VH	X(YZ)Z	0	$\gamma^2/15$
HV	X(ZY)Z	0	$\gamma^2/15$
HH	X(ZX)Z	0	$\gamma^2/15$

The contribution to the spectrum for three-dimensional isotropic systems is shown in Table 9.1. These results show that the intensities of the *VH, HV,* and *HH* experiments should be identical.

Theoretically, the depolarization ratio can have values ranging from 0 to 3/4, depending on the nature and symmetry of the vibrations. Nonsymmetric vibrations give depolarization ratios of 3/4. Symmetric vibrations have depolarization ratios ranging from 0 to 3/4, depending on the polarizability changes and symmetry of the bonds in the molecule. Accurate values of the depolarization ratio are valuable for determining the assignments of Raman lines, and in conjunction with dichroic measurements in the IR they represent a powerful structural tool for polymers.

One of the major problems is that the polarization can be scrambled whether by optical inhomogeneities in the sample or by the inherent birefringence of the samples. Many spectra show evidence of extensive depolarization of both the incident and scattered light, and thus each spectrum is an indeterminate mixture of four different polarizations: the intended polarization, the one resulting from depolarization of the incident light, the one resulting from depolarization of the scattered light, and (to a lesser extent) the one resulting from depolarization of both the incident and scattered light. When this occurs, little useful information is available from the individual polarized spectra. However, one can approximate the isotropic and anisotropic parts of the polarizability tensor (*18*). The spectra added for the isotropic spectrum are Y(XX)Z, Z(YY)X, and Y(ZZ)X, where Z is the long axis of the polymer sheet, X is the shorter axis, and Y is perpendicular to the sheet (*18*). For the anisotropic spectrum the Y(ZY)Z, X(ZX)Z, and Y(XY)X spectra are added.

Raman polarization measurements are complicated in microscopic measurements as the microoptics also contribute to the polarization. The assumption is normally made that the illumination is collimated, but in the microscope the light is collected through a small angle and the polarization ratio measured must account for this effect (*18, 19*) in the following manner:

$$\rho = (A + B)\gamma^2/\{15A\delta^2 + [(4/3)A + B]\gamma^2\} \qquad (9.19)$$

where A and B are given by

$$A = \pi^2[(4/3) - \cos\theta - (1/3)\cos\theta^3] \qquad (9.20)$$

$$B = [(2/3) - \cos\theta - (1/3)\cos\theta^3] \qquad (9.21)$$

and θ is the gathering half-angle of the objective. If $B \ll A$, the usual polarization equations are obtained. Even this correction may not be adequate for totally polarized bands because the polarizing effects of the microscope objective are not considered.

Fluorescence Interference with Raman Spectroscopy

One of the principal problems in Raman spectroscopy is *fluorescence*. Fluorescence occurs when the excitation line is partially absorbed and remitted. *Phosphorescence* is luminescence that is emitted for a significant period of time (in the millisecond to second time regime) after excitation. Many compounds that do not fluoresce do exhibit phosphorescence.

The basic problem in Raman spectroscopy is the inefficiency of the Raman scattering, since roughly only 1 out of 10^8 incident photons is Raman-scattered. This, coupled with the fact that most industrial systems give rise to an interfering fluorescent background, leads to a problem of detectability. For example, if a system contains one part per million of an interfering system that fluoresces under visible excitation, the same incident flux of 10^8 photons will produce 100 fluorescent photons that will completely mask the Raman signal. A number of methods have been developed in an attempt to minimize fluorescence interference including time-based discrimination, UV excitation, heavy-atom quenching, and surface-enhanced techniques. No universal method exists for the removal of fluorescence. However, the most successful and practical approach has been using near-IR excitation for the Raman spectrum (*20*). Near-IR excitation is below the electronic absorption process that leads to fluorescence for most organic substances.

Raman Instrumentation

A Raman instrument has three basic components:

- a signal generator, usually a laser;
- a signal analyzer, either a spectrometer or interferometer; and
- a signal detector, either a monochannel or two-dimensional array.

A Raman microprobe also has a focusing component, usually a microscope, and a mapping unit such as a computer-controlled micromanipulator. Each of these components plays an important role in the nature of the Raman mapping experiment.

Raman Illumination Sources

Although initially the sun was used as a source by C. V. Raman in 1926, he realized that it was not practical because of the inefficient nature of the Raman scattering process. Intense arc systems were subsequently utilized, but the S/N ratio and the quality of the spectra left much to be desired. Then the laser was introduced as a source, and Raman spectroscopy became a more useful tool as high-quality spectra could be obtained. The laser provides a coherent, single-frequency (with single-mode operation), high-power, small-beam source (~100 μm) that is nearly ideal for Raman spectroscopy.

Typically, the focused laser spot produces a high power density of the order of 10^4 to 10^6 W cm^{-2}. Power densities of this magnitude can produce heating and decomposition of sensitive, especially absorbing, samples. The laser can heat up the sample substantially. If the incident laser beam of radiance I_0 (W/cm^2) radiates a sample of radius r (cm) in contact with a substrate of thermal conductivity K (W cm^{-1} K^{-1}) with efficiency E, then the temperature rise, ΔT, is given by

$$\Delta T = EI_0 r/4K \qquad (9.22)$$

For a typical transparent material, E is approximately 1×10^{-4} so if the sample is in contact with air having thermal conductivity $K = 2.6 \times 10^{-4}$ W cm^{-1} K^{-1}, the temperature of the sample will rise ~200 °C. With the Raman microprobe, the microsample is more likely to be in contact with a glass surface, which is a better thermal conductor (with $K = 0.1$ W cm^{-1} K^{-1}), so the temperature rise will be 10–20 °C if the laser power is limited to a few milliwatts (2–10 mW) (*21*). Defocusing of the beam or the use of less laser power can minimize these effects but at the expense of a lower S/N ratio.

Improvements in lasers as the technology has evolved have resulted in a large array of available frequencies ranging from the UV to the IR, and Raman spectroscopy has been the beneficiary of these advances, particularly in the reduction of fluorescence where near-IR lasers can be used with energies below the electronic energy levels of most molecules.

The choice of laser excitation frequency, ν, depends on the type of sample being examined. In most cases, the laser wavelength is chosen to avoid any absorption by the sample as it may be destroyed by photodecomposition. Visible excitation may produce a large background of fluorescence. Sometimes, for particular samples, the fluorescence can be decreased by exciting at longer wavelengths. Occasionally, a particular exciting wavelength may reasonably enhance certain features of the spectrum. In a resonance Raman experiment where the laser is absorbed by the sample, certain vibrational modes that are coupled to the absorbing chromophore are enhanced in the Raman. In some cases the Raman scattering cross-section can be orders of magnitude larger, making the Raman signal greatly enhanced. However, there is also a high risk of the sample being destroyed by photochemical effects or thermal heating under resonance Raman conditions.

Since the Raman scattering cross section varies as ν^4 the wavelength of the source should be as short as possible to increase the probability of Raman-scattered photons. If, however, fluorescence is a problem, a frequency of excitation that is below the threshold for excitation of fluorescence must be used. Presently, the most practical laser of choice is the Nd:YAG system, lasing at 1.064 μm (9395 cm^{-1}). On the horizon are diode-pumped Nd:YAG lasers, which have a significantly lower source fluctuation noise than the traditional flash-lamp-pumped Nd:YAG lasers.

Dispersion Raman Spectrometers

In the dispersive mode of Raman spectroscopy, a monochromatic laser beam (excitation radiation) is radiated on the sample and the scattered light is focused on the entrance slit of a monochromator. The dispersion monochromator (often double or triple) discriminates between the strong elastic scattering (Rayleigh scattering) and the weak inelastically scattered light (Raman scattering) with different frequencies and scans the desired frequency interval. The spectral resolution is determined by the width of the entrance slit, the number of rulings of the gratings, and the focal length of the spectrometer. The bandwidth of the monochromator limits the resolution in terms of data points per wavenumber.

The Raman spectrum is given by the detection of the intensity of the scattered, frequency-shifted light by a photoelectric system like a photomultiplier tube. The resulting signal of the detector is amplified and converted to a form appropriate for plotting as a function of the shifted frequency. The line shape measured is the result of the convolution of the spectral image in the focal plane and the rectangular exit slit.

The dispersive Raman spectrometer requires a high-dispersion, high-throughput, low-stray-light monochromator system. Typical single monochromators provide stray light rejections of 10^{-5}–10^{-6} (as a fraction of the Rayleigh light that enters the spectrometer) limited by the imperfections on the optical surfaces such as gratings. Double and triple monochromators are often required to obtain adequate stray light rejection. However, the 1–5% transmission efficiency due to reflection losses and grating efficiencies of triple monochromators reduces the Raman signal to extremely low levels for detection.

The stray light can be reduced by decreasing the amount of Rayleigh-scattered light entering the monochromator. This can be accomplished by using notch or edge filters in front of the spectrometer entrance slit. The ideal filter has a frequency cutoff close to the exciting line, a large optical density in the cutoff region, and high transmission outside the cutoff with a sharp slope during transition from peak cutoff to full transmission. No such ideal filter exists, but Rayleigh filtering has been accomplished practically by using narrow-band rejection filters including holographic edge and notch filters, interference filters, atomic line vapor cells, and crystalline colloidal Bragg diffraction filters.

To use single-stage spectrographs with a multichannel detector, one needs sharp cutoff filters. The insertion of holographic filters at the entrance reduces the Rayleigh-scattered light so that a throughput approaching 50% can be achieved. A dispersion Raman system based entirely on holographic elements has been reported (*22*).

One of the major problems involved with dispersive Raman spectrometers is the poor reproducibility of the frequencies (± 0.2 cm^{-1}) due to backlash in the grating positioning. A frequency error of this magnitude is sufficient to distort spectral subtractions to the point where they are not useful. The largest of these frequency errors arises from the inability to reposition the spectrometer precisely due to backlash. Additional errors are introduced by temperature fluctuations and by variations in the sample position relative to the sampling optics. However, such errors in the frequency can be corrected using least-squares and Fourier-domain phase shifting (*23*).

Multiplexed Raman Spectrometers

One of the primary weaknesses of conventional scanning Raman spectrometers is that only a very small fraction of the total scattered light strikes the detector at any one time; the rest is discarded. One partial solution is to use multiplex Fourier transform Raman(FT-Raman) instrumentation or multichannel detectors. First we will discuss the multiplex approach, which utilizes an interferometric system.

The motivation for carrying out FT-Raman spectroscopy is

- high etendue,
- intrinsic spectral accuracy,
- spectral resolving power independent of input aperture diameter, and
- spectral resolution under computer control.

A figure of merit for optical measurements is the *etendue*. The etendue measures the ability of an optical system to accept light. The high etendue or throughput of an interferometer due to the lack of spectral slits relative to a monochromator is a major advantage for Raman spectroscopy. The large throughput implies that the interferometer can capture most of the light scattered from the illuminated volume of sample. This advantage allows the use of lower laser power, which could be particularly important for measurements on photofragile materials where sample damage and nonlinear optical processes could occur.

The improved frequency precision of FT-Raman relative to dispersive Raman allows spectral subtraction to be used to minimize background and "spectrally purify" the spectra by removal of the undesirable background and overlap contributions.

A visible light FT-Raman spectrometer using high-optical-density Rayleigh line filters has been designed (*24*) coupled with a low-noise, high-quantum-efficiency photodetector. The movable mirror was mounted on a linear electromagnetic motor, and submicrometer position control was based on a compact noninterferometric position sensor. The scattered light autocorrelation function was obtained over a range of mirror path differences. Subsequently, the Raman signal was obtained by applying a user-selected window function to the waveform, and then the fast Fourier transform was performed on the data.

Near-IR FT-Raman spectroscopy became practical due to the commercial availability of the cw Nd:YAG laser at 9395 cm^{-1} (*20,25*). Typically, FT-Raman is carried out using near-IR excitation, which gives additional advantages:

- Near-IR excitation minimizes the fluorescent background.

- Raman and IR measurements can be carried out on the same instrument.

An optical diagram of an FT-Raman spectrometer is shown in Figure 9.6. The Stokes-shifted Raman scattering occurs in the 5000–9000 cm^{-1} spectral range, coincident with the near-IR region. The laser photon energy at 1076 nm is below the electronic energy levels of most organic molecules and therefore does not cause fluorescence, but the intensity of the Raman scattering at 1076 nm is diminished by a factor of ~22 due to the fourth-power dependence on the frequency of the incident radiation in comparison to excitation at 488.0 nm. This factor can be compensated for by increasing the excitation power.

There is an additional limitation due to the lack of good detectors (shot noise limited) in the near IR relative to the photomultiplier tube used in the visible. Improvements in detectors are on the horizon so this

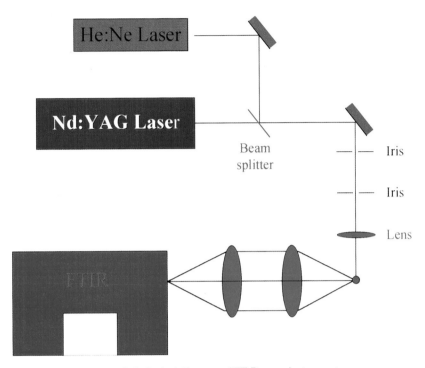

Figure 9.6. Optical diagram of FT-Raman instrument.

problem may be overcome with improved technology. Initially, FT-Raman instruments used either lead sulfide or germanium detectors, which yielded limited instrumental performance. Currently, indium–gallium–arsenide (InGaAs) detectors have extended the performance. The current indium–gallium–arsenide detectors have a noise equivalent power (NEP) of 10–15 $(W/Hz)^{1/2}$ or better (*26*).

Because of the optical configuration of the Michelson interferometer both Rayleigh- and Raman-scattered light enter the system. To obtain Raman spectra with high S/N ratios using the FT-Raman method, the unwanted Rayleigh scattering must be removed. The intense Rayleigh line is from 10^6 to 10^8 times more intense than the Raman scattering. A signal of this intensity drives the detector into the distributive noise regime, resulting in high-frequency noise spread throughout the transformed spectrum. However, spectra with useful S/N ratios can be obtained using notch filters and other devices to filter out the Rayleigh line. Holographic filters are nearly ideal notch filters, which provide high optical density (>6) at the Rayleigh line and have transmittances approaching $90\%T$ at other wavelengths (*27*).

Raman spectroscopy is always photon-limited, and the energy analyzer must have a high-throughput efficiency. Recent developments suggest that an increase of a factor of 10 in sensitivity has occurred relative to the initial FT-Raman instrumentation (26). A recent review of the instrumental techniques for micro-FT-Raman indicates the power of the technique for analysis of a variety of samples (28).

Unfortunately although fluorescence is substantially reduced, changes in the sample due to laser IR heating can occur if the sample absorbs at the frequency of the laser. Thus Raman signal attenuation varies with sample absorptive or scattering properties. Little Raman scattering can be collected through a sample thickness of more than one extinction path length, and the relative amount of heating for two different excitation wavelengths (or specimens) may be estimated as the inverse of their extinction path lengths. The thermal conduction for heat removal depends on the thermal properties of the sample and the sample cell. In order for one to obtain less than 1 °C heating, which is important for biological systems, the laser power needs to be less than 740 mW (29).

Raman Spectroscopy using Two-Dimensional Array Detectors

To improve the performance of the dispersive Raman spectrometer, the system can be coupled to a multichannel array detector, and the advantages of such coupling have been discussed (30). In this new generation of dispersive Raman instruments, a polychromator or spectrograph and a multichannel detector are used. Two-dimensional detectors such as vidicons, intensified diode-arrays, or charge-coupled devices (CCDs) are used rather than the single-channel photomultiplier. In these instruments, the photoactive detector elements lie in the spectrograph image plane. The entire spectrum is collected simultaneously using an array of detectors in which each element of the multichannel detector can have comparable sensitivity to a single photomultiplier.

The assumption is that all light incident on the detector is correctly positioned optically upon each single detector. In this way the time needed to record a spectrum is reduced markedly. Consequently, the S/N ratio can be considerably improved and measurements can be made that are not possible with single-channel spectrometers. The line shape measured with a two-dimensional detector is a convolution of the spectral image and the geometrical response function of the pixels. The spectral width of the two-dimensional detector is different at each pixel since the pixel spacing is equidistant only with respect to geometric spacing; it is not equidistant in wavelength or wavenumber. Regardless of the spectrometer spectral width, the total spectral resolution will be limited by the width of the pixel column (31). The density of measurement points, in terms of wavenumbers, is restricted by the dimensions of a single detector element. The two-dimensional array has a particular detector function since each detector element has its own sensitivity and noise characteristics. The ultimate limitation of any array detector is the resolution–bandwidth tradeoff. It is necessary to select either high resolution or large spectral bandwidth.

Vidicons and diode arrays were the first array detectors used for Raman spectroscopy (*32,33*). The two-dimensional photodiode consists of an array of individual photodiodes, each with an associated capacitor to store the signal. The readout region is separate and consists of a videoline connected to each photodiode's capacitor through a switch. Switch control is provided by a shift register that serially turns each switch on, connecting one pixel's signal to the video line at a time.

Microchannel plate-intensified photodiode arrays provide a 10^3 to 10^4 gain prior to detection by a Si photodiode array (*34*). A photocathode is positioned in close proximity to the face of the microchannel plate. Photons striking the photocathode cause photoelectron emission. The photoelectrons enter the hollow glass fibers striking a resistive coating and produce a secondary emission of additional electrons. These electrons after exiting the microchannel plate are accelerated toward a phosphor screen and ultimately coupled with a fiber-optic bundle to the photodiode array.

CCD detectors have a high quantum efficiency over a wide wavelength range and are particularly useful for detection at very low light levels, which is a requirement for Raman spectroscopy (*35*). CCD detectors have a unique capability that distinguishes them from other detectors. This is the readout mode, in which charge from more than one detector element is combined on the chip before being read out (*35*). This process, called *binning*, involves moving the charge from a number of neighboring detector elements (e.g., a single column or row of the CCD chip) into a single "bin" and then transferring the charge from the bin to the output mode. To do this, each pixel of the CCD performs two functions: collecting the photoelectrons during the integration period and transferring the charge out of the chip during the readout period. An area CCD consists of multiple columns of pixels placed next to each other to form rows. At the end of the columns and running parallel to the rows is a readout register. To read out the photoelectrons, each row of pixels is simultaneously transferred down the columns to the next row of pixels. The photoelectrons in the last row, which is parallel to the readout register, are transferred into the register. Once the readout register transfers the signal to the output, the next row is transferred into the readout register and the process is repeated until the entire array has been read out.

The use of silicon as the photosensitive element in CCDs permits operation in the near-IR region with 632–830 nm source lasers, thus reducing fluorescence interference (*36, 37*). Germanium detectors can also be used and commercial detectors are available with 128×128 elements with dimensions of 0.1×2.5 mm (*38*).

Photonic noise is equal to the square root of the number of photoelectrons in a pixel. The detector noise of a CCD can be as low as 3 electrons per pixel compared to a photodiode array, which is approximately 1500 electrons per pixel (*39*). By comparison, it has been estimated that the CCD detector is roughly 3400 times more sensitive than the InGaAs detector (*40*). Thus CCD Raman spectrometers generally yield a situation in which the only noise source is sample shot noise and background or internal scatter from the surface of the detector.

In principle, the dynamic range for intensified photodiode arrays could be 10^4 and 10^6 for CCD. However, the observed ranges are lower because of internal scattering or reflection of light at the interfaces of the detector components (*41*).

The instrumental response must be determined for CCDs because of the problem of wavelength inaccuracies introduced due to variable detector position and nonlinearity across a wide flat field. It is desirable to calibrate the filter transmission, spectrograph throughput, and detector response so proper instrumental corrections can be made (*42*).

Multiscanning Two-Dimensional Raman Spectroscopy

One of the obvious extensions of the use of two-dimensional detectors is to combine them with a scanning system. Indeed, a scanning multichannel method has been demonstrated as a technique for improving the *S/N* ratio (*43*).

Raman Microscopes

Conventional laser-excited Raman spectroscopy is inherently a microprobe technique due to the small diameter of the laser beam (~100 μm) (*44,45*). However, attaching a microscope to the system allows higher spatial resolution in the *x–y* plane. The depth dimension, *z*, of the sample can be obtained by using a confocal microscope in conjunction with the Raman spectrograph.

Traditionally, Raman microspectra are obtained by illuminating the whole microscopic field of view with laser light, wavelength selecting the scattered light, passing it through the spectrometer, and imaging it onto a vidicon-type TV detector.

For Raman microscopes, an optical microscope is used to focus the laser beam into a small spot (~2–5 μm) on the specimen, and the scattered light is collected from the focus point. The scattered light is sent on the entrance slit of the optical filter, which can work as a Raman microspectrometer when the detector is a photomultiplier followed by an amplifier and a chart recorder (monochannel detection) or a Raman microspectrograph when the detector is a two-dimensional detector. In this latter case the spectrum is visualized on a monitor (*46*).

Micro-Raman Mapping and Imaging

There are three different illumination methods of obtaining Raman maps (*48*):

- point-by-point methods,
- line-scanning methods, and
- wide-field or global illumination.

In the point illumination mode, individual pixels are recorded; in the line mode the illumination takes the form of a line and the illumination line is scanned over the sample; in global illumination, the entire sample is illuminated and the scattered light is recorded on a two-dimensional detector.

In addition to these general mapping methods, the type of the detection system plays a role. With signal detection at a single wavenumber shift and global illumination, the mapping must be repeated for each wavelength desired. With simultaneous frequency and point illumination detection of the complete Raman spectrum, all the pixels must be recorded individually. Finally, global illumination with complete recording of the Raman frequency range can occur. In this case, Hadamard encoding of the spatial information at N different settings of the encoding mask occurs (*48*).

Polymeric systems are generally weak Raman scatterers, which makes the volumetric resolution limited as the volume must be sufficient to generate a measureable Raman signal. Also, polymer systems are notorious for their fluorescent capability, and this interfering radiation leads to substantial loss of S/N ratio when visible excitation lasers are used.

Point-by-Point Raman Imaging Instrumentation

In the *point illumination mode,* an area of interest on the sample is selected using the light microscope. The laser beam is positioned to this point and a complete Raman spectrum is obtained of the tightly focused area or point. To obtain a map, the sample is moved under the illuminated spot using a computer-controlled stage. The point illumination method is useful in that a complete spectrum is obtained of the selected area.

The point-by-point Raman mapping method, although the simplest, is limited by

- low signal-to-noise ratio, low spatial resolution,

- high laser power densities, and

- long experimental measurement times.

In order to obtain high spatial resolution, a small pixel area is selected. A low signal results from the small volume of sample. Signal averaging at each spatial position is often required, which requires a considerable amount of experimental time to obtain high-fidelity images. The spatial resolution is determined by the focused laser spot size, which can be as low as 1 μm (in theory). Blurring is a function of the repositioning error of the translation stage. To obtain high spatial resolution, high local laser power can be used but may induce thermal decomposition of the sample. Total incident power densities can only be reduced below the threshold decomposition levels at the sacrifice of spatial resolution.

The point illumination approach was the basis of the first commercial Raman microscope (46). Figure 9.7 is an optical diagram of the molecular optic laser examiner (MOLE) system (47). The MOLE was commercialized by Jobin–Yvon Optical systems. The system consisted of a double monochromator, containing master plane holographic gratings, which is optically and mechanically coupled to a microscope. Either 40× or 80× objectives with respective numerical apertures of 0.85 or 0.90 are used to focus the laser beam and collect the Raman-scattered radiation. The system can have monochannel or multichannel detection so two modes of operation are possible.

As previously discussed, the first Raman microscope operated on a point-by-point basis although the global radiation mode is possible. Recently a point-by-point data collection microscope has been developed that utilizes an intensified linear diode array detector with a spectrograph and subtractive double monochromator as a Rayleigh line filter and a stepper-motor-driven X–Y microscope stage (35).

Since this first commercial Raman microscope, several other Raman microscopic systems have been developed and commercialized. A particularly useful approach is to use fiber-optic coupling to the Raman microscope (49). This approach eliminates many of the alignment problems encountered with ordinary Raman microscopes. Fiber coupling to the microscope requires the incorporation of a filter module. This module must

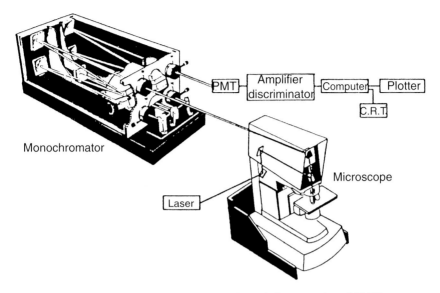

Figure 9.7. An optical diagram of the molecular optic laser examiner (MOLE) system commercialized by Jobin–Yvon Optical systems. (Reproduced with permission from ref 47. Copyright 1985 Science & Technology Press (Polymer) Elsevier.)

remove the Raman spectra of the silica generated in the excitation fiber and transmit Raman scattering efficiently. A holographic optical transmission grating can be used to remove the silica contribution to the Raman spectra, and a holographic notch filter prevents the laser line from returning to the spectrograph. Both the local power density and spatial resolution are lower than obtained with a conventional microscope. Raman spectra from 6 μm polystyrene have been observed without interference from the embedding epoxy using single-mode optical fibers coupled to the Raman microscope (*49*).

The point-by-point method is simple and straightforward but is experimentally demanding for obtaining high-fidelity images.

Raman Microline Imaging

The problem of photoinduced degradation of samples by the use of high laser power density can be reduced by using the line-illumination method, which defocuses the excitation source in one dimension to a slit geometry and systematically translates the sample through the laser slit illumination (*27,50–53*). The line focusing allows the measurements to be made with a laser power density 320 times lower than point focusing with equivalent signal to noise (*50*). Linear illumination makes better use of the CCD

detector. Both spectral and spatial information can be simultaneously recorded. Raman maps can be built up from the profiles by moving the
sample perpendicularly to the laser line.

The line-scanning method is more efficient than the point-by-point
method so the experimental time required is less. The maximum attainable spatial resolution is a convolution of the laser linewidth and the image step size. Positioning errors have a particularly damaging role in reducing the quality of the line-scanning image.

Raman microline focus spectroscopy (MiFS) uses a line-focused Raman microscope coupled to a nonastigmatic spectrograph (*50*). The
method uses interference filters to remove the laser line and cylindrical
optics coupled to the objective to generate the microline focus. The resultant spatial resolution is determined by the objective used and the detector element structure. Movement of the sample in a single direction
results in data collection from the whole analysis area. By this method, a
spatial resolution of 3.5 μm per CCD pixel was obtained at a 1.06 cm^{-1}
spectral resolution with an exposure time of 5 s (*50*). The advantages of
the line-scattering compared to point-by-point method are

- improved sensitivity;

- reduced risk of sample damage due to the distribution of the laser
  power;

- simultaneous collection of a number of spatially related Raman
  data, obviating the need to monitor laser power and other experimental variations; and

- faster speed of data collection.

The line illumination mode appears to be a suitable tradeoff between
point and global illumination for polymers as it acquires all of the spectral
and spatial data in a reasonable measurement time without sacrificing the
illumination power density to the point of low Raman signal generation.

This approach has been used to study carbon fibers and polymer degradation (*54*).

Raman Global Illumination Imaging

In the *global illumination mode*, the spectrometer is set to a particular band-
pass frequency range characteristic of a species of interest, a widened
beam (typically 20–400 μm) is scanned over the total sample area, and
an intensity map of the sample for the measuring frequency range is obtained. The scattered light from the imaged area is then passed through
an energy (band-pass) filter before the image is formed. For direct global

Raman imaging, a number of band-pass filters have been used including diffraction gratings, Hadamard transforms, multilayer-dielectric filters, and acousto-optic and liquid crystal tunable filters (*see* following sections). The wide-field or direct imaging is mechanically and optically simple as well.

Global illumination has the following advantages relative to point and line illumination:

- shorter overall exposure times,
- high fidelity of images due to simultaneous exposure of the entire sampled region,
- higher spatial resolution, and
- reduced blurring due to greater simplicity of measurement (no moving parts).

The disadvantages of global illumination relative to point and line methods are that

- it is difficult to use with fluorescent samples,
- it is generally not confocal,
- only one Raman line can be imaged at a time,
- it has low spectral resolution($\sim$20 cm^{-1}),
- it has poor reduction in scattered light at low frequencies, and
- it is vulnerable to artifacts due to sample geometry.

For polymeric systems, the low power density may not allow the detection of sufficient signal for good spatial resolution in Raman imaging as polymers are generally weak Raman scatterers. The fluorescence problem is particularly bothersome with global imaging of polymers. Typically, the spectral resolution for global illumination is low ($\sim$20 cm^{-1}), and this low spectral resolution may not be sufficient to resolve different structural components in complex polymeric systems.

The success of global imaging depends on the frequency filtering system. The filter must (55)

- have high transmission,
- discriminate against the scattered laser light,
- provide spectral selectivity,
- maintain image fidelity,
- have high tuning reproducibility and speed,

- have narrow band pass for high spectral resolution, and
- have large aperture.

Several technologies are being used as optical filter devices for global Raman imaging.

Hadamard Transform Raman Methods

Hadamard Raman microscopy makes use of wide-field illumination and allows the collection of images in very short experimental times (56–61). In Hadamard imaging, a CCD detector is not employed to directly view the magnified image of the Raman-scattered sample. Rather, the x and y spatial dimensions of the image are collected independently with Hadamard masks and the CCD vertical detector elements, respectively. The result is that high incident laser powers can be used, which has been termed the *power distribution advantage* (21). Using this approach images can be obtained in 3–10 min. While the y spatial dimension is reproduced directly with the CCD camera, problems associated with Hadamard multiplexing can limit the quality of the x spatial dimension image.

The principal limitations of the Hadamard method are associated with mask alignment and positioning defects. Attempts to remove these problems have been made by developing numerical methods of correcting the Hadamard imaging defects (60).

Global Illumination Raman using Holographic Filters

Holographic Raman filters are spectroscopic devices for Rayleigh line rejection and are designed for a variety of laser wavelengths. In general, the filters provide between 5 and 6 orders of laser rejection while uniformly transmitting 75–80% of the Raman emission within 75 cm^{-1} of the exciting line. Placing these filters in front of a CCD detector results in very efficient Rayleigh line rejection (62, 63). The use of holographic Raman filters can improve the performance of conventional Raman systems at a minimum cost. An improvement of ~3.5 has been reported using a holographic filter relative to the normal subtractive dispersion spectrometer approach (86).

A Raman spectrograph has been reported that uses a volume holographic transmission grating as a dispersing element (65). Volume holographic gratings are based on the refractive index modulation of dichromated gelatin. The modulation produces diffraction by the light wave and behaves like a reflection grating.

By using a holographic grating, a collection efficiency for the Raman scatter >25% can be achieved, compared to <1% for many of the earlier Raman microscope systems. For microprobe experiments, the low-aperture holographic filters can be used as the diameter of the incident laser beam is small. One immediate advantage is the ability to use low-power laser densities as a result of the higher throughput efficiency, thus reducing the potential of sample damage.

A direct Raman imaging system can be developed using dual matched holographic gratings as a tunable filter (*83*). The filter system allows 72% throughput over the visible tuning range with a spectral bandwidth of 250–300 cm^{-1}.

Global Illumination Raman using Acousto-Optic Filters

An *acousto-optic tunable filter* (AOTF) is another useful electronic light-dispersive device for use in Raman instrumentation. These AOTFs are computer-controlled notch filters that can provide random wavelength access, wide spectral coverage, and moderate spectral resolution. AOTFs are prepared from optically transparent birefringent crystals such as TeO_2 to which an array of piezoelectric transducers are bonded.

In Raman imaging applications, the AOTF is used to spectrally filter the source. Under computer control for collection of point-to-point images, the AOTF is swept through a wavelength range, and at prescribed intervals images are recorded. Alternately, the AOTF may be digitally tuned to specific frequency bandwidths where global images may be collected.

An AOTF operates on the principle of acousto-optic interaction in an isotropic medium. Incident white light is diffracted by the AOTF into a specific wavelength when a particular radio frequency (rf) is applied. The propagating acoustic waves produce a periodically moving grating that will diffract portions of an incident beam. Therefore, the spectral bandpass of the filter can be tuned over large optical regions by simply changing the frequency of the applied rf. The wavelength of the light diffracted decreases with increasing frequency of the rf signal. The efficiency of the diffraction is proportional to the power of the rf signal. Thus, the wavelength and intensity of the diffracted light can be controlled by controlling the frequency and the power of the applied rf signal.

Apertures range from 2 to 10 mm^2 with an acceptance angle up to 5° off the optical axis. The bandwidths are typically 1–10 nm across the visible, yielding a spectral bandpass of approximately 10 cm^{-1}. AOTFs provide out-of-band rejections of 10^5 and a diffraction efficiency greater than 90% (*67*).

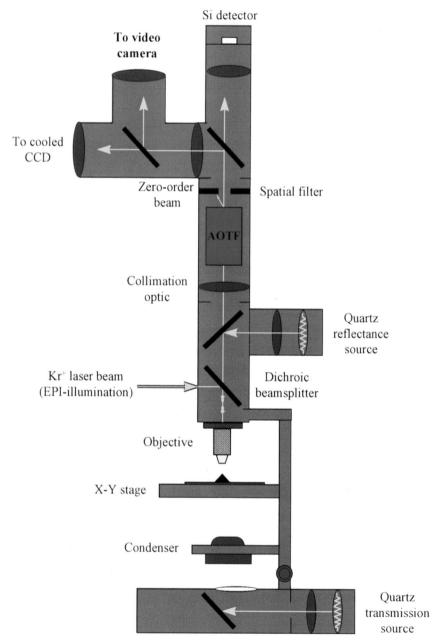

Figure 9.8. Schematic diagram of an AOTF microscope. (Reproduced with permission fromref 70 Copyright 1993 Spectroscopy.)

In practice, the AOTF is placed between the Raman scatterer and a Si focal-plane array detector yielding a no-moving-parts Raman spectroscopic imager (Figure 9.8) (*68*). With an AOTF, it is possible to tune the spectrometer from one wavelength to another in 5 µs by the use of an rf generator. The AOTF has been called "the new generation" monochromator (*69*). The AOTF can replace the dispersive monochromator and is operable between 400 and 1900 nm. One of the advantages for imaging of the AOTF as an electronic polychromator compared to the conventional grating system is the fact that the light rays from the different wavelengths are not deviated by the AOTF by the chromaticity of the light. With conventional gratings, the chromaticity produces different light ray paths (*69*). As a result, an AOTF can transmit two-dimensional images, and their use has been implemented in an AOTF-based Raman imaging spectrometer. At each discrete wavelength a CCD frame is collected, digitized, and stored. The size, wavelength range, wavelength increment, and exposure time for each image frame may be modified under computer control (*71*).

The advantages of an AOTF are its (*69*)

- compactness (it is all solid state, rugged, and contains no moving parts);
- wide angular field;
- high throughput (diffraction efficiencies of the filter are generally greater than 85%);
- wide tuning range (from UV through visible to IR);
- high spectral resolution (bandwidth of light transmitted by the filter is about 2–6);
- rapid scanning ability (order of a few microseconds);
- adjustable-intensity light;
- high-speed random or sequential wavelength access; and
- imaging capability.

AOTFs show relatively strong side lobes outside their bandpass. These side lobes can be reduced to approximately 3% of peak transmission by amplitude apodization of the rf drive signal. The AOTF is an isotropic crystal, and this causes image blur at high magnification (*67*).

Coupled with a fiber-optic microimaging probe, high-spatial-resolution Raman images can be obtained from samples located meters away. This fiber-optic method has the potential for remote-monitoring imaging (*72*). The described system has a 350-µm-diameter microfiber consisting of a bundle of 6000 4-µm-diameter fused-silica fibers. The spatial resolution of the system is 4 μm, and the spectral resolution is of the order of 4.5 nm full-width at half-maximum (FWHM). A microscope focuses the source into the coherent fiber that carries the illumination to the sample. The same bundle collects the reflected Raman image and collimates it by sending it back through the microscope objective. A 121-camera lens focuses the beam on the CCD detector (72).

Global Illumination Raman using Dielectric Filters

A recent approach has been to do away with the diffraction grating of the traditional Raman dispersive spectrometers. Wavelength selectivity and laser line rejection are achieved with specially designed multilayer dielectric filters. These dielectric filters, which have a resolution of 2 cm^{-1} for Raman spectra and 20 cm^{-1} for Raman imaging, are tuned by rotating the filter with respect to the incident beam. The center wavelength of the filter band-pass is given by

$$\lambda(\theta) = \lambda(0)(n^2 - \sin^2 \theta)/n^2)^{1/2}$$

where θ is the angle of incidence (to the normal) and n is the refractive index. Thus for a He–Ne laser (633 nm), the band pass-band filter with refractive index $n = 2$ centered at 700 nm for normal incidence can cover Raman shifts between 900 and 1500 cm^{-1} for angles between 0 and 35°.

A Raman microscope has been developed using multilayer dielectric filters to select the Raman line width from which the image is formed (Figure 9.9) (67). The width of the intermediate slit is matched to the

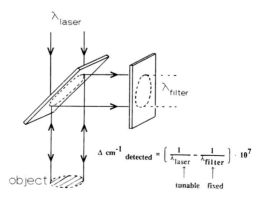

Figure 9.9. Schematic diagram of a filter Raman microscope system. (Reproduced with permission from ref 73. Copyright 1993 Spectroscopy (Eugene, OR).)

size of the focal spot. The result is a high degree of spatial filtering in the direction perpendicular to the slit.

The optical quality of the filters is sufficiently high not to degrade the image-forming capabilities of the microscope. The specimen is mounted on the stage of a conventional microscope, which is used both to deliver the laser beam and to collect the back-scattered Raman light. In the microprobe mode, the parallel laser beam is brought to ~1 μm diameter focus at the specimen, and the Raman-scattered light is focused on a few pixels at the center of the CCD detector. When the Raman imaging mode is selected, the laser beam is defocused to illuminate an area of the specimen about 200 μm in diameter. The characteristic Raman band (which is determined during the microprobe mode) is selected with the filter and is then used to form a functional-group image on the CCD detector (*74, 75*).

There are some optical problems inherent in this design. First, light coming from different parts of the object passes through the dielectric narrow-pass filter under slightly different angles, causing vignetting. Thus different parts of the object are imaged at different wavenumber shifts, leading to a spectrally inhomogeneous image (*76*). The transmission bands for *p*- and *s*-polarized light diverge so that different wavenumber intervals are transmitted for polarized Raman-scattered light (*76*).

Another disadvantage of filter-based Raman microscopies is that the bandwidth of the filter cannot be adjusted for optimum detection conditions for each specimen. Access to the low-wavenumber Raman shifts is also difficult with filter-based systems.

However, filter-based Raman microscopic instruments are easy to use, fast, and yield reproducible results that can be used to compare different samples. Efforts are under way to extend the illumination wavelength into the near IR to reduce fluorescence interference.

The effects of sampling parameters on Raman line images have been studied (*77*) including the spectral resolution as determined by the spectrometer slit width, spatial sampling resolution, and the number of wavelength variables measured. Studies were made of 10-μm spheres of polystyrene in an epoxy matrix. The spectrometer slit widths affect the spectral resolution, the spatial resolution, and the degree of confocal filtering. The spectra of pixels are substantially different when the spectral resolution is low and had a large impact on the effectiveness of PLC analysis. The effect of spatial resolution was reflected in a substantial change in baseline offset, and the higher the spatial resolution (more spectra) the less serious was the baseline offset. The number of wavelength variables measured influences the spectral resolution but is not equivalent to reduction of the spectral resolution by increasing the slit width. Overall, it was determined that both the number of wavelength variables and the specific wavelengths themselves were important.

Global Illumination Raman using Liquid-Crystal Filters

Liquid-crystal tunable Lyot filters (LCTF) using nematic liquid crystals have been used for Raman imaging (55, 78, 79). Potentially, liquid-crystal (LC) filters have higher attainable spectroscopic resolution than the other electronically tunable filters. LCTFs are electronically tunable over a broad frequency range. The LCTF provides a large optical aperture and rapid voltage-controlled frequency tuning with only slight distortion of the images. However, the LCTF has a low transmission. The LCTFs have been used with Fabry–Perot interferometers (55, 78) and via an optical fiber to an infinity-corrected microscope (79).

The Lyot birefringent filter consists of a series of cascaded components with birefringent elements positioned between parallel linear polarizers. The output from a previous stage is the input for the subsequent stage. The retardation increases in powers of 2 for each successive stage. Each successive component exhibits a transmission spectrum with half the free spectral range and half of the bandpass of the previous one. Only frequencies that are in-phase are transmitted to the next cascaded stage. The transmittance of the nth stage is given by

$$T_{\mathrm{n}}(\lambda) = \cos^2 [\Gamma_n(\lambda)/2] \qquad (9.23)$$

where $\Gamma_n(\lambda)$ is the wavelength-dependent retardation given by

$$\Gamma_n(\lambda) = 2\pi(\Delta nd)/\lambda \qquad (9.24)$$

where Δn is the birefringence of the material and d is the thickness of the LC.

To vary the retardation of the LC elements, one applies an electric field that is parallel to the incident light. Spectral tuning by the electric field is equivalent to rotating the crystal axis of a uniaxial crystal, changing the birefringence and the retardance. Varying the electric field applied to the LC gives control over the band pass wavelength, λ_{B}.

A comparison has been given of the performance of LCTFs and AOTFs (55), and the results are shown in Table 9.2.

Table 9.2. Comparison of Tunable Filter Performance

Parameter	LCTF	AOTF
Peak Transmittance	13.6%	40.0%
Amplitude Stability	±0.46%	±0.80%
Tuning reproducibility	±0.06%	±0.05%
Free Spectral Range	500–741 nm	560–1120 nm
Tunability (minimum increment size)	0.05 nm	0.10 nm
Bandpass (647.1 nm)	11.1 nm	2.1 nm
Aperture	20-mm diameter (circular)	7 × 7 mm (square)
Out-of-band rejection	10^4	10^3
Tuning speed	50 ms	25 ms

NOTE: LCTF is liquid-crystal tunable Lyot filter; AOTF is acousto-optic tunable filter.

Confocal Raman Microscopy

The advent of the confocal technique in Raman spectroscopy(*80–82, 76*) has greatly extended the potential of the technique. The major advantage of the confocal arrangement over conventional microscopy is the introduction of depth resolution, which allows optical sectioning. A Raman confocal microscope offers the possibility of obtaining spectral information to create a three-dimensional image from a volume element of a sample rather than the two-dimensional spatial array ordinarily obtained.

The key elements of a Raman confocal microscope are the two optically conjugated "confocal pinholes" placed in the light path (Figure 9.10) (*83*). Light passing through the first pinhole is focused by the objective onto the sample so that it converges to a small (diffraction-limited) spot. Light emerging from this illuminated volume is focused onto the second pinhole, which is located in front of the detector. One pinhole is used to filter the laser beam, and the second is placed in an enlarged image of the specimen and isolates an adjustable region coincident with the illuminated spot. This configuration rejects the photons that arise from outside this focused region and provides the axial (depth) resolution needed for three-dimensional imaging. This configuration also improves both axial and lateral resolutions and enhances the contrast. An image is obtained by mechanically scanning through the microscope focal volume of the sample and recording the map at each new focal position. The depth resolution is proportional to the diameter of the confocal hole and the optical properties of the microscope objective.

A Raman confocal microscope allows optical sectioning of the specimen in the axial dimension and provides an efficient way to virtually eliminate contributions of the surrounding material to stray light and fluores-

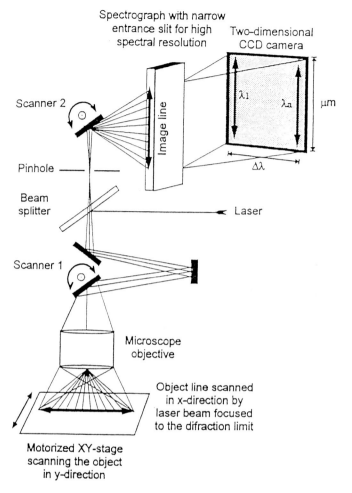

Figure 9.10. Schematic representation of a confocal laser line-scanning Raman imaging spectrometer. (Reproduced with permission from ref 83. Copyright 1995 Society of Spectroscopy.)

cence. Such an instrument has been developed by DILOR under the European patent 92400141.5 (1992) to M. Delhay et al. An improved confocal instrument has been reported (*84*). An instrument has also been described that will generate three-dimensional images (*85*). This instrument takes full advantage of the CCD to collect spectral and spatial information of the scanned line simultaneously (*83*). In theory, the lateral depth spatial resolution is different when the laser beam is scanned over the sample. A study of the depth profiling of confocal arrangement (100× magnification, 100-μm-diameter pinhole) is shown in Figure 9.11.

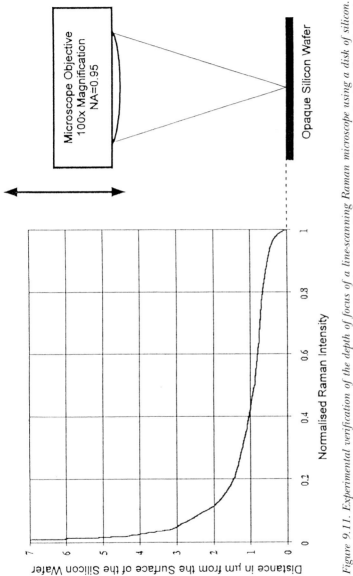

Figure 9.11. Experimental verification of the depth of focus of a line-scanning Raman microscope using a disk of silicon. The intensity of the 512 cm^{-1} Raman line is plotted against the relative displacement of the sample along the optical axis toward the focal plane of the microscope. (Reproduced with permission from ref. 83. Copyright 1995 Society of Spectroscopy.)

The *axial resolution* (r_z) measures the ability of the confocal optical system to discriminate signals from adjacent layers of the sample in the direction of the optical axis of the objective. A method of defining the depth resolution of a confocal Raman microscope is to measure the FWHM of the depth scan of the Raman signal through an infinitely thin sample. A depth resolution of 1.8 μm was obtained for a static optical laser beam. It was also concluded, from other experiments, that the depth resolution is not seriously affected by the line-scanning Raman imaging technique (*87*).

The critical portion of the Raman confocal microscope is the selection of the pinhole diameter, which must be optimized for the experimental conditions. A large pinhole diameter yields a high signal but low axial resolution, while a small-diameter pinhole increases the axial resolution but also increases the measurement time to obtain a sufficient *S/N* ratio (*24*). A theoretical model has been described that demonstrates the relationship between pinhole diameter and depth of focus. The theoretical results are shown in Figure 9.12.

The attainable contrast in the image is highly dependent on the pin-

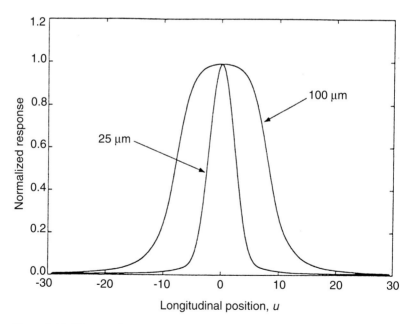

Figure 9.12. Theoretical longitudinal (depth) response of a confocal Raman microscope for pinhole diameters of 25 and 100 μm. The spectral response from the target falls on the detector only in a narrow region centered on the geometric focus, while light outside this region is rejected. Increasing the aperture diameter to 100 μm and keeping the other parameters constant increases the microscopic depth response. (Reproduced with permission from ref 24. Copyright 1995 Society of Spectroscopy.)

hole's rejection of out-of-focus Raman scattered light. Other factors can contribute to the loss of S/N ratio, including optical and geometrical aberrations in the system. For a pinhole diameter of 25 μm, an experimental axial resolution of ~5 μm with a spectral resolution of 22 cm^{-1} is attainable (*24*). Another approach for Raman confocal microscopy is to use an optical fiber as the confocal pinhole. Using a fiber-optic bundle, the signal from the central fibers yields the conventional confocal image, and the signal from the entire fiber bundle is integrated to yield information about a larger volume of the sample (*88*).

One approach to confocal micro-Raman images is the combination of the scanned line focus system with stage scanning in the depth dimension. A Raman line-imaging system yielding confocal Raman images has been described in which the spectrograph entrance slit functions as a spatial filter. The instrument uses a scanning galvanometer mirror to generate uniform intensity line illumination (*86*).

A line scanning three-dimensional confocal instrument has been described (*83, 90*). The diagram of this instrument is shown in Figure 9.13. The confocal optical system has five elements: two optically coupled adjustable confocal diaphragms, a beam splitter, and two synchronized mir-

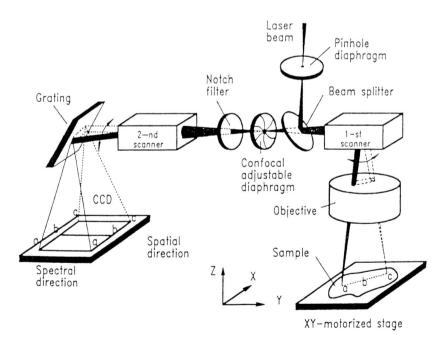

Figure 9.13. Principal elements of the confocal optical scheme and a general scheme for the acquisition of the confocal spectral image. (Reproduced with permission from ref. 89. Copyright 1995 Society of Spectroscopy.)

ror scanners. Light scattered by the scanning line is collected by the objective and transformed into a coaxial light beam. After the first scanner, 50% of the collected light passes through the beam condenser and is focused on the plane of the adjustable confocal diaphragm, which provides the lateral and axial resolution. The Raman signal passes through the notch filter and reaches the second mirror scanner. The second mirror scanner decodes the signal into a spatially resolved signal by symmetric vibration in-phase with the first scanner. This signal is dispersed by the spectrograph into a spectrum and is directed onto the two-dimensional CCD. The spectra from the different positions in the sample are detected by the different rows of pixels along the CCD. Then the next line of spatial data is examined to construct the three-dimensional image.

Confocal Raman spectroscopy has high potential for polymer studies since it provides an optically slicing technique for depth profiling and fluorescence rejection. The three-dimensional imaging capability is also a valuable result as visual interpretation is considerably simplified.

Applications of Raman Microscopy and Mapping

Recent reviews report many applications of Raman microscopy to polymers (*91–93*). Raman microscopy and mapping has great potential for studying the structural and spatial distributions of polymer systems based on its high spatial resolution and short image measurement times, particularly with global illumination. The problem has been the omnipresent fluorescence for commercial polymer samples. Near-IR laser excitation minimizes this effect, but the development of such microscopes has not been as rapid as optical laser systems.

Additionally, FTIR instrumentation and microscopes are more accessible in most laboratories, and this instrumentation is used in preference to the Raman microscopes because of the greater familiarity with IR compared with the Raman. There are specific structural and spatial problems in which Raman spectroscopy plays a dominant and important role based on higher sensitivity (due to resonance enhancement) and higher spatial resolution. Not all of the applications of Raman can be covered here, but an attempt will be made to demonstrate the various approaches using isolated examples.

Contaminant Analysis

The earliest applications of the Raman microprobe were contaminant analysis (*94*). This early work entitled "Identification of Inclusions and

Particles by Raman Microprobe" lists a number of examples including analysis of geological samples, biological samples, and synthetic materials like polyethylene terephthalate (PET) In this early paper we read:

> *Formation of defects is a problem currently encountered in the commercial production of materials. Characterization of bubblelike and hazelike defects inside synthetic fibers or films which are at the origin of their breaking have been successfully solved by Raman microprobe technique.*

Unfortunately, contaminant analysis of most commercial polymer systems using the classical visible laser source was limited by interfering fluorescence, but the FT-Raman techniques have helped minimize this problem.

A particularly useful application of FT-Raman microprobe analysis, demonstrating the utility of the higher spatial resolution of Raman, is in the identification of the source of pinholes and craters in coating systems (*95*). In general, the pinhole sizes in the coatings are smaller than the spatial resolution of FTIR (i.e., <10 μm), making FT-Raman viable. With the laser probe, the material was identified in the bottom of a pinhole (less than 5 μm in diameter at its base). The Raman spectrum (Figure 9.14) clearly shows the material to be polydimethylsiloxane (bands at 488 and 710 cm^{-1}); this type of antiflow agent is used in the film formation

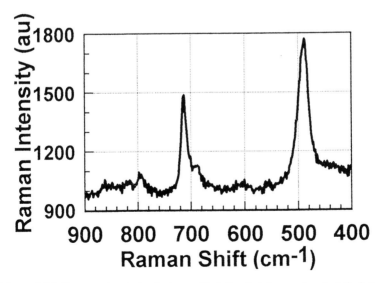

Figure 9.14. Raman microprobe spectrum of polydimethylsiloxane found at the bottom of a pinhole coating. (Reproduced with permission from ref 95. Copyright 1994 American Chemical Society). (ACS National Meeting).

process (*94*) and apparently can lead to pinhole contamination. Control of the antiflow agent should lead to improved coating appearance and performance due to the decrease in the formation pinholes.

Pathological Tissue Specimens

In optical microscopy, it is usually necessary to stain tissue specimens to achieve optical contrast. The primary advantage of the Raman microprobe is that tissue samples can be studied without having to extract the material or to decompose the tissue. However, unstained tissue specimens can be sectioned and mounted on ordinary microscope glass slides and studied directly using Raman microscopy as the characteristic Raman lines serve as a unique "stain" for different components of tissue. The Raman procedure provides a rapid and precise in situ identification of components and may be useful for the diagnosis of metabolic disorders. In this fashion, cystine crystals in liver and spleen specimens are identified by the enhancement of Raman lines at 501 and 2916–2967 cm^{-1} arising from the disulfide (S–S) and the C–H stretching motions, respectively (*96*).

Raman imaging has been applied to the histopathological characterization of biopsied human breast tissue containing foreign components (*97*). The Raman images revealed that polyester inclusions occur in the human breast implant capsular tissue of the specimens examined.

FT-Raman microscopic studies have been made of the drug distribution in a skin patch. The patch is a transdermal drug delivery device. The patch provides drug delivery by a controlled release mechanism directly into the systemic circulation. It has been used to administer estradiol, which replaces the natural estrogen that the female body stops producing at menopause. The estradiol is contained within a reservoir in the patch from which it passes through a permeable membrane to the skin surface. The patch is held in place by an outer adhesive layer. Raman microscopic mapping was used to analyze the homogeneity of the estradiol distribution in the patch. The advantage of the Raman technique is the nondestructive nature of the method, and the spatial resolution could be reduced to 20 μm. Little or no sample preparation is required. The mapping is obtained with a Nd:YAG laser power of 450 mW for 2000 scans at 4 cm^{-1}. A 40× objective is used yielding a sample spot 25–30 μm in diameter. A crystal in the patch approximately 250 μm in diameter is identified as estradiol. The mapping profile showed the presence of an inhomogeneous inclusion in the drug delivery device. The Raman spectra suggest that the crystal inclusion is modified in terms of the degree of crystallinity relative to the pure drug. It was not possible to conclude from this study when the inhomogeneous inclusion is formed (*98*). The impact of the inclusion on the drug release performance of the patch could not be ascertained.

Resonance Raman microspectroscopy is a valuable tool for probing local chemical interactions and secondary structural changes in biological systems. The resonance Raman spectra "can provide unique fingerprints reflecting the unique combinations of chemotaxonomic markers present in each type of organism" (*99*). UV resonance Raman spectra have been obtained for microsamples of dAMP deoxymononucleotide and DNA of a single living T47 D cultured cell (*99*). These results represent the ultimate in specific sensitivity for living tissue analysis.

A Raman image (not resonance enhanced) of a human lymphocyte (isolated from peripheral blood by standard density centrifugation methods) using a laser (660 nm) microprobe system with a dielectric narrow band-pass filter is shown in Figure 9.15 and compared with the bright-field image. The Raman image was obtained with the microscope tuned to 1515 cm^{-1} (carotenoid $-C = C-$ in-phase stretching vibration) from which a background image obtained at 1500 cm^{-1} was subtracted (*76*). Clearly, the subcellular distribution of carotenoids in human lymphocyte subpopulations can be studied (*100*).

The sensitivity of Raman imaging of cell biological samples can be enhanced by using extrinsic Raman labels (*101*). That is, the Raman signal is measured from labeling molecules exhibiting extraordinary Raman intensities, which act as reporter molecules. The extrinsic Raman labels require molecules with a Raman spectrum that can be easily distinguished from the background scattering of the sample. These labels must have a large Raman cross section to be used to detect cell species that are too weak to be detected with the intrinsic Raman signal. Two extrinsic labels have been evaluated including filipin, which is a cholesterol-binding mol-

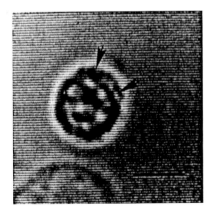 

Figure 9.15. A: Bright field image of human lymphocyte. B: Raman image obtained with microscope tuned to 1515 cm^{-1} (carotenoid $-C = C-$ in-phase stretching vibration) from which a background image obtained at 1500 cm^{-1} was subtracted. (Reproduced with permission from ref 76. Copyright 1993 Society of Spectroscopy.)

ecule with a high Raman cross section. The Raman spectrum of filipin has an intense band at 1586 cm^{-1} that can be used for imaging. For the study of rat eye lens, the lens constituents themselves only contribute a weak background signal. The Raman imaging results using filipin suggest that the cholesterol is concentrated in the membranes of the eye lens fibers. Carboxylated polystyrene microspheres were covalently coupled to rabbit anti-mouse protein. Polystyrene has a strong band at 999 cm^{-1}, which is then used as an extrinsic label for the recognition of the phenotype of peripheral blood lymphocytes (101). A recent report has been made of surface-enhanced Raman microscopic spectral analysis of intact zebra fish embryos using coated microelectrodes with 1–3 μm active silver tip diameters (102).

Laminate Analysis

Polymer laminates are often required for applications in packaging where diffusion of oxygen or water is detrimental to the packaged food product. Packaging laminates are made up of multilayer structures in which the polymer layers have the required resistance to diffusion or the strength required for the packaging. These laminates can be examined with Raman microscopy by scanning directly on-edge of the laminates if the thicknesses of the internal layers are sufficient or by confocal depth profiling methods through the laminate for thinner layers.

A polymer film laminate with a film thickness of 120 μm from the lid of a packaged microwave meal was examined directly on-edge using FT-Raman microscopy (103). The Bruker FT-Raman microscope was used with fiber-optical interconnects. Four main layers were identified: the outer layer was PET, the second layer was of poly(vinylidene chloride), the next layer was nylon 6, and the innermost layer was polyethylene (PE). The innermost PE layer has a thickness over ~60 μm, and the other three layers are ~20 μm in thickness.

The depth profiling through the laminate of PE and polypropylene (PP) was accomplished with a confocal Raman microscope focused at different depths of the sample (104). Figure 9.16 shows the Raman spectra obtained with different focusing levels and with a 300-μm pinhole. The appearance of the PP bands with penetration corresponds to the decrease in the PE bands of the upper layer.

This same group examined a multilayer foil with three layers: PE (26 μm), polyamide (PA) (30 μm), and PE (26 μm) (105). In Figure 9.17, the spectra are shown with and without the 300-μm pinhole. As can be seen there is a clear separation between the PE and PA layer contributions in the spectra. The increase in the intensities of the lines around 1080

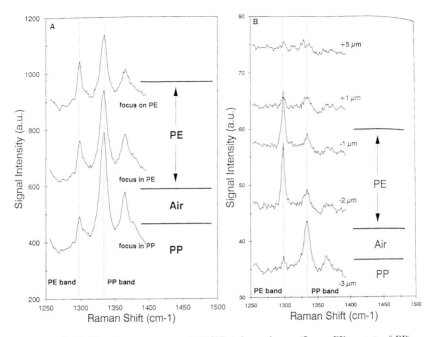

Figure 9.16. The Raman spectra of PE/PP polymer layers (2 µm PE on top of PP). A: Recorded without a pinhole. B: Same system as A with the × 100 objective and a 300-µm pinhole. (Reproduced with permission from ref 105. Copyright 1992 Society of Spectroscopy.)

and 1280 cm^{-1} are associated with the PA component, and all of the bands in the middle spectrum are from PA.

A similar result has been obtained for a PET sample that was recorded through a relatively thick covering layer of PE (2.3 mm) (*28*). Figure 9.18 demonstrates the results for PE layers of different thicknesses. Using confocal Raman microspectroscopy in the depth profiling mode it is possible to detect the hydrogen-bonding interactions between the ester and alcohol groups near the interfacial regions for polymethyl methacrylate (PMMA) poly(vinyl alcohol) (PVOH) laminates (*106*). The thicknesses of the PMMA and PVOH layers were 4.5 ± 0.5 µm and 9.5 ± 0.5 µm, respectively. The depth resolution of the confocal arrangement was 2.0 ± 0.3 µm, and depth profiles were obtained of the laminates. Initially, the spectra show no bands due to PMMA and only PVOH, but subsequent spectra at greater depth show the presence of PMMA, reflecting the interpenetration of the two polymers. Considerable broadening of the carbonyl band occurs at the interfacial region, suggesting the presence of a hydrogen-bonding interaction.

It is apparent that Raman microspectroscopy, particularly in the confocal mode, is a powerful and useful technique for laminate analysis. No

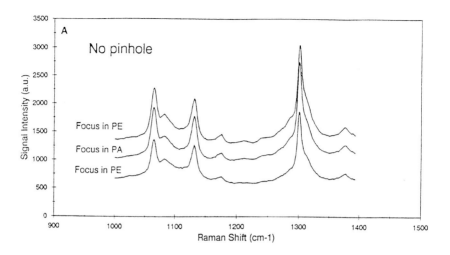

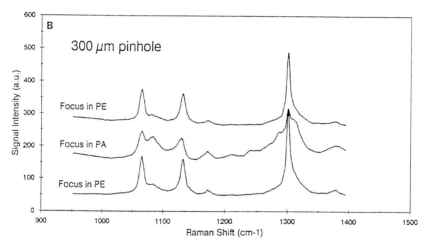

Figure 9.17. Raman spectra of a scan through a PE/PA/PE multilayer foil (26 μm/ 30 μm/26 μm) with a × 50 objective. A: No pinhole. B: 300-μm pinhole. (Reproduced with permission from ref 105. Copyright 1992 Society of Spectroscopy.)

sample preparation is required, and the measurements are relatively rapid and sensitive to the polymer composition in the inner layers as well as interactions that occur across the interfaces.

Analysis of Fibers and Stressed Samples

Typically, commercial fibers are too thick optically for FTIR measurements, making it difficult to use the usual sampling methods. However, the micro-Raman spectra of fibers are particularly easy to obtain by simple

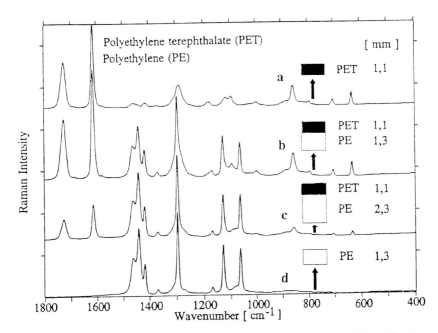

Figure 9.18. Raman spectra of polymer layers. (a): Polyethylene terephthalate. (b): the same layer as (a) but with 1.3 mm PE in front of it; (c): same as for (b) but with a PE layer of 2.3 mm; (d): PE only, 1.3 mm thickness. (Reproduced with permission from ref 28. Copyright 1994 American Chemical Society.)

reflection of the focused laser beam (*104, 107*). In addition to the identification of the chemical composition of the fiber and the presence of finishes on the fiber, Raman microscopy can monitor the degree of molecular stretching (i.e., the molecular elongation). The contribution of molecular stretching increases with fiber modulus (*108*). Similar results have been observed for rigid-rod polymer fibers (*109–111*), carbon fibers (*112*), and oriented polymer films (*113*). Raman microscopy can also be used to measure the degree of molecular orientation in the fiber.

Carbon fibers have been studied by micro-Raman techniques with the purpose of determining the level of microcrystalline order in pristine and brominated pitch-based carbon fibers and the chemical state of the bromine in these fibers as a function of bromination time (*114*).

It is possible to detect the presence of macroscopic residual stress in gel-spun ultradrawn high-*E*-modulus PE fibers by observing the frequency shift of the C–C stretching modes at 1060 and 1130 cm^{-1} (*105*). The fibers were stressed in the Raman machine, and the Raman signal was obtained from the outer 3 μm of the fiber. The two observed Raman lines showed two components in the stressed samples indicative of a bimodal distribution of stress in the fibers.

It has been demonstrated that some Raman bands in aramids shift to lower wavenumber on the application of stress (*108*). For aramid fibers this shift in frequency $d\Delta v$ is proportional to the applied stress. The penetration of the laser beam is only of the order of 1–2 μm so the Raman signal comes essentially from the skin regions of the fibers (*108*).

Micro-Raman has been used to study the distribution of crystalline phase transformation along the neck of stretched poly(vinylidene fluoride) (PVF$_2$) (*115*). The Raman spectra were recorded along the stretching direction at intervals of 20 μm and along the perpendicular direction at intervals of 40 μm. The crystal transformation of PVF$_2$ from form II(α) to form I(β) can be followed by measuring the intensity of the two Raman bands at 799 and 840 cm^{-1}, which are assigned to the nonplanar (TGTG′) conformation of the II(α) phase and to the planar zigzag (TTTT) conformation of the I(β) phase, respectively. The relative evolution of the crystalline modification along the neck was expressed by the determining of the coefficient of the crystalline modification, *R*, given by

$$R = 100I(480 \text{ cm}^{-1})/[I(480 \text{ cm}^{-1}) + I(799 \text{ cm}^{-1})] \quad (9.25)$$

at intervals of 40 μm. Figure 9.19 gives a three-dimensional map representing the coefficient of the crystalline modification *R* along the neck of the sample stretched at 1 mm min^{-1}.

Polymer Degradation

Poly(vinyl chloride) (PVC) dehydrochlorinates when exposed to heat. The process involves the elimination of HCl and the formation of a double bond in the polymer chain. The double bond acts as a site for subsequent rapid elimination of HCl, and a conjugated polyene is formed via the so-called unzipping reaction. The presence of these polyenes causes the PVC to darken in color. This observed color change is due to the polyenes exhibiting a UV–visible absorption, and Raman spectroscopy makes use of this property.

Dehydrochlorination of PVC results in the formation of all trans-conjugated polyene chains in the polymer backbone. The polyene chains, of varying chain length, have a broad electronic absorption in the visible region, which supports a strong resonance enhancement of vibrations associated with the $-C-C-C-$ backbone. In particular the $C = C$ stretch near 1500 cm^{-1} and the C–C stretch near 1200 cm^{-1} are enhanced sufficiently to allow detection of PVC degradation at $10^{-5}\%$ levels. No resonance Raman spectra are observed for uncolored PVC films. The degree of resonance enhancement is determined by the proximity of the laser excitation wavelength to the electronic absorption band profile maximum. This is in turn determined by the polyene chain length distribution in the

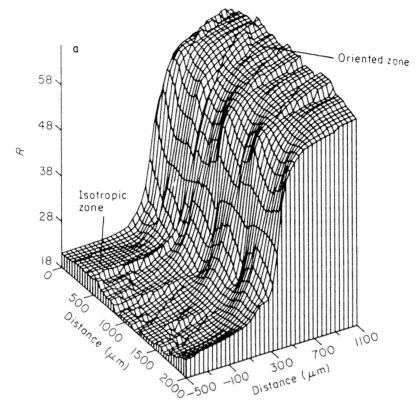

Figure 9.19. A three-dimensional map representing the coefficient of the crystalline modification R along the neck of the sample stretched at 1 mm min⁻¹. (Reproduced with permission from ref 115. Copyright 1992 Science & Technology Press (Polymer) Elsevier.)

sample (*116, 117*). Less than 0.0001% dehydrochlorination can be detected. The relationship between the conjugated polyene sequence length (*n*) and the value of v_2 is (*118*)

$$v_2 = 1461 + 141.6 \exp\ (-0.0925n) \tag{9.26}$$

The use of resonance Raman spectra in this application is complicated by a number of experimental difficulties: the photooxidation and photodestruction of molecules, distortion of spectral information due to optical saturation phenomena and photoinduced transients, and fluctuations of the scattered light intensity in inhomogeneous samples. It was found that examination of the solid samples of PVC were complicated because the laser power introduced changes during the measurement if power den-

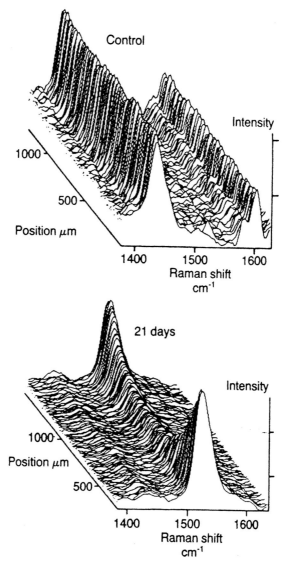

Figure 9.20. Raman profiles of PVC samples exposed to triethylenediamine for 21 days. (Reproduced with permission from ref 121. Copyright 1991 American Chemical Society.)

sities above 20 W/mm^2 were used (*119*). If the samples are dissolved in tetrahydrofuran, no effects were observed; however, the solution process results in an altered polyene distribution. Solutions must be stored in a cooled state and in the dark to avoid disappearance of the polyenes.

Using the MiFS technique the laser power is considerably reduced so Raman profiles can be obtained for samples exposed to triethylenedi-amine for 21 days (Figure 9.20). Bands at 1436 and 1602 cm^{-1} with a weak shoulder at 1574 cm^{-1} are due to plasticizer and serve as an internal standard (*120, 121*).

There is a fine line in the evolving degradation process of the polymer between when resonance enhancement of the Raman signal occurs and when laser destruction of the degrading polymer samples occurs. However, since most of the interest commercially is in the lower levels of degradation, resonance Raman is applicable as is elegantly demonstrated for PVC.

Analysis of Polymer Blends

The attractiveness of polymer blending lies in the potential for tailoring the properties to produce commercial articles to meet specific physical and mechanical requirements. Blending of available polymers offers lower costs and faster development times when compared to the time required to develop new polymers. However, most binary mixtures of polymers are not miscible on the molecular level because the entropy of mixing is not favorable for high-molecular-weight polymers.

Miscible blends of two polymers can be engineered through specific interactions such as hydrogen bonding or van der Waals forces. The advantage of miscible systems is that they are characterized by single glass transition temperatures (thermodynamically single homogeneous phases) allowing the modification of processing temperatures. Other properties of miscible blends tend to follow the rule of mixtures behavior although both positive and negative synergies have been observed. In principle, miscible blends can be processed as a single-phase system and then phase-separated in a controlled manner to achieve a desirable morphology.

Compatible blends are two-phase materials with properties controlled by the properties and geometry of each phase and the nature of the connectivity between phases (compatibilizers modify/improve the interface).

The miscibility of two polymers can have either energetic or entropic origin. In the former case A and B links simply attract each other; in the latter, A and B links like to be in the neighborhood of each other because in this way, greater freedom for internal rotations in the chains is achieved. This may be, for example, due to steric reasons; if A links are more bulky than B links, their separation with the formation of the pure

A phase may restrict the set of possible conformations of A chains that are otherwise realized in the presence of some fraction of B chains.

The phase separation process for blends is understandable. Consider the case in which the glass transition temperatures for pure poly(A) and poly(B) are T_g^A and T_g^B ($T_g^A > T_g^B$) and that these polymers are miscible at high temperatures. For a given composition of the blend poly(A)–poly(B), the phase separation begins at a certain temperature T_c located in the interval $T_g^A > T_c > T_g^B$. In this case the process of the separation into macroscopic phases cannot be completed: as soon as large enough domains of the A-rich phase appear, they become glassy by virtue of the inequality $T_g^A > T_c$ and the process stops. As a final result, microdomains appear and are frozen in.

Information on the domain sizes and distribution in polymer blends can be obtained from micro-Raman imaging (84). Raman frequencies exist that are unique to each component of the blend providing image contrast. Figure 9.21 shows the Raman image at 2 μm resolution of a rubber-toughened epoxy in the form of a thin film. The Raman image was obtained using a frequency of 1665 cm^{-1}, characteristic of the C = C stretching mode of the rubber with a background image using the 1720 cm^{-1} line subtracted. The light features in the image represent the dispersed rubbery domains and appear as spherical domains dispersed in the epoxy matrix (84). Similar results were obtained for blends of PE–PP and a polyester [PET/poly (butylene terepthalate) (PBT)] blend (84).

A confocal Raman imaging study has been made of a PE/PP/ethylene–propylene copolymer (EPM) at a resolution of 1 μm in the lateral direction and 3 μm in depth. The Raman images indicate that segregated particles of PE/EPM occur in a matrix predominantly PP. Furthermore, a small part of the PE must be dissolved in the PP matrix. The PP in the vicinity of the segregated PE/EPM particles has an increased crystallinity suggesting trans-crystallization. Trans-crystallization is the crystallization that is induced near a surface or interface. Additionally, the molecular orientation of the individual components in the blend could be studied independently (122).

Interdiffusion in Polymers

Interdiffusion of polymers has been studied using surface-enhanced Raman scattering (SERS) (115). For studies of the interdiffusion of polystyrene (PS) and deuterated polystyrene (DPS), the SERS spectrum was obtained from a bilayer consisting of a film of PS having a thickness of ~1100 overlaid with a film of DPS of similar thickness. SERS spectra were obtained as a function of heating time at 170 °C. Initially, no bands due to

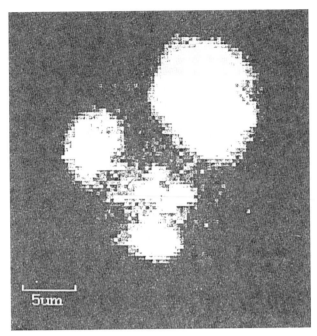

Figure 9.21. The Raman image at 1-μm resolution of a rubber-toughened epoxy in the form of a thin film. The Raman image was obtained using a frequency of 1665 cm⁻¹, characteristic of the C = C stretching mode of the rubber with a background image of 1720 cm⁻¹ subtracted. The light features in the image represent the dispersed rubbery domains. (Reproduced with permission from ref 73. Copyright 1993 Society of Spectroscopy.)

DPS were observed. The SERS intensities of the lines near 1014 and 976 cm^{-1} should be proportional to the amounts of PS and DPS located in the near-surface region. After several minutes, lines due to DPD were observed indicating the DPS had diffused to the surface regions of the silver. In this experiment, the average diffusion distance was around 1100 Å and the diffusion time was around 4 min. The diffusion constant of DPS obtained was 0.87 × 10^{-12} cm^2 s^{-1} (which was lower by a factor of 2 than previously reported).

Interfaces in Composites

Micro-Raman spectroscopy has been used to study the interfacial regions in fiber–epoxy composites (*124–128*). The compositional differences between the interface, the interfacial region, and the bulk can be detected. Thus it is possible to investigate the bonding chemistry using Raman mapping. Direct monitoring using remote spectroscopic measurements may be helpful in developing high-performance composites.

Raman imaging of glass-reinforced composites has been reported. It is possible to distinguish between areas of different chemical composition (126). A protocol has been developed for monitoring chemical and physical spatial variations in glass-reinforced composites via multivariate statistics (128). The multivariate images highlight differences between the fiber and surrounding epoxy and the presence of stress cracks and bubbles. Using PCA analysis over the frequency range from 1200 to 1290 cm^{-1}, which includes the epoxide band at 1243 cm^{-1} related to the cure of the epoxy, the image of the cure percentage variance of the epoxy can be observed. The images suggest that the fiber acts as a nucleation site for regions of lower cure percentage of the epoxy.

Cure in Network Polymers

The homogeneity of melamine–formaldehyde (MF) resins has been studied using confocal Raman imaging. In particular, a Raman spectroscopic method has been developed that allows for the determination of free melamine content, not only in a soluble resin but also in a cured resin (125). Practically, MF resins are used as pressed plates, and it is desirable to have local information about the free melamine content. The Raman spectra of MF adducts and resins with varying formaldehyde/melamine (F/M) ratio have a 676 cm^{-1} line that vanishes as the melamine ring is substituted. For MF resins that contain free melamine, the 676 cm^{-1} line is still present indicating that the 676 cm^{-1} line is useful for determining the free melamine content in MF resins. The line at 975 cm^{-1} is assigned to the triazine ring-breathing vibration and can be used for determination of the total melamine content of an MF resin. Therefore, the ratio of the intensities of the 676/975 lines can be used to quantify the amount of melamine present in the unsubstituted form. This Raman method was applied to different MF resins. For the cured resin with F/M = 1.7 no free melamine was detected; for the cured resin with F/M = 1.0 a free amine content of ~4% was found.

Raman imaging was carried out on an area of 44 × 44 μm divided in 3.7 × 3.7 μm area Raman pixels. The images show that there are regions of high free melamine content of ~10 μm area. A line intensity profile was generated (125). This result is important as the mechanical properties of the cured resin can be influenced by heterogeneities on the micrometer level.

Identification of Phases in Polymers

Semicrystalline polymers have a number of phases associated with their internal structure. These phases include a crystalline phase, an amorphous phase, and a possible third phase of interfacial material. The interfacial region between the crystalline and the amorphous phase is considered to consist of crystalline-like structures that have lost their lateral order and so cannot be characterized as another crystalline modification. Raman spectroscopy has been used to make a determination of the nature of the interfacial region and the fraction of this phase in PE (*129*). An early determination was based on the assumption that if PE was simply a two-phase system, it should be possible to synthesize the Raman spectrum of a sample of semicrystalline PE from a weighted sum of the spectrum of a model crystalline compound and of a model amorphous compound. The Raman band at 1415 cm^{-1} is only observed in the orthorhombic crystalline phase, and the band at 1440 cm^{-1} is present only in the melt (amorphous) phase. The fraction of interfacial phase is obtained from the differences between unity and the sum of the amorphous and crystalline fractions.

More recently, it has been suggested that there is a fourth phase in PE. The extra fourth phase was characterized as another intermediate phase (*130*). Finally, recent results suggest that the Raman methods previously employed have some basic problems, and the use of Raman for the determination of the intermediate phases is suspect (*131*).

A fiber-coupled Raman microprobe has been used to study the opposite surfaces of unaxially drawn PET films (draw ratio 3.5). Differences in crystallinity were observed by focusing the laser beam at various points through the film thickness (*132*).

Lubrication Analysis

Microscopic Raman spectroscopy can be used to determine fluid film thickness for lubricating systems (*132*). The basis of measuring the film thickness relies on the linearity of the Raman scattering intensity with the number of molecules in an optical collection volume, which in this case is the thickness of the fluid sample. To quote the authors (*133*) "the excitation laser beam acts like a 'dip stick' in determining the amount of lubricant present." In the experiments described, a one-inch-diameter stainless steel bearing (grade 10) was one of the lubricated surfaces and a Drukker cut diamond was the other. The fluid used was a polyphenyl-ether consisting of a mixture of isomers of bis-(bisphenoxyphen-oxy)benzene. The Raman band used for the thickness measurement was the 1002 cm^{-1} band.

The inherent advantage of the Raman method for lubrication applications is its utility for curved surfaces. The scattering of the laser is from a very small area of the lubricated surface so the curvature of the surface is large in comparison to the probed optical area.

Examination of Historic Art Works

In the materials analysis of irreplaceable works of art, it is important that the analytical technique be noninvasive and nondestructive. Raman microspectroscopy has been used to identify and characterize small paint samples of artistic and historic pieces of art (*134–136*). In such cases, very small paint samples, with surface areas of 0.5 mm² or less, are examined that have between 5 and 15 layers of varnish, glaze, paint, and glue containing various paint pigments. Raman microspectroscopy can be used to study pigments in situ for those cases where the paint layers are too thin to remove from the painting without causing considerable damage (*137, 138*). Raman microspectroscopy is rapid and easy to perform and provides rapid and reliable means of identifying the materials components from pigments, minerals, or resins.

These advantages of Raman microspectroscopy have been demonstrated in the analysis of a sample from Tintoretto's *The Dreams of Men* (*134*). The Raman microspectra gave new insights into the light-induced transformation of realgar. Studies of this type offer potential for generating methods of influencing the preservation of paintings using this and other pigments.

Summary and Speculations About the Future

Great progress has been made in Raman microspectroscopy. Raman spectroscopy is well-suited for imaging because

- the excitation and scattered radiation (visible or near-IR) can be readily guided for mapping purposes;
- it is a scattering technique and so can be readily applied irrespective of sample form;
- the instrumentation is compact and mechanically simple;
- the use of fiber-optic bundles opens the possibility of remote analysis; and
- the rich information content of the spectra yields high specificity of the analysis.

The primary limitations are the inherent weakness of the Raman signal, the ever-present threat of fluorescence, and the relative lack of experience of analysts with Raman spectroscopy relative to IR spectroscopy. Additionally, Raman spectroscopic imaging has yet to find the "big" application that demands the presence of a Raman imaging device in every polymer laboratory.

The biggest barrier for the analysis of industrial polymer samples by Raman imaging is the fluorescence problem. The use of near-IR illumination minimizes this problem but enhances the potential for thermal-induced effects in the samples.

The use of Raman imaging is becoming increasingly important in the fields of inorganic materials and biological chemistry. An increasing role in polymeric materials can be expected.

References

1. Ferraro, J. R.; Nakamoto, K. *Introductory Raman Spectroscopy;* Academic : Orlando, FL, 1994.
2. Grasselli, J.; Bulkin, B. J. *Analytical Raman Spectroscopy;* Wiley: New York, 1991.
3. Long, D. A. *Raman Spectroscopy;* McGraw-Hill: New York, 1977.
4. Strommen, D.; Nakamoto, K. *Laboratory Raman Spectroscopy;* Wiley: New York, 1984.
5. *Practical Raman Spectroscopy;* Gardiner, D. J.; Graves, P. R., Eds.; Springer Verlag: Berlin, Germany, 1989.
6. Asher, S. *Anal. Chem.* **1993,** *65,* 59.
7. Chang, R.; Furtak, T. E. *Surface Enhanced Raman Spectroscopy;* Plenum: New York, 1982.
8. Laserna, J. J.; Cabalin, L. M.; Montes, R. *Anal. Chem.* **1992,** *64,* 2006.
9. Boerio, F. J.; Tsai, W. H.; Montaudo, G. *J. Polym. Sci. Part B: Polym. Phys.* **1989,** *27,* 2477.
10. Koenig, J. L. *Spectroscopy of Polymers;* American Chemical Society: Washington, D.C., 1992.
11. Hecht, L. *J. Raman Spectrosc.* **1991,** *22,* 473.
12. Pigeon, M.; Prud'homme, R.; Pezolet, M. *Macromolecules* **1991,** *24,* 5687.
13. Bower, D. I. *J. Polym. Sci. Polym. Phys. Ed.* **1972,** *10,* 2135.
14. Bower, D. I. *J. Phys. B: At. Mol. Phys.* **1976,** *9,* 3275.
15. Mariani, L.; Morresi, A.; Giorgini, M. G.; Paliani, G.; Cataliotti, R. S. *Appl. Spectrosc.* **1993,** *47,* 1227.
16. Mariani, L.; Morresi, A.; Giorgini, M. G.; Paliani, G.; Cataliotti, R. S. *Appl. Spectrosc.* **1993,** *47,* 1232.
17. Nyquist, R. A.; Putzig, C. L.; Leugers, M. A.; McLachlan, R. D.; Thill, B. *Appl. Spectrosc.* **1992,** *46,* 981.
18. Turrell, G. *J. Raman Spectrosc.* **1984,** *15,* 1103.
19. Bernard, C.; Dhamelincourt, P.; Laureyns, J.; Turell, G. *Appl. Spectrosc.* **1985,** *39* 1036.
20. *Fourier Transform Raman Spectroscopy;* Chase, D. B.; Rabolt, J. F., Eds.; Academic: Orlando, FL, 1994.

21. Treado, P. J.; Morris, M. D. *Appl. Spectrosc. Rev.* **1994**, *29*, 1.
22. Battery, D. E.; Slater, J. B.; Wludyka, R.; Owen, H.; Pallister, D. M.; Morris, M. D. *Appl. Spectrosc.* **1993**, *47*, 1913.
23. Shen, C.; Vickers, T. J.; Mann, C. K. *Appl. Spectrosc.* **1992**, *46*, 772.
24. Brenan, C. J. H.; Hunter, I. W. *Appl. Spectrosc.* **1995**, *49*, 971.
25. Hendra, P.; Jones, C.; Warnes, G. *Fourier Transform Raman Spectroscopy: Instrumentation and Chemical Applications;* Prentice Hall: New York, 1991.
26. Chase, D. B. *Appl. Spectrosc.* **1994**, *48*, 14.
27. Tedesco, J. M.; Owen, H.; Pallister, D. M.; Morris, M. D. *Anal. Chem.* **1992**, *64*, 1154.
28. Schrader, B.; Baranovic, G.; Keller, S.; Saawatzki, J. *Fresenius' J. Anal. Chem.* **1994**, *349*, 4.
29. Archibald, D. D.; Yager, P. *Appl. Spectrosc.* **1992**, *46*, 1613.
30. Bilhorn, R. B.; Swedler, J. V.; Epperson, P. M.; Denton, M. B. *Appl. Spectrosc.* **1987**, *41*, 1125.
31. Deckert, V.; Kiefer, W. *Appl. Spectrosc.* **1992**, *46*, 322.
32. Campion, A.; Brown, J.; Grizzle, W. H. *Surf. Sci.* **1982**, *113*, 153.
33. Campion, A.; Woodruff, W. H. *Anal. Chem.* **1987**, *59*, 1301.
34. Bormett, R. W.; Asher, S. A. *Appl. Spectrosc.* **1994**, *48*, 1.
35. Pelletier, M. J. *Appl. Spectrosc.* **1990**, *44*, 1699.
36. Pemberton, J.; Sobocinski, S. *J. Am. Chem. Soc.* **1989**, *111*, 432.
37. Pemberton, J.; Sobocinski, S. *Appl. Spectrosc.* **1990**, *44*, 328.
38. Engert, C.; Eckert, V.; Kiefer, W.; Umapathy, S.; Hamaguchi, H. *Appl. Spectrosc.* **1994**, *48*, 933.
39. Gallagher, P. *Photonics Spectra* **1994**, 109.
40. Baraga, J. J.; Feld, M. S.; Rava, R. P. *Appl. Spectrosc.* **1992**, *46*, 187.
41. Bilhorn, R. B.; Sweedler, J. V.; Epperson, P. M.; Denton, M. B. *Appl. Spectrosc.* **1987**, *41*, 1114.
42. Deckert, V.; Kiefer, W. *Appl. Spectrosc.* **1993**, *46*, 322.
43. Fryling, M.; Frank, C. J.; McCreery, R. L. *Appl. Spectrosc.* **1993**, *47*, 1965.
44. Rosacso, G. J. *In Advances in Infrared and Raman Spectroscopy;* Clark, R. J. H.; Hester, R. E., Eds.; Heydon: London, 1980; Vol. 7.
45. Louden, G. D. *In Practical Raman Spectroscopy;* Gardiner; Graves, Eds.; Springer-Verlag: Berlin, Germany, 1989.
46. Delhaye, M.; Dhamelincourt, P. *J. Raman Spectrosc.* **1975**, *3*, 33.
47. Adar, F.; Noether, H. *Polymer* **1985**, *26*, 1935–1943.
48. Puppels, G.; Grond, M.; Greve, J. *Appl. Spectrosc.* **1993**, *47*, 1258.
49. Davis, K. L.; Slater, J. B. Presented at the Microbeam Society Meeting, Breckenridge, CO, July 1995; p 115.
50. Ivanda, M.; Furic, K. *Appl. Opt.* **1992**, *31*, 6371.
51. Bowden, M.; Gardiner, D. J.; Rice, G.; Gerrard, D. L. J. *J. Raman Spectrosc.* **1990**, *21*, 37.
52. Bowden, M.; Gardiner, D. J.; Southall, J. M. *J. Appl. Phys.* **1992**, *71*, 521.
53. Bowden, M.; Dickson, G. D.; Gardiner, D. J.; Wood, D. J. *Appl. Spectrosc.* **1990**, *44*, 1679.
54. Bowden, M.; Bradley, J. W.; Dix, L. R.; Gardiner, D. J.; Dixon, N. M.; Gerrard, D. L. *Polymer* **1994**, *35*, 1654.
55. Morris, H. R.; Hoyt, C. C.; Treado, P. J. *Appl. Spectrosc.* **1994**, *48*, 857.
56. Treado, P. J.; Morris, M. D. *Appl. Spectrosc.* **1990**, *44*, 1.
57. Treado, P. J.; Govil, A.; Morris, M. D.; Sternitzke, K. D.; McCreery, R. L. *Appl. Spectrosc.* **1990**, *44*, 1270.
58. Govil, A.; Pallister, D. M.; Chen, L.; Morris, M. D. *Appl. Spectrosc.* **1991**, *45*, 1604.

59. Chen, K.-L.; Sheng, R.-S.; Morris, M. D. *Appl. Spectrosc.* **1991,** *45,* 1717.
60. Treado, P. J.; Morris, M. D. *Spectrochim. Acta Rev.* **1990,** *13,* 355.
61. Morris, M. D.; Govil, A.; Liu, K. L.; Sheng, R. *Proc. SPIE Int. Soc. Opt. Eng.* **1992,** *1439,* 95.
62. Carabba, M. M.; Spencer, K. M.; Rich, C.; Rauh, D. *Appl. Spectrosc.* **1990,** *44,* 1558.
63. Pelletier, M. J.; Reeder, R. C. *Appl. Spectrosc.* **1991,** *45,* 765.
64. Everall, N. *Appl. Spectrosc.* **1992,** *46,* 746.
65. Battey, D. E.; Slater, J. B.; Wludyka, R.; Owen, H.; Pallister, D. M.; Morris, M. D. *Appl. Spectrosc.* **1993,** *47,* 1913.
66. Pallister, D. M.; Govil, A.; Morris, M. D.; Colburn, W. S. *Appl. Spectrosc.* **1994,** *46,* 746.
67. Morris, H. R.; Hoyt, C. C.; Treado, P. J. *Appl. Spectrosc.* **1994,** *48,* 857.
68. Lewis, E. N.; Treado, P. J.; Levin, I. W. *Appl. Spectrosc.* **1993,** *47,* 539.
69. Tran, C. D.; Furlan, R. J. *Anal. Chem.* **1992,** *64,* 2775.
70. Ciurczak, E. W. *Spectroscopy* **1993,** *8,* 12.
71. Treado, P. J.; Levin, I. W.; Lewis, E. N. *Appl. Spectrosc.* **1992,** *46,* 1211.
72. Angel, S. M. et al. Presented at PittCon, Chicago, IL, March 1966; paper 754.
73. Garton, A.; Batchelder, D. N.; Cheng, C. *Appl. Spectrosc.* **1993,** *47,* 922.
74. Batchelder, D. N.; Cheng, C.; Pitt, G. D. *Adv. Mat.* **1991,** *3,* 566.
75. Batchelder, D. N.; Cheng, C.; Muller, W.; Smith, B. J. E. *Makromol. Chem., Macromol. Symp.* **1991,** *46,* 171.
76. Puppels, G. J.; Grond, M.; Greve, J. *Appl. Spectrosc.* **1993,** *47,* 1256.
77. Hayden, C. A.; Morris, M. D. *Appl. Spectrosc.* **1996,** *50,* 708.
78. Christensen, K. A.; Bradley, N. L.; Morris, M. D.; Morrison, R. V. *Appl. Spectrosc.* **1995,** *49,* 1120.
79. Morris, H. R.; Hoyt, C. C.; Miller, P.; Treado, P. J. *Appl. Spectrosc.* **1996,** *50,* 805.
80. Puppels, G. J.; Colier, W.; Olmikhof, J. H. F.; Otto, C.; de Mul, F. F. M.; Greve, J. *J. Raman Spectrosc.* **1991,** *22,* 217.
81. Barbillat, J.; Dhamelincourt, P.; Delhaye, M.; Da Silva, E. *J. Raman Spectrosc.* **1994,** *25,* 3.
82. Sharonov, S.; Nobiev, A.; Chourpa, I.; Trofanov, A.; Valisa, P.; Manfait, M. *J. Raman Spectrosc.* **1994,** *25,* 699.
83. Markwort, L.; Kip, B.; DaSilva, E.; Roussel, B. *Appl. Spectrosc.* **1995,** *49,* 1411.
84. Puppels, G. J.; Colier, W.; Olmikhof, J. H. F.; Otto, C.; de Mul, F. F. M.; Greve, J. *J. Raman Spectrosc.* **1991,** *22,* 217.
85. Govil, A.; Pallister, D. M.; Morris, M. D. *Appl. Spectrosc.* **1993,** *47,* 75.
86. Bradley, N. L.; Morris, M. D. Presented at FACSS XII, Cincinnati, OH, October 1995; paper 594.
87. Tabaksblat, R.; Meier, R. J.; Kip, B. J. *Appl. Spectrosc.* **1992,** *46,* 60.
88. Schrum, K. F.; Ko, S.; Ben-Amotz, D. Presented at the Federationof Analytical Chemists and Spectroscopy Societies, XII, Cincinnati, OH, October 1995; paper 685.
89. Sharonov, S.; Nabiev, I.; Chourpa, I.; Feofanov, A.; Valisa, P.; Manfait, M. *J. Raman Spectrosc.* **1994,** *25,* 699.
90. Sharonov, S.; Chourpa, I.; Morjani, H.; Nabiev, I.; Manfait, M.; Feofanov, A. *Anal. Chim. Acta* **1994,** *290,* 40.
91. Meir, R.; Kip, B. *Microbeam Anal.* **1994,** *3,* 61.
92. Williams, K. P. J.; Wilcock, I. C.; Hayward, I. P.; Whitley, A. *Spectroscopy* **1996,** *11,* 45.
93. Huong, P. V. *Vib. Spectrosc.* **1996,** *11,* 17.

94. Delhaye, M.; Barbillat, J.; Dhamelincourt, P. *In Analytical Techniques in Enviornmental Chemistry;* Albaiges, J., Ed.; Pergamon: Oxford, England, 1980.
95. Claybourn, M. *Abstracts of Papers,* 208th National Meeting of the American Chemical Society, Washington, DC; American Chemical Society: Washington, DC, 1994; PMSE 76.
96. Centeno, J. A.; Ishak, K. G.; Mullick, F. G.; Gahl, W. A.; O'Leary, T. J. *Appl. Spectrosc.* **1994,** *48,* 569.
97. Schaeberie, M. D.; Kaiasinsky, V. F.; Luke, J. L.; Lewis, E. N.; Levin, I. W.; Treado, P. J. *Anal. Chem.* **1966,** *68,* 1829.
98. Armstrong, C. L.; Edwards, H. G. M.; Farwell, D. W.; Williams, A. C. *Vib. Spectrosc.* **1996,** *11,* 105–113.
99. Sureau, F.; Chinsky, L.; Amirand, C.; Ballni, J. P.; Duquesne, M.; Laigle, A.; Turpin, P. Y.; Vigny, P. *Appl. Spectrosc.* **1990,** *44,* 1047.
100. Puppels, G. J.; Garritsen, H. S. P.; Kummer, J. A.; Greve, J. *Cytometry* **1993,** *14,* 251.
101. Sijtsema, N. M.; Duindam, J. J.; Puppels, G. J.; Otto, C.; Greve, J. *Appl. Spectrosc.* **1996,** *50,* 545.
102. Todd, E. A.; Morris, M. D. *Appl. Spectrosc.* **1994,** *48,* 545.
103. Turner, P. H. *Bruker Appl. Notes;* 1994.
104. van der Zwagg, S.; Northolt, M. G.; Young, R. J.; Robinson, I. M.; Galiotis, C.; Batchelder, D. N. *Polym. Commun.* **1987,** *28,* 276.
105. Tabaksblat, R.; Meier, R. J.; Kip, B. J. *Appl. Spectrosc.* **1992,** *46,* 60.
106. Hajatdoost, S.; Yarwodd, J. *Appl. Spectrosc.* **1996,** *50,* 558.
107. Young, R. J.; Lu, D.; Day, R. J. *Polym. Int.* **1991,** *24,* 71.
108. Young, R. J.; Lu, D.; Day, R. J.; Knoff, W. F.; Davis, H. A. *J. Mater. Sci.* **1992,** *27,* 5431.
109. Day, R. J.; Robinson, I. M.; Zakikhani, M.; Young, R. J. *Polymer* **1987,** *28,* 1833.
110. Young, R. J.; Day, R. J.; Zakikhani, M. *J. Mater. Sci.* **1990,** *25,* 127.
111. Young, R. J.; Ang, P. P. *Polymer* **1990,** *31,* 47.
112. Robinson, I. M.; Zakikhani, M.; Day, R. J.; Young, R. J.; Galiotis, C. *J. Mater. Sci. Lett.* **1987,** *6,* 1212.
113. Fina, L. J.; Bower, D. I.; Ward, I. M. *Polymer* **1988,** *29,* 2146.
114. Afanasyeva, N. I.; Jawhari, T.; Klimenko, I. V.; Zhuravleva, T. S. *Vib. Spectrosc.* **1996,** *11,* 79.
115. Jawhari, T.; Merion, J. C.; Rodriguez-Cabello, J. C.; Pastor, J. M. *Polymer* **1992,** *33,* 4199.
116. Gerrard, D. L.; Maddams, W. F. *Macromolecules* **1981,** *14,* 1356.
117. Gerrard, D. L.; Maddams, W. F. *Macromolecules* **1975,** *8,* 54.
118. Baruya, A.; Gerrard, D. L.; Maddams, W. F. *Macromolecules* **1983,** *16,* 578.
119. Kip, B. J.; van Aaken, S. M.; Meier, R. J.; Williams, K. P. J.; Gerrard, D. L. *Macromolecules* **1992,** *25,* 4290.
120. Bowden, M.; Gardiner, D. J.; Rice, G.; Gerrard, D. L. *J. Raman Spectrosc.* **1990,** *21,* 37.
121. Bowden, M.; Donaldson, P.; Gardiner, D. J. *Anal. Chem.* **1991,** *63,* 2915.
122. Markwort, L.; Kip, B. *Proceedings of the 29th Annual Meeting of MAS;* Editz, E., Ed.; VCH Publishers: Breckenridge, CO, 1995; p 109.
123. Hong, P. P.; Boerio, F. J.; Clarson, S. J.; Smith, S. D. *Macromolecules* **1991,** *24,* 4770.
124. Stellman, C. M.; Booksh, K. S.; Myrick, M. L. Presented at PITTCON, March,1995, New Orleans, LA, 1995; paper 413.
125. Scheepers, M. L.; Meier, R. J.; Markwort, L.; Gelan, J. M.; Vanderzande, D. J.; Kip, B. J. *Vib. Spectrosc.* **1995,** *9,* 139.

126. Stellman, C. M.; Booksh, K. S.; Myrick, M. I. Presented at the Federation of Analytical Chemists and Spectroscopy Societies, XII, Cincinnati, OH, October 1995; paper 681.

127. Myrick, M. L.; Angel, S. M.; Lyon, R. E.; Vess, T. M. *SAMPE J.* **1993**, *28*, 37.

128. Stellman, C. M.; Booksh, K. S.; Myrick, M. L. *Appl. Spectrosc.* **1996**, *50*, 552.

129. Strobel, G. R.; Hagedorn, W. J. *J. Polym. Sci. Polym. Phys. Ed.* **1978**, *16*, 1181.

130. Mutter, R.; Stille, W.; Strobel, G. *J. Polym. Sci. Polym. Phys. Ed.* **1993**, *31*, 99.

131. Naylor, C. C.; Meier, R. J.; Kip, B. J.; Williams, K. P. J.; Mason, S. M.; Conroy, N. L.; Gerrard, D. *Macromolecules* **1995**, *28*, 2969.

132. Everall, N.; Davis, K.; Owen, H.; Pelletier, M. J.; Slater, J. *Appl. Spectrosc.* **1996**, *50*, 388.

133. Hutchinson, E. J.; Shu, D.; LaPlant, F.; Ben-Amotz, D. *Appl. Spectrosc.* **1995**, *49*, 1274.

134. Trentelman, K.; Stodulski, L.; Paviosky, M. *Anal. Chem.* **1996**, *68*, 1756.

135. Huong, P. V. *Physica C: (Amsterdam)* **1991**, *180*, 128.

136. Clark, R. J. H.; Cooksey, C. J.; Daniels, M. A. M.; Withnall, R. *Endeavour* **1993**, *17*, 191.

137. Best, S. P.; Clark, R. H. J.; Daniels, M. A. M.; Withnall, R. *Chem Br.* **1993**, *29*, 118.

138. Best, S. P.; Clark, R. H. J.; Daniels, M. A. M.; Porter, C. A.; Withnall, R. *Stud. Conserv.* **1995**, *40*, 31.

10

Nuclear Magnetic Resonance Imaging

"MRI (NMR imaging) is, without question, the most significant innovation in diagnostic imaging since the discovery of X-rays."

—Felix W. Wehrli

The basis of nuclear magnetic resonance (NMR) is that atomic nuclei oscillate in the presence of a magnetic field—like tiny gyroscopes, and are forced into alignment with the magnetic field. Because of thermal motion, perfect alignment is not achieved, and these spinning nuclei precess around the direction of the magnetic field. The rate of precession (cycles per second) is proportional to the strength of the magnetic field. There is no signal when the nuclei are at equilibrium. In order to activate the nuclei so that they emit a signal, energy must be transmitted to the sample. On bombardment with a radio wave from a radio-frequency (rf) transmitter, the nuclei absorb the rf energy and topple out of alignment with the magnetic field. The nuclei emit rf energy as they realign. The emitted energy is the NMR signal. By measuring the specific radio frequencies that are emitted by the nuclei and the rate at which the realignment occurs, spectroscopists can obtain detailed information about the molecular structure and dynamics of the samples they are studying (*1–4*).

The nuclear magnetization induced in the sample by the magnetic field is the local sum of the magnetic fields of the nuclei. The magnetization is a bulk property of the sample rather than a property of the individual protons. The magnitude of the induced magnetization is a function of the *gyromagnetic ratio, γ*, which is a physical property of the nuclei. Different chemical elements have different gyromagnetic ratios.

The different gyromagnetic ratios result in different energies and allow the study of the different nuclei individually. The gyromagnetic ratio of protons is 4.26×10^7 Hz T^{-1} (or 2.68×10^8) rad s^{-1} T^{-1}. The magnetization, **M**, is given by a simplified Block equation (neglecting relaxation processes):

$$d\mathbf{M}/dt = \gamma \; \mathbf{M} \times \mathbf{H} \qquad (10.1)$$

where t is time and **H** is the strength of the applied magnetic field. The nuclear magnetization, **M** (as a function of time), is the source of the NMR signal. Initially, the magnetization has components M_x^0, M_y^0, M_z^0. Letting **H** = H_0**k**, where **k** is the wave vector, we can write in coordinate form:

$$dM_x/dt = \gamma \; H_0 \; M_y \qquad (10.2)$$

$$dM_y/dt = \gamma \; H_0 \; M_x \qquad (10.3)$$

$$dM_z/dt = 0 \qquad (10.4)$$

These equations have the solution:

$$M_x(t) = M_x^0 \cos \omega_0 t - M_y^0 \sin \omega_0 t \qquad (10.5)$$

$$M_y(t) = M_x^0 \sin \omega_0 t + M_y^0 \cos \omega_0 t \qquad (10.6)$$

$$M_z(t) = M_z^0 \qquad (10.7)$$

where ω_0 is the *Larmor frequency* of the spin system, given by

$$\omega_0 = -\gamma H_0. \qquad (10.8)$$

This is the *resonant* or *natural frequency*, which characterizes each nucleus in a static field.

In addition to the magnitude of the gyromagnetic ratio and the applied static field, individual nuclei of the same type can exhibit different magnetization because of another phenomenon. The chemical electrons surrounding a given nucleus can shield the nucleus from the externally applied magnetic field by a small amount. This shielding changes the induced magnetization and shifts the resonance frequency by a small amount, termed the "chemical shift." These chemical shifts, which for protons cover a range of about 10 ppm and for ^{13}C a range of 250 ppm, are the substance of NMR.

The measurements of chemical shifts in NMR are used to identify chemically shielded nuclei, and correlation tables between chemical shift and chemical structure have been established (5). The *chemical shift* is the resonance NMR frequency of a chemically shielded nucleus measured relative to that of a suitable reference compound. The chemical shift values, δ, are typically on the order of 10^{-6} and therefore commonly specified in parts per million relative to a reference. These δ units are independent of the strength of the magnetic fields, thereby allowing direct comparison of chemical shift measurements from different instruments. In absolute frequency terms, the separation between nuclei with different chemical shifts increases with increasing magnetic field, yielding better dispersion of resonances at high field.

The resonance intensities are also influenced by "relaxation times" T_1 and T_2, which are measures of the relaxation processes involved. The constant T_1 is the spin-lattice relaxation time, and it governs the evolution of the longitudinal magnetization, M_z, toward the equilibrium value, M_0. The physical process involved in this relaxation is the dissipation of energy from the collection of nuclei, the "spin system," to the atomic and molecular environment of the nuclei, the "lattice." The constant T_2 is the spin–spin relaxation time, and it governs the evolution of the transverse magnetization toward its equilibrium value, which is zero. The physical processes for T_2 include the same processes as for T_1 as well as the magnetic coupling between the neighboring nuclei. This process of relaxation can be thought of as the randomization of the individual spins so that the sum of the fields of the nuclei in the collection goes to zero.

Conventional NMR spectroscopy can be used to determine the type of chemical structure on the basis of resonance frequency, but not the spatial position of the stimulated nuclei in a heterogeneous rigid sample. If the magnitude of the field is made nonuniform in a controlled manner, then protons at different points in space will precess at different frequencies, that is, the gradient system spatially encodes the spins.

One might not suspect that NMR has any potential for imaging because of the magnitude of the radiation's wavelength. At 60 MHz, for example, the wavelength is 5 m, which, in terms of optical diffraction, hardly merits consideration (6) as such resolution has little if any utility in materials science. However, polarization coupling effects were discovered that mitigate the expected optical diffraction limit imposed on the wavelength of the radiation, and spatial resolutions in the range of 6–40 μm are obtainable by NMR imaging (NMRI). MRI is the term used for medical nuclear magnetic imaging. This term is preferred in the medical profession because the public has a highly suspect view of the term "nuclear," although obviously the term as used here has a considerably different connotation than radioactivity.

NMRI is a method in which the stimulating signal is spatially encoded so an image can be reconstructed showing the spatial distribution of nuclei in a heterogeneous sample (7–12). Other than spatially encoding the signal, NMRI works on the same principles as standard NMR.

The fundamental element in NMRI is the spatial dependence of the spins' precession frequency, which equates to the their phase in the presence of a magnetic field gradient, **G**:

$$G = (\partial H_0/\partial x, \partial H_0/\partial y, \partial H_0/\partial H_z) \qquad (10.9)$$

which is superimposed on the static polarizing field $\mathbf{H}_0 = H_z$. The resonance frequency as a function of the spatial position, **r**, in the presence of a magnetic field gradient is given by

$$\omega(\mathbf{r}) = \gamma(H_0 + \mathbf{G}\cdot\mathbf{r}) \qquad (10.10)$$

The incremental field at location **r**, resulting from the superimposed field gradient **G**, is given by the scalar product:

$$\mathbf{G}\cdot\mathbf{r} = G_x x + G_y y + G_z z \qquad (10.11)$$

When we ignore γH_0 (i.e., we operate in the rotating frame of reference), the NMR signal evolves as

$$\exp(i\gamma\omega t) = \exp(i\gamma\, \mathbf{G}\cdot\mathbf{r}t) \qquad (10.12)$$

In a volume element $dxdydz$, the signal $dS(t)$ is given by

$$dS(t) \propto M_{xy}\,(\mathbf{r},\, t)\, \exp\,[i\gamma(G_x x + G_y y + G_z z)t]\,dxdydz \qquad (10.13)$$

where $M_{xy}\,(\mathbf{r},\, t)$ is the magnetization density at location **r**.

We can now define the wave vector (i.e., the spatial frequency vector) as follows:

$$\mathbf{k}(t) = (2\pi)^{-1}\, \gamma\!\int G(t')\,dt' \qquad (10.14)$$

where $\mathbf{G}(t')$ is the time-dependent spatial encoding gradient. The wave vector has the units of cm^{-1} or m^{-1} and measures "reciprocal space." If we ignore the effects of relaxation, we can write:

$$S(\mathbf{k}) = \iiint_{object}\, \rho(\mathbf{r})\, e^{j2\rho k(t)\cdot\mathbf{r}}d\mathbf{r} \qquad (10.15)$$

where $\rho(\mathbf{r})$ represents the spin density of the object within the range Δx, Δy, Δz, over which the integration occurs. Practically, the spatial frequency is the phase rotation per unit length of the object that the magnetization undergoes after being exposed to a gradient $G(t)$ for a time t. Observe that

$$\rho(\mathbf{r}) = \iiint_{\mathrm{kspace}} S(\mathbf{k})\ e^{-i2\pi\mathbf{k}(t)\cdot\mathbf{r}}d\mathbf{k} \qquad (10.16)$$

so $S(\mathbf{k})$ and $\rho(\mathbf{r})$ are Fourier transform pairs. Therefore, the spatial spin density at any one location can be obtained as the inverse Fourier transform of the spatial frequency signal.

In the simple one-dimensional case, the absorption signal recorded in the presence of a static magnetic field represents the one-dimensional projection of the object's spin density perpendicular to the gradient direction. This result is illustrated in Figure 10.1.

Consider two-dimensional imaging using the Fourier zeugmatography method, which is based on the successive application of orthogonal pulsed field gradients prior to and during sampling of the free induction decay (fid). Following the excitation by a radio-frequency exciting pulse, a gradient G_x is activated. During the initial period, when the G_x gradient is active, a single-frequency fid is obtained. During readout, a G_z gradient is

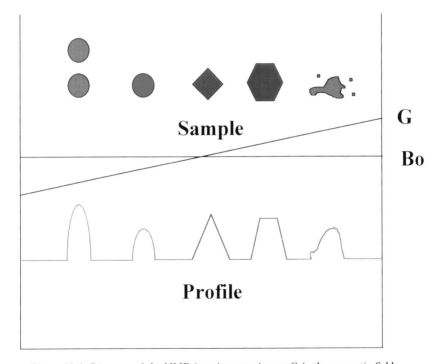

Figure 10.1. Diagram of the NMR imaging experiment: G is the magnetic field gradient, and B_0 is the magnetic field.

switched on, producing two fids containing the frequencies given by the gradient amplitude and the samples' location on the z axis. A Fourier transform with respect to t_z generates a "spectrum", reflecting the phase modulation being caused by the spins' evolution during the time period t_x. The second Fourier transformation with respect to t_x then yields the two-dimensional zeugmatogram.

Now consider the Fourier zeugmatography method for the simple case of 3-D imaging with the gradients having constant amplitude. Following the excitation by a radio-frequency exciting pulse, the spins in the sample are subjected to variable periods of field gradients G_x and G_y. While these gradients are active, spins at locations x and y resonate at frequencies $\omega(x) = \gamma G_x x$ and $\omega(y) = \gamma G_x y$. Their relative phase at time $t_x + t_y$ is then given by $\omega(x) t_x + \omega (y) t_y$. Hence, the phase of the resulting fid, recorded in the presence of a third gradient, G_z, is modulated, with the modulation being a function of the spins' location. By stepping the periods, t_x and t_y, in equal increments, a three-dimensional array of time domain data is obtained whose Fourier transform yields a three-dimensional spectrum.

There are several ways of spatially encoding the NMR signal (13). One is to apply a linear magnetic field gradient on the original static field (Figure 10.1). The purpose of the linear gradient magnetic field is to label, or encode, different regions of the sample linearly with different NMR frequencies. Since the magnetic field is varied in a linear manner across the sample, the recorded frequency of the NMR signal indicates the spatial position of the resonating nuclei (Figure 10.1). In one dimension, the position of the sample is related to a frequency by the relationship:

$$\Delta\omega_z = \omega_z - \omega_0 = \gamma G_z z \qquad (10.17)$$

where γ is the gyromagnetic ratio, the magnetic field gradient, G_z, is given by $\delta B_z / \delta z$, and z is the Cartesian coordinate. A tailored rf pulse with a narrow frequency range is used to excite only those nuclei at corresponding positions in the z dimension. The amplitude of the NMR signal received from the z axis line is a measure of the number of resonant nuclei on that line, and so the NMR spectrum represents a graph of spin density versus distance (neglecting relaxation effects). The signal may be accumulated and Fourier-transformed in the usual way to obtain a measure of the nuclear magnetization present in the plane at z. In this way, nuclei on one side of the sample will be exposed to a weaker total magnetic field than those at the other side. This produces a signal that, after the Fourier transform, takes on the shape of the sample (Figure 10.1).

Conventional NMR imaging methods rely on the use of static field gradients and generally involve spin echoes or gradient echoes for refocusing nuclear magnetization and/or avoiding artifacts due to gradient switching. Any echo has some magnetization loss due to the transverse relaxation.

The use of static field gradients results in effects due to the switching process, including eddy currents created in the materials surrounding the probe, and spectrometer perturbations due to finite rise and fall times of H_0 gradients.

In another technique, called *projection–reconstruction imaging*, a magnetic field gradient is rotated to obtain snapshots of the sample at many different angles covering an arc of at least 180° (*14*). From such a set of data, a computation can reconstruct a cross-sectional image of the sample.

Resolution in NMRI

Spatial resolution is limited by the amount of sample that can be detected by NMR. Spatially resolving a given volume in an NMR image is equivalent to making NMR spectroscopic measurements on that volume. To resolve two spatially distinct volume elements, one must apply a magnetic field gradient of sufficient strength that the elements one wishes to resolve are shifted in resonance frequency from each other by an amount greater than the natural line width. Stated more simply,

$$\text{resolution} = \text{line width/gradient strength} \qquad (10.18)$$

The highest resolution reached so far is $10 \times 10 \times 100$ μm. This corresponds to an observable volume element of 10^{-5} mm^3.

The attainable resolution is limited by spectroscopic and hardware factors. Spectroscopic factors are the line width and the spread of the chemical shift of an NMR signal, diffusion processes, and susceptibility gradients, both within the object and at its boundaries. Hardware factors may be the magnetic field inhomogeneity or instability, nonlinearity of the magnetic gradient field, and the achievable S/N ratio.

To attain a spatial resolution of 10 μm with a gradient strength of 150 mT/m (15 G/cm), in samples like water, the line widths must lie in a range of 1 to 50 Hz. In solids where the line width can reach 50 kHz because of dipolar interactions of protons, a resolution of 1 mm requires a gradient strength of 1100 mT/m (*15*). In addition, in order to avoid the interfering effects of eddy currents, these very strong gradients must be switched within 20 μs.

The difficulties of solid-state imaging arise from the fact that the solid-state line width is approximately 1000 times broader than its solution counterpart (*16–19*). Increasing the gradient by three or four orders of magnitude to maintain spatial resolution in solids imaging is a formidable task from a hardware point of view. Much effort has gone into finding pulse sequence alternatives to such a brute force approach (*20*).

Utility of NMRI

Most of the optical spectroscopic imaging techniques presently used are two-dimensional techniques based on reflection from surfaces or transmission through thin optical slices. NMRI, on the other hand, is an "internal" technique, and the portion of the sample to be examined can be internal and "selectively excited." A three-dimensional image is reconstructed from a collection of two-dimensional images obtained in a series of sequential steps without destruction of the sample. NMRI is a means of detecting and imaging previously invisible internal material heterogeneities. Its potential applications in the field of polymeric materials are many and diverse (*4*). These applications are based on one of the following: the detection and imaging of subsurface defects, including interfacial flaws and microcracks, or detection and characterization of areas modified through the introduction of foreign substances, such as additives, degradation products, and contaminants. The NMR imaging technique is noninvasive, so multiple measurements can be made on the same sample under different conditions, such as after different exposures to external environmental fields. No special sample preparation is required, which makes in situ studies of materials possible, including superposition of images obtained before and after application of stresses, and exposure to environmental factors including stress, fatigue, temperature, and penetrants. The underlying purpose of NMR imaging is to detect the presence or absence of inhomogeneities in situ. By using computer enhancement techniques, it is possible to compare a control sample with a modified sample and determine the differences between samples. Defects such as voids and inclusions are represented by very small image discontinuities and are easily detected.

Contrast in NMRI

The spin densities and the molecular environments of the nuclei are reflected in the time variation of the amplitude of the measured rf signal, and hence are reflected in the intensity of each voxel in the image. When the spin densities, T_1, and T_2 are different in the voxels of a heterogeneous sample, these differences result in changes in resonance intensities, which can be exploited to develop contrast in the NMR images.

In imaging, the first requirement is to define the slice to be imaged.

This is the basis of the *tomographic* process. This is accomplished by a method called *selective excitation*. A field gradient is applied along the z direction, and the object is irradiated with a 90° pulse having a narrow spectral width, corresponding to a narrow range of magnetic field values. Only those nuclei that fall within this narrow range of field values will be excited in a slice of the object perpendicular to z. Only a slice of the object is excited, and the remainder of the object does not respond. The field gradient is switched on in the defined slice (xy plane), and the fid evolves and is recorded. The Fourier transform of this fid yields the NMR spectrum of the selected slice of the object. This method has the advantage that the NMR response is measured immediately after excitation, yielding maximum intensity, but the disadvantage that the rf pulse may excite the entire range of frequencies produced by the gradient.

An extension of this procedure is termed two-dimensional Fourier transform imaging. In this method, after the 90° pulse, a field gradient is applied that defines the slice. This is followed by a pulse gradient that introduces phase-encoding of the spins along one dimension in the slice. A read gradient is applied during NMR signal acquisition that defines the second dimension of the slice. This sequence is repeated for a set of pulsed gradients for a specified repetition rate (TR). The number of different gradient pulses used defines the resolution along the phase-encoding dimension. A two-dimensional Fourier transform is used to process the data into an image, that is, a set of Fourier transformations generates spectral information along the read-gradient dimension, and a second set of Fourier transforms along the phase-encoding dimension creates a two-dimensional image.

Unfortunately, the use of a read gradient results in a suboptimal NMR image. A decrease in spectral resolution results because a read gradient is in fact an intentional distortion of the homogeneity of the magnetic field, and the resonance peaks are broadened, leading to blurring. To avoid the necessity of applying rf pulses at times when the gradients are on, a spin echo pulse sequence can be used. The sequence begins with the 90° pulse that generates magnetization transverse to the direction of the magnetic field. Following this pulse, a magnetic field gradient is turned on. This gradient is called the dephasing gradient because it causes the various spins to precess at different rates; this causes the spins to lose phase coherence. After the dephasing gradient is turned off, a 180° rf pulse is applied to the sample followed by a second gradient pulse with magnitude and direction identical to the first. This readout gradient pulse causes the spins to rephase to form a spin echo during the acquisition window.

A second useful NMR imaging technique is known as *spin-warp* imaging. The effects of the rf pulses and the G_x gradient are the same as found above, but a second orthogonal gradient, G_y, is applied during the first half of the pulse sequence. The amplitude of this gradient is incremented systematically for each row of data collected. The sequence that is usually used to measure the T_2 relaxation phenomena in images is called multiple spin-echo. At a given repetition time, TR, the NMR signal is measured at several different echo times, TE. These echoes provide a measure of the T_2 relaxation. By repeating the process at different TR values, the T_2 relaxation can also be measured.

Because differences in relaxation times and spin densities determine image contrast, data on relaxation times are important in the selection of the optimal rf pulse sequence for imaging a selected sample. The ability to accurately quantify relaxation rates is important in understanding and optimizing image contrast. Spin density and T_1 and T_2 images can be computed from measurements made using pulse sequences with predetermined variations. These fundamental images represent the inherent data in the system and can be recombined to reconstitute computed images for a given pulse sequence.

Contrast in NMRI depends on both material-specific and operator-selected parameters. The material-specific parameters include the spin density and the relaxation times T_1, $T_{1\rho}$ and T_2. The operator-selected parameters include the pulse sequence (inversion recovery, spin echo, etc.) and the pulse delay and repetition times (timing parameters). For a given imaging system and pulse sequence, it is the delay and repetition times in conjunction with the intrinsic material parameters that dictate the appearance of the final image. If the correct pulse sequence is employed and the relaxation times of the two materials are known, it is possible to calculate the delay and/or repetition times that will produce the maximum difference in signal intensity between those materials.

The spin echo (SE) technique is the most common pulse sequence applied in MRI today. As shown in Figure 10.2, the spin echo method consists of a series of rf pulses that are repeated many times in order to achieve a sufficient signal-to-noise ratio. Images are constructed by acquiring a multitude of projections (typically 256 per image), each with an identical setting of a readout gradient during which the sequence is sampled. Each projection is differentiated from the others by a phase difference that is produced by advancing the phase encoding gradient. Each projection is produced by a 90° pulse, followed by a 180° pulse for induction of the spin echo. The 90° rf pulse tips the magnetization into the xy plane, where it begins dephasing. The 180° rf pulse is applied after a time, t, and forces the magnetization to refocus at a time $2t$ (also known as the echo time, TE) after the 90° rf pulse, at which time the data is collected. The frequency encoding gradient, G_x, causes the spins to precess at different frequencies depending on their position in the static magnetic field. The phase encoding gradient, G_y, is orthogonal to G_x. Varying the

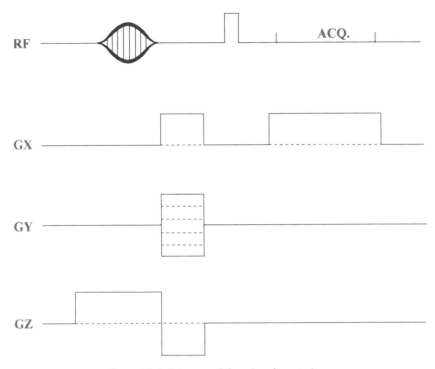

Figure 10.2. Diagram of the spin echo experiment.

intensity of G_y causes the spins to dephase at different rates, providing the second dimension of a two-dimensional image. The slice selection gradient, G_z, and the Gaussian-shaped 90° rf pulse determine the position and thickness of the region of interest. The data is Fourier-transformed in two dimensions to produce the image of the selected slice. The time delay between the observation pulse and the observation is called the echo time (TE). The time between two consecutive pulse sequences is labeled the repetition time (TR), and it usually ranges from 250 to 2500 ms.

Spin echo techniques have a unique position in NMR applications. The main problem with NMR imaging is the long data collection time, mainly due to the long spin-lattice relaxation time, T_1. Each measurement necessitates a time period on the order of T_1 (which is approximately 0.5 s for aqueous systems) for the system to return to equilibrium magnetization. By using spin echo repetition, a large number of spin echoes can be repeated within a T_1 or T_2 decay period.

Chemical-Shift Imaging

NMRI usually assumes that the measured nuclear spins, usually protons, precess at the same frequency. However, due to chemical shift differences

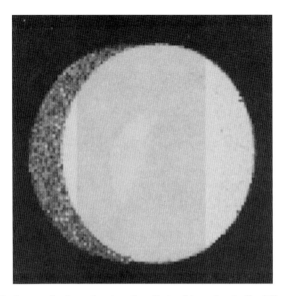

Figure 10.3. Image of xylene showing the effects of two chemically different protons. The separate images are due to the aromatic and methyl protons, and they are separated by ca. 4.8 ppm from each other. Each image is centered at its resonant frequency in the absence of a magnetic field gradient, and the resulting image is therefore smeared. (Reproduced with permission from ref 48. Copyright 1989 Gordon & Breach.)

arising from different chemical environments of protons in substances, some of the spins are naturally exposed to slightly different local fields, and hence process at slightly different frequencies, that is, aliphatic versus aromatic. The local field change is written as σH_0, where H_0 is the static field and σ is the chemical shift in parts per million (ppm). In imaging, the presence of two different types of resonating nuclei can lead to overlapping images and artifacts, as shown in Figure 10.3.

Figure 10.3 (48) shows the results of an image of xylene. The two observed images are due to the aromatic and methyl protons, which are separated by ca. 4.8 ppm from each other. Since the read or frequency encoding gradient spreads out resonance frequencies according to positions along the gradient direction, the observed image actually consists of partially overlapping sets of data (one corresponding to each type of nucleus). If the resonances are due to different species, two or more different images will be obtained. If all resonances arise from the same molecule, they will have identical spatial distributions and images.

The usual imaging schemes apply a linear gradient, G, to frequency-encode the data. Applying an inverse Fourier transform maps the spin density, as a function of frequency, linearly to spatial location. The linear relation between frequency, ω, and position, x, is given by

$$\omega = \gamma Gx \qquad (10.19)$$

where γ is the gyromagnetic ratio for hydrogen. In a gradient-free environment, the decrease in the precessional frequency of the proton of molecule a is given by

$$\Delta\omega_a = \gamma\sigma_a H_0 \qquad (10.20)$$

This leads to a shift in the image position of molecule a, with respect to that of the protons of water, by an amount given by

$$\Delta x_a = \Delta\omega_a/\gamma G_r \qquad (10.21)$$

Consequently, the image of molecule a would overlap water in the region of interest and cause an artifact in the image, which might be incorrectly interpreted as actual spatial features. By increasing G_r, the pixel shift due to chemical differences is reduced. However, much valuable information is contained in the image if the chemical shifts can be sorted out correctly. It is possible to form an image from only a selected portion of the total NMR spectrum. This process is called chemical shift imaging.

A particular resonance peak can be selectively excited by rf irradiation to the exclusion of others in the chemical shift spectrum. A long, low-power, amplitude-shaped rf pulse can be used to excite a narrow range of resonant frequencies distributed about a particular frequency. Such a "soft" pulse is more frequency sensitive than a short, square "hard" pulse.

High-resolution NMR spectra displaying chemically shifted resonances provide information on the chemical species present in the system and their relative concentrations. The magnetic resonance response can be simultaneously obtained from all regions of a heterogeneous sample by using a four-dimensional Fourier transform technique, where the high-resolution spectrum obtained during data acquisition defines one dimension and the other three dimensions form a Cartesian coordinate system.

The application of various spatially resolved NMRI techniques for the observation of high-resolution, chemically resolved spectra has been limited. This is largely due to the mutually exclusive requirements of both the highly homogeneous magnetic field, which is necessary for the observation of chemical shift information, and the inhomogeneous field, which is applied as a linear magnetic field gradient and is necessary to obtain spatially resolved data. Chemical shift imaging techniques use pulsed magnetic field gradients, which in the standard configuration of superconducting magnets generate sufficiently large eddy currents upon gradient removal to temporarily degrade the field homogeneity. This is one of the reasons for the difficulty of high-resolution, chemically resolved spatial spectroscopy.

Currently, there are several approaches to observing chemical shift effects in NMRI. One may attempt to construct an image corresponding to a preselected chemical shift of a sample, either locally or globally. When the chemical shifts originate from different chemical species, an image taken at a specific chemical shift will provide information on the spatial distribution of the corresponding species while excluding the interference of other species in the image. A local method assumes knowledge of the chemical shift and usually produces an image of the chemical species under consideration. The in-phase and out-of-phase experiments can be used for this purpose (21). In addition, chemical-shift selective suppression of an unwanted species or selective excitation of the species to be imaged (22, 23), as well as a method based on chemical-shift-specific slice selection (24), have been proposed as local methods. A global method produces essentially a chemical-shift spectrum for each localized region or volume element and thus creates a stack of chemical-shift images. A global deconvolution calculation technique has been proposed (25) utilizing a combination of the Wiener filter and an apodization function.

A method of convolution has also been suggested whereby the image is deconvoluted by the NMR spectrum of the sample (26). This latter method shifts the image of each individual resonance such that it is centered about the carrier frequency. No totally adequate method of suppressing the chemical-shift artifacts has yet been developed, but all of the methods improve the quality of the images when multiple chemical shifts are present.

On the other hand, chemical-shift imaging is highly desirable. Selective-excitation chemical-shift imaging is possible only if the spectrum of the sample is resolvable for the entire imaging volume. It has been suggested that chemical-shift-sensitive NMR images can be obtained using spectral simplification by tailoring the excitation pulses (26). An example of a chemical-shift image for an adhesive system is shown in Figure 10.4. The image was obtained by selective excitation of the proton chemical shifts associated with each component of the two-component adhesive system polyether polyol with an isocyanate curing agent (27).

Chemical-shift images have also been reported for two rubbery polymers, polybutadiene and poly(dimethylsiloxane) (28).

Application of Measurements of Spin Density by NMRI to Polymers

The number of applications of NMRI to polymers is growing at a rapid rate as the instrumentation becomes more available and the interest and

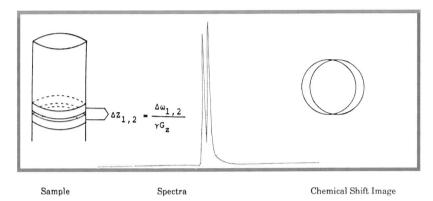

| Sample | Spectra | Chemical Shift Image |

Figure 10.4. Chemical-shift images of an adhesive system. (Reproduced with permission from ref 25. Copyright 1989 Society of Spectroscopy.)

knowledge of the technique grows. A number of reviews have been written documenting this progress (*29–37*).

Detection of Voids

One of the applications of NMRI measurements of spin density is the determination of the number and distribution of voids. Because NMR imaging allows one to obtain the image of a slice of a polymeric sample, internal defects can be measured if they are larger than the resolution of the technique (currently >20 μm).

Voids in Composites

The performance of composites depends primarily on the interfacial adhesive bond strength between the resin and the reinforcing fiber. A number of variables influence the ultimate interfacial bonding characteristics, including fiber surface functionality, the amount and structure of the adhesion promoter, the type and distribution of interfacial bonds, drying and curing temperatures, humidity, and various degradation and aging effects. We consider a number of defects or flaws that can contribute to composite failure.

Nonbonded Regions. Nonbonded regions are where the resin and fiber are locally separated from each other. The extent and distribution of the nonbonded regions determine the properties in shear.

Voids. A void is a local lack of resin or porosity of the resin resulting from incomplete flow of the matrix in the fabrication process. Voids are regions with no load-bearing capacity, and they can contribute to failure under tensile load.

Fiber-Rich Areas.   Fiber-rich areas are domains where the fiber distribution is excessive compared to the norm. Fiber-rich areas do not allow a distribution of the applied load over the sample, and they generate a highly stressed region subject to internal failure.

Resin-Rich Areas.   Resin-rich areas are domains where the resin concentration is excessive compared to the norm. All composite matrices are in a liquid or liquidlike state before the final curing process. The liquid state is a requirement for good flow, leveling, and wetting, and it is typically achieved by dispersing the resin prior to fabrication. The liquid state is obtained by using low-molecular-weight, low-viscosity resins—which cure by a chemical reaction caused by a curing agent, heat, or radiation. Resin-rich areas represent domains of low modulus and strength compared to the normal fiber–resin regions, and can become foci for failure under stress.

Resin Structural Defects.   Foam (from air bubbles) is present in some degree in most composite matrices, and it is undesirable because it affects bond strengths. A high content of foam is a major problem and must be eliminated. Composites can also contain various types of physical and chemical inhomogeneities. Poor mixing can result in separate phases in the matrices. The chemical composition and structure of the resin is a function of a number of process parameters, including cure temperature, pressure during curing, time of curing, distribution of hardener, and inhomogeneous mixing of components. Manufacturing and fabrication can produce gradients in the resin structure, which introduce gradients in the mechanical behavior of the composite.

Fiber Alignment.   The reinforcing fibers can be misaligned as a result of the fabrication process. Improper fiber alignment leads to a nonuniform distribution of stress among the fibers and can lead to failure in tension.

The void content of pultruded composite rods has been studied using NMRI (*38*). Glass-fiber-reinforced nylon rods with fiber contents of 51% by volume were first mixed with different amounts of catalyst following the reaction injection molding (RIM) process and then pultruded. The diameter of the die in the pultruder was 0.9 cm. The rods were then soaked in 80 °C water for 25 weeks before imaging. The uptake in water was 3.7% as measured by the increase in weight of the composite rods. The images were recorded on a Bruker MSL 300 spectrometer using a spin echo pulse sequence. The slice thickness was 1 mm, and the slices

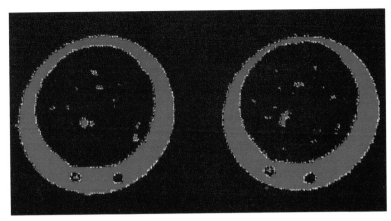

Figure 10.5. Two NMR images taken 0.5 cm apart through the pultruded rod. The light areas in the images represent void areas filled with water. The marker in the upper left-hand portion of the image is 1 mm in diameter. Comparison of the sizes of the voids in the pultruded rod indicates that some of the voids approach the magnitude of the marker. (Reproduced with permission from ref 38. Copyright 1989 Gordon & Breach.)

were taken transverse with a resolution of 100 μm × 100 μm to the fiber axis. The rods were standing in 1.5-cm-diameter vials containing water. In Figure 10.5, two images were taken 0.5 cm apart through the pultruded rod (*38*). The light areas in the image represent void areas filled with water. The marker in the lower left-hand portion of the image is 1 mm in diameter. Comparison of the sizes of the voids in the pultruded rod indicates that some of the voids approach the magnitude of the marker. Comparison of the corresponding edge-enhanced images shows that some of the voids in the images occur in the same location, which indicates that the voids are connected or tubular in shape. Thus, a channel-like void region is suggested over a length of 0.5 cm. From the computer comparison of the two images taken 0.5 cm apart, it is possible to identify a tubular-shaped void running from one image to the other within the nylon rod. Such a void could be obtained if an air bubble were trapped in the matrix during the pultrusion process. It appears that water diffuses by following the fibers in the composite.

Cell Structure of Foams

Foams represent the ultimate in void content, and NMRI has been used to study the distribution of pores and their connectivity. The experiment entails the filling of the foam with water and examining the proton image. An example is shown in Figure 10.6 for a polyurethane foam filled with water (*39*). The light portions of the image represent pores filled with

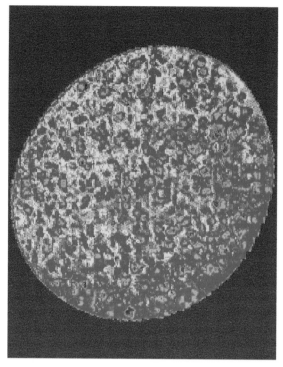

Figure 10.6. Image of polyurethane foam. (Reproduced with permission from ref. 39. Copyright 1989 John Wiley & Sons.)

water, whereas the dark areas arise either from solid foam or an absence of material (air). With an edge detection algorithm, one can observe the outline of the pores, as shown in Figure 10.7. Obviously, it is possible to observe the distribution of the pore sizes. A histogram of the pore sizes versus the number of pores of a selected size can be constructed. Such a histogram can be correlated with the chemical foaming process variables, which could lead to an understanding of the mechanism. The connectivity between the pores can be determined by injecting water into a specific site in the foam and following the flow pattern (*39*). Such knowledge could promote better process control and an improved foamed product.

Voids and Debondings in Adhesives

The performance of adhesives depends primarily on the interfacial adhesive bond strength between the adhesive and the adherent. A number of variables influence the ultimate interfacial bonding characteristics, including adherent surface functionality, amount and structure of the adhesion promoter, type and distribution of interfacial bonds, drying and

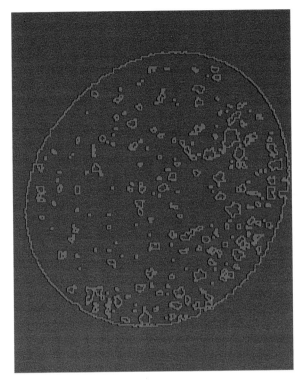

Figure 10.7. Image of polyurethane foam after application of an edge dislocation algorithm. (Reproduced with permission from ref. 39. Copyright 1989 John Wiley & Sons.)

curing temperatures, humidity, and various degradation and aging effects. The failure mode of an adhesive bond generally falls into one of two categories: adhesive and cohesive. A failure is adhesive in nature if it occurs along one of the interfaces between the adhesive and the substrate. A cohesive failure exists if it occurs entirely within one of the three components making up the bond system—the two substrates and the adhesive layer itself (*40, 41*).

The mechanical properties of adhesive-bonded systems depend on the nature, number, and distribution of defects such as voids, resin-rich areas, poor surface bonding sites, debonding sites, etc. Until now, it has not been possible to determine the relative role of these defects in the performance of adhesives, since no techniques have been available to monitor the nature, number, and distribution of defects and also allow mechanical testing of the evaluated samples (*42*). NMRI has the capability of measuring imperfections in adhesive systems while being noninvasive and nondestructive.

The principal advantage of NMRI is the viewing of the internal structure of the adhesive bond (*43, 44*) which yields the potential of determining the strength of the adhesive bond without destroying the bond by testing. The nondestructive nature of the NMRI experiment allows one to expose samples to environmental agents such as water and evaluate the same samples over different time periods. In addition, the NMR relaxation parameters can be used to characterize the mobility and state of cure of the adhesive system. Modulating either T_1 or T_2 relaxation times by varying the NMR pulse sequence allows differentiation between mobile and immobile regions of the adhesive bond (*45–49*).

The ultimate expression of "chemical-shift-resolved tomography" would be "four-dimensional" data that encodes chemical shift as an additional dimension along with the three spatial coordinates (*50, 51*).

Using NMRI, it may be possible to develop methods of assessing the informational content of NMR images of adhesive bonds by manufacturing known defects in bonded regions. It is necessary to determine the information content of the images in terms of the nature of the defect, that is, voids, porosity, and debondings. To establish the correlation between the image patterns and the nature of the defect, samples can be fabricated under conditions maximizing a particular type of defect. Voids arising from inadequate flow can be manufactured by reducing mold pressure and temperature and the corresponding flow between the adherents. Defects due to adhesive porosity can be obtained by including foaming agents in the resin that will decompose under the curing conditions, producing high-porosity regions. Debonding can be produced by protecting the adherent surface rather than activating it as is usually the case. In this manner, the contribution of the different defects to the image can be ascertained and utilized in evaluating the actual bonding processes. Furthermore, it will be possible to evaluate adhesive-bond formation variables by studying their role in property determination with materials qualification through NMRI.

NMRI of the Curing of Water-Based Adhesives

All adhesives exist in a liquid or liquidlike state before the final cure process. The liquid state, a requirement for good flow, leveling, and wetting, is typically achieved by dissolving or dispersing the actual binding substances in water or solvents. This liquid state is ideal for NMRI . In the formation of adhesive bonds, the adhesive goes from a viscous liquid layer in the glue line to a bonded solid layer by drying or cross-linking. Both of these processes result in a decrease in the volume of the adhesive. In water-based systems such as poly(vinyl acetate) (PVAc, as used in Elmer's glue), the solids content of the adhesive is on the order of 50–60%, so

Figure 10.8. Drying of a plasticized PVAc emulsion between two wood blocks. A trough in the wood block has a diameter of 11 mm and a depth of 3 mm. Adhesive less than 100 μm thick is visible in the joint in (a). The black space in the middle of the trough in (c) is caused by water absorbed from the adhesive into the wood. (a) 10 min; (b) 4 h; and (c) 16 h. (Reproduced with permission from ref 42. Copyright 1987 VSP B. V.)

the adhesive at set occupies only 50% of the available volume in the glue line. For adhesive bond integrity, one must know how this non-load-bearing volume or voids are distributed with respect to the adhesive joint. It would be desirable to know how the adhesive curing can be controlled so that the excluded volume of the adhesive does not disrupt the bond integrity of the adhesive joint.

Adhesives that have a solids content less than 100% have volatile substances such as water that leave voids in the adhesive, if no external pressure is applied to force their adherents together. Voids in the adhesive decrease the strength of the structure. Defects, voids, or damaged areas of the adhesive materials are clearly shown in the NMR image.

NMRI measurements have been made using the water-based PVAc adhesive system to join wood blocks (52). The changes occurring in adhesive during bond formation can be observed in Figure 10.8. The image in Figure 10.8a was taken 10 min after application of the adhesive. It shows a very thin glue line between the wood pieces and a 3-mm-deep, 11 mm-wide trough made in the wood piece. After drying for 4 h (Figure 10.8b), there is no sign of wet adhesive in the thin glue line, although the adhesive in the trough is still wet and mobile. Even after 16 h (Figure 10.8c), the PVAc emulsion in the trough has not dried totally, although most of the water has penetrated into the wood. The image in Figure 10.8c shows an empty space or void (the large black area) in the adhesive.

NMRI is a sensitive method for finding imperfections because the technique allows detection of even minor differences in physical and chemical properties of adhesives (53, 54). Figure 10.9a shows a homogeneous glue line of a PVAc emulsion No. 1 (solids content 50%, water content 50%, and viscosity 35 Pa s) between a wood block and a ceramic tile.

Initially, the intensity of the signal from the adhesive is uniform

Figure 10.9. (A) Proton image of a homogeneous glue line of PVAc emulsion No. 1 (solids content 50%, water content 50%, and viscosity 35 Pa s) between a wood block and a ceramic tile. The intensity of the signal from the adhesive is uniform throughout the glue line. Small empty spaces within the adhesive layer are due to the bottom pattern of the ceramic tile. (B) Proton image of a glue line made of emulsions No. 1 and No. 2 (solids content 60%, water content 40%, and viscosity 35 Pa s). These emulsions have the same viscosity but different solids contents. Because this image was acquired with a relatively long repetition time and a short echo time, the observed contrast is mainly from differences in proton spin densities, i.e., water concentration. The higher signal intensity corresponds to emulsion No. 1 and the weaker intensity to emulsion No. 2 (C). Proton image obtained from emulsion No. 1 and emulsion No. 3 (solids content 50%, water content 50%, and viscosity 2 Pa s), which differ in their viscosities. For this image, the stronger signals are due to emulsion No. 3. (Reproduced with permission from ref 53. Copyright 1989 VSP B. V.)

throughout the glue line. Small empty spaces within the adhesive layer are due to the bottom pattern of the ceramic tile. Figure 10.9b represents a glue line made of emulsions Nos. 1 and 2 (solids content 60%, water content 40%, and viscosity 35 Pa s). These emulsions have the same viscosity but different levels of solids. Because this image has been acquired with a relatively long repetition time and a short echo time, the observed contrast is mainly due to differences in proton spin densities, that is, water concentration. The higher signal intensity corresponds to emulsion No. 1, and the weaker intensity to emulsion No. 2. Figure 10.9c shows a comparison of images obtained from emulsions Nos. 1 and 3 (solids content 50%, water content 50%, and viscosity 2 Pa s), which differ in their viscosities. For this image, the stronger signals are due to emulsion No. 3. Hence, NMRI imaging can be used to detect differences in the physical properties of the different adhesives.

Voids that result from absorption or evaporation of water from water-based adhesives are abundant in thick joints. The location of voids can be very critical to final joint strength. NMRI has been used to study the structural variations found in single lap joints made of ceramic tiles and water-based styrene acrylate adhesives (55). Figure 10.10 shows a diagram of the experimental lap joining in the NMRI instrument and the interface.

The loss of water during curing of these adhesives creates voids that represent approximately 20% of the total joint volume. The location of the voids can be critical to the final joint strength. Figure 10.11 schematically depicts typical changes in the joint during the curing process.

Studies of the void formation of bonded systems have been made (55). To detect the voids, the bonded adhesive samples were immersed in water at 30 °C before the images were collected. Several samples were imaged, and these same samples were tested mechanically. Each image represents a 1-mm slice of the sample in the joint area, and successive images were taken 3 mm apart. Typical images are shown in Figure 10.12. The images show (by the indented regions at the interface) that setting of the adhesive caused the edges at the interface to shrink ca. 500 μm. This volume corresponds to about 5% of the total joint volume, which indicates that the voids are located internally in the joint. However, since water cannot penetrate these adhesives, the voids in the central area of the joint cannot be detected. The images in Figure 10.13 are shown for two samples with different compressive strengths. The image in Figure 10.13a shows a large void and a crack in the image. The image in Figure 10.13b was obtained at a position where a visible hole ca. 300 mm in diameter was observed.

These and other NMR images were correlated with wet strength measurements to show that void content and location of the voids are critical to strength. The major mechanism for water intrusion into the joints is penetration through voids and cracks. This is particularly fortunate, since cracks that reach the surface are the most damaging to compressive strength:

Environmental failure of the adhesive bonded joint of a metal can be caused by displacement of the adhesive (or coating–primer) from the adherent by ingress of moisture, which causes:

- weakening or rupture of secondary bonds,

- transformation of the oxide into a weaker type, and

- hydrolysis of the adhesive (or coating–primer) layer near the adherent surface.

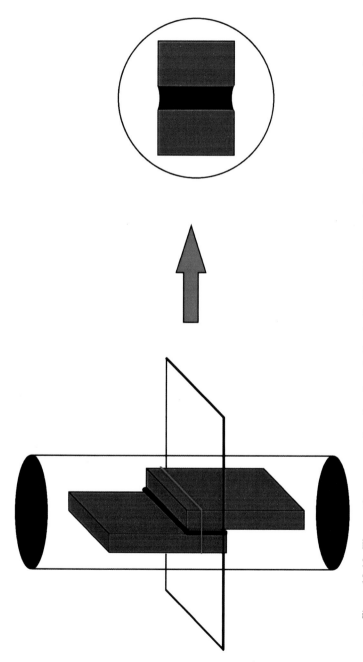

Figure 10.10. The sample arrangement and a slice selection in the NMRI instrument. (Reproduced with permission from ref 55. Copyright 1990 VSP B. V.)

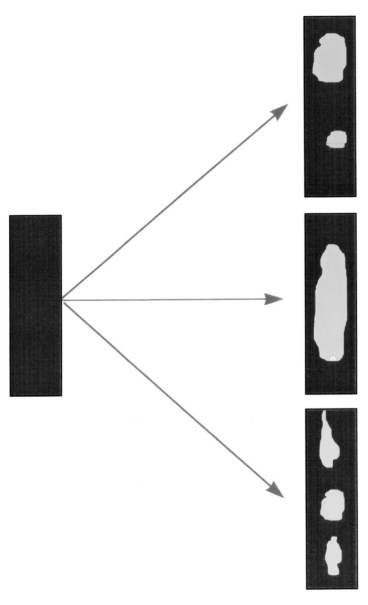

Figure 10.11. Schematic of typical void locations in a joint made of a water-based adhesive. (Reproduced with permission from ref 55. Copyright 1990 VSP B. V.)

Figure 10.12. Three images of a typical joint geometry after setting of the adhesive. Three slices were collected at different positions in the sample. (Reproduced with permission from ref 55. Copyright 1990 VSP B. V.)

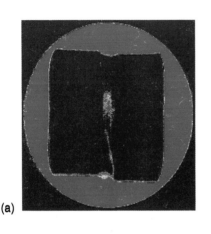

0.063 N/mm²

(a)

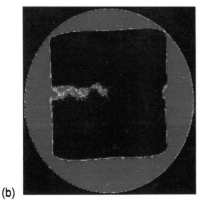

0.097 N/mm²

(b)

Figure 10.13. Two images of samples with lower compressive strengths than the samples in Figure 10.12. The image in (a) shows a large void and a crack. The image in (b) was obtained at a position where a visible hole was observable in the sample. (Reproduced with permission from ref 55. Copyright 1990 VSP B. V.)

Generally, in the absence of moisture, interfacial bonds are stable and do not fail under normal stresses on the bonded components. However, water has the ability to compete with the chemical functional groups on the adhesive molecules for interfacial bonding sites on the substrate surfaces. This weakens the adhesive bonds. In the worst case, water can completely displace the adhesive from the substrate surface. From a practical point of view, the stability of the interfacial adhesive bond is directly related to its resistance to the deteriorating effects of water. The successful use of long-lived adhesive systems requires the matching of the adhesive to the environment. More information is needed about the chemistry of the interfacial bonds formed, the importance of moisture adsorption, and the possible effects of field aging on equilibria involving competitive absorption and bond stability as a function of time.

NMRI of the Curing of Epoxy Adhesives on Metal Surfaces

The curing behavior of adhesive systems has been studied extensively by a number of surface techniques, and considerable insight has been obtained into the interfacial reactions of epoxy and metal substrates. However, all of these techniques are limited to thin films deposited on the metal surface. NMRI is a technique which allows the study of the in situ curing of, for example, an epoxy adhesive between metal substrates. NMRI can reveal the spatial aspects of cured adhesives in situ on metals that were subjected to different chemical treatments to enhance the adhesive strengths. NMRI images of glue lines of epoxy adhesive joints of aluminum–epoxy–aluminum have been observed (56) and voids are also detected. Figure 10.14 shows an example of such a void by the bright area in the center of the bonded Al system. There appears to be only one large void, suggesting that sufficient mobility is present during curing for the smaller voids to coalesce into a single large void.

To obtain a strong and durable bond, it has been found necessary for the aluminum substrates to go through an extensive and costly pretreatment procedure. Generally, aircraft manufacturers use a combination of etching (in a chromic–sulfuric acid solution) and chromic acid anodizing. These pretreatments are followed by applying a thermosetting epoxy primer to pretreat the adherents of adhesive-bonded aluminum joints. This pretreatment process results in a high initial strength and good long-term durability of the adhesive-bonded joints.

Modern surface pretreatment processes (cleaning, etching, anodizing, and priming) have the following effects on the adhesion and failure mechanism of adhesive-bonded aluminum joints:

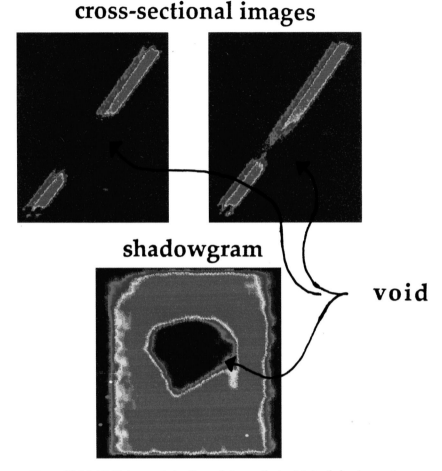

Figure 10.14. NMR image of glue lines of epoxy adhesive joints of aluminum–epoxy–aluminum. Voids are detected as the dark areas in the image. A (A, B) Interfacial region; and (C) shadow image, showing the large void in the center of the bonded sample. (Reproduced with permission from ref 48. Copyright 1989 Gordon & Breach.)

- improvement of the wetability and reactivity of the surface, making the formation of secondary bonds possible,

- creation of a surface structure that is suitable for mechanical interlocking,

- increase in the long-term durability of the joints by the creation of stable oxides and the use of corrosion-inhibiting primers, and

- decrease in the fatigue life of aluminum adherents due to the formation of a fatigue-sensitive oxide layer.

The surface pretreatments and primers can be divided into the following parts:

- cleaning methods (e.g., by hand with methanol, or by alkaline bath pretreatments),

- etching methods (e.g., etching in a chromic–sulfuric acid solution, or a sodium hydroxide etch),

- mechanical methods (e.g., brushing, grinding, and grit blasting),

- anodizing (e.g., in phosphoric acid, chromic acid, or sulfuric acid),

- adhesion promoters (e.g., silanes),

- water-based primers, and

- primers using inhibitors other than chromates.

Differences in the nature of the epoxy cure have been observed using NMRI to study the images of treated and untreated surfaces of Al. Figure 10.15 shows the cross-sectional images for the untreated aluminum-epoxy sample spatially segmented into uncured and cured regions. The red area

Figure 10.15. Cross-sectional images of an untreated aluminum–epoxy sample segmented into uncured and cured regions. The red area corresponds to uncured epoxy, and the green area corresponds to cured epoxy. The cure time was 0.6 h at 175 °C. The image has been magnified 2.5 times (pixel size 0.0016 cm²). (Reproduced with permission from ref 49. Copyright 1993 Gordon & Breach.)

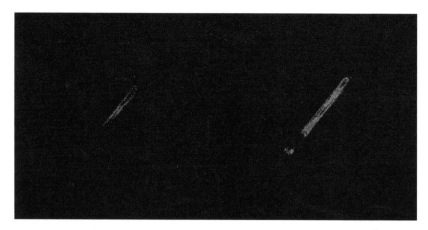

Figure 10.16. Cross-sectional NMR images of a CAA–epoxy sample. The red areas are uncured domains, and the green areas are cured domains. (Reproduced with permission from ref 49. Copyright 1993 Gordon & Breach.)

corresponds to uncured epoxy, and the green areas correspond to cured epoxy. The image, which has been magnified 2.5 times, shows the uncured epoxy segregates at the center of the sample and away from the surface.

Figure 10.16 shows a segmented image of the uncured and cured domains in a chromic acid anodized (CAA) aluminum–epoxy sample. In this case, it shows that most of the uncured epoxy (in red) is at the interface of the sample, although some uncured epoxy appears in the interior as well. Comparison of the images in Figures 10.15 and 10.16 shows the contrasting behavior between the treated and untreated Al surfaces. The uncured epoxy in joints of epoxy adhesive and anodized aluminum is at the surface, whereas for untreated aluminum joints it is away from the surface (center of the bond). Figure 10.17 shows a schematic of the epoxy structures in the interface for treated and untreated samples, where the dark regions are the uncured epoxy regions. These images suggest that the chemically treated surfaces influence the curing processes of the epoxy resin. The chemically treated surface yields less harmful voids and therefore stronger and more durable adhesive bonds.

Images can also be obtained through the bonded metal–epoxy–metal samples, thus showing a shadowgram or total projection of the samples. NMRI images of untreated aluminum–epoxy samples obtained with cures of 0.2, 0.4, 0.6, and 0.8 h at 175 °C are shown in Figure 10.18. The images display a random distribution of the cured and uncured domains, showing no chemical correlation between the neighbors during curing, leading to rough images. Contrasting behavior is observed for the anodized samples. As shown in Figure 10.19, the chromic acid anodized aluminum–epoxy samples show simply connected regions for the uncured domains. The

UNTREATED ALUMINUM ADHESIVE

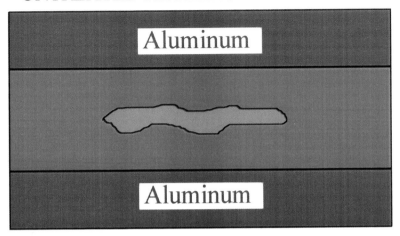

ANODIZED ALUMINUM ADHESIVE

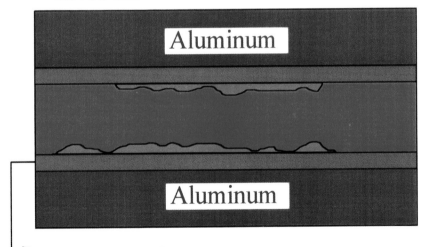

ANODIC OXIDE

Figure 10.17. Schematic of the epoxy structures in the interface for treated and untreated samples, where the dark regions are the uncured epoxy regions. (Reproduced with permission from ref 49. Copyright 1993 Gordon & Breach.)

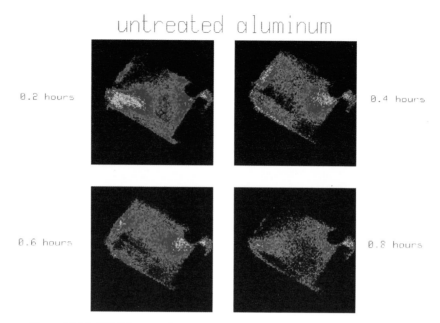

Figure 10.18. NMR images of untreated aluminum–epoxy samples obtained at 0.2, 0.4, 0.6, and 0.8 h cure at 175 °C. (Reproduced with permission from ref 49. Copyright 1993 Gordon & Breach.)

results can be interpreted as in Figure 10.20, where the diagrams show the expected behavior for random (left) and correlated polymerization (right). The trend for the data is a steady increase in kurtosis and skewness values as a function of cure time for the anodized samples. Contrasting behavior is seen for the untreated and chemically treated (Iridite-treated) aluminum samples, where the kurtosis as well as the skewness goes through a minimum. These minima are due to redistribution of signal intensities within the images.

Applications of NMRI Utilizing Changes in Molecular Mobility in Polymers

In contrast to the other spectroscopic mapping techniques such as infrared and Raman spectroscopies, NMRI generates information about the relative mobility of chains in a polymer sample through the contrast in the image due to contributions of T_2 and T_1. The T_2 relaxation time is sensitive to local motion of the nuclei. Generally, freely mobile molecules, with short correlation times, have long T_2 times, whereas motionally restricted or immobile molecules have long correlation times and short T_2

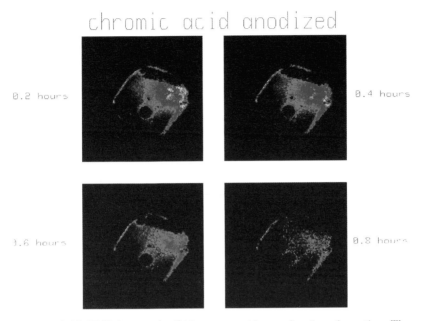

Figure 10.19. NMR images of a CAA–epoxy sample as a function of cure time. The red regions correspond to uncured domains, and the green regions correspond to cured domains. (Reproduced with permission from ref 49. Copyright 1993 Gordon & Breach.)

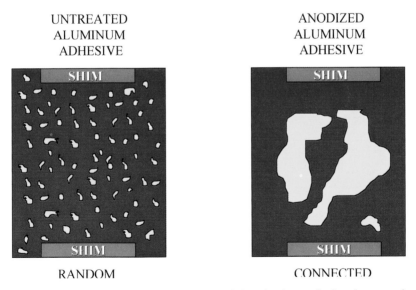

Figure 10.20. Schematic of cured and uncured domains in anodized and untreated aluminum–epoxy joints. Dark regions are uncured regions. (Reproduced with permission from ref 49. Copyright 1993 Gordon & Breach.)

times. Thus, NMRI is particularly useful for processes in which large changes in molecular mobility occur. Several examples of the utility of this unique aspect of NMRI are given in this section.

NMRI Studies of the Polymerization Process

A number of studies of the polymerization process have been made utilizing the NMRI technique. Polymerization leads to a characteristic change in local motion, which induces a marked drop in T_2 as the polymer is formed (56). It is possible to observe and quantify the rate of polymer propagation through a solution by measuring the velocity of the front (57). The reaction zone approximates a delta function, and the front propagation is easily detected.

In particular, NMRI has been used to examine the traveling waves of benzoyl peroxide initiated methacrylic acid polymerization (57). The polymerization of methacrylic acid is autocatalytic, as the exothermic free-radical polymerization is coupled to the thermal decomposition of the initiator, benzoyl peroxide. An autocatalytic reaction in an unstirred vessel supports a constant-velocity wave front resulting from the coupling of molecular diffusion to the chemical reaction. Figure 10.21 shows typical profile intensity (horizontal dimension) of a traveling wave as a function of time (vertical dimension). The sharp drop in intensity in the profiles marks the point at which the polymer front occurs. The image shows an initially uniform distribution of monomer solution of high signal intensity. The reaction front is represented by the discontinuity in signal intensity, which moves forward rapidly but then settles into a constant-velocity front that propagates through the rest of the methacrylic acid solution. The plot of the interfacial position is linear as a function of time and yields a measure of the front velocity. The front velocity is 0.61 cm/min with an uncertainty of 1%. NMRI was used to follow the highly anisotropic polymerization induced by a photoinitiator (58). Spatial variations are easily observed.

The curing of epoxy in a composite can be imaged. With curing, the T_2 decreases, and the spatial distribution of the degree of cure is reflected in the contrast in the image. Using a calibration obtained from the dependence of the epoxy T_2 and viscosity on temperature, a viscosity map of the sample can be obtained (59), which corresponds to the state of cure of the epoxy.

NMRI Studies of the Vulcanization Process

Useful rubber articles are made by vulcanization, as raw rubbers are generally not strong, do not maintain their shape after a large deformation,

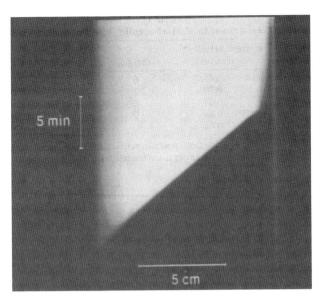

Figure 10.21. Image of the traveling wave of benzoyl peroxide initiated methacrylic acid polymerization. The typical profile intensity (horizontal dimension) is shown as a function of time (vertical dimension). The image shows an initially uniform distribution of monomer solution of high signal intensity. The reaction front is represented by the discontinuity in signal intensity, which moves forward rapidly but then settles into a constant-velocity front that propagates through the rest of the methacrylic acid solution. (Reproduced from ref 57. Copyright 1992 American Chemical Society.)

and can be sticky (raw rubber has about the same consistency as chewing gum). In order to produce a proper elastomer, the polymer chains must be very flexible and chemically cross-linked. *Vulcanization* can be defined as a chemical cross-linking process that generates a rubber material with useful properties.

In the rubber industry, a sulfur cross-linking system is generally used for vulcanization. Since the vulcanization rate with sulfur alone is slow, chemical accelerators and activators are necessary. The vulcanization process may occur via different reaction mechanisms, depending in particular on the chemicals–fillers used (60).

The exact mechanism under which accelerated sulfur vulcanization occurs changes as the class of accelerators and activators changes. However, a generally accepted sequence of reactions is as follows:

- An interaction of the curatives occurs to form what is termed the active sulfurating agent;

- The polymer chains interact with the sulfurating agent to form polysulfidic pendant groups terminated by accelerator groups;

- Polysulfidic cross-links are formed;

- Numerous parallel reactions occur, such as continuing cross–link formation, cross–link destruction (reversion) by thermal degradation, formation of inefficient cyclic cross–links (sulfides), and numerous other network-modifying reactions.

Typically, benzothiazole or sulfenamide accelerators are used. Zinc oxide is the activator, and a fatty acid such as stearic acid is the coactivator (61).

The elastomeric networks are difficult to characterize because of the low concentration of cross-links and the insolubility of the networks. However, high-resolution solid-state NMR has made considerable contributions to our understanding of the structure, particularly in terms of the chemical nature and concentration of the cross-links (62–69). An additional utility of high-resolution solid-state NMR is an evaluation of the relative mobility of the chains as influenced by the cross-links (70–73).

NMRI is particularly useful for the study of elastomer networks, as the line widths of the proton resonances are narrow as the polymer is well above its glass transition temperature. For elastomers, the proton–NMR line widths are not excessively broad, T_2 is typically on the order of 10 ms, and imaging is possible with reasonably fast gradient switching (~ 250 μs or less), which can easily be achieved with most NMR microimaging equipment. As a result of the narrow line width, the resolution of the images is high (down to a resolution of 20 μm). In addition, the molecular sensitivity of the mobility enhances the contrast of the images. Consequently, NMRI has been useful in the determination of internal inhomogeneities arising from inadequate mixing, gradients in crosslinking chemistry, filler distribution, and impurities (74–76).

The ordinary industrial accelerator systems generate a broad distribution of cross-link densities in rubber articles because of the poor mixing of the solid accelerator/sulfur recipe in the raw rubber (77). In particular, the variation in cross-link density leads to substantial differences in the coefficients of thermal expansion and high internal stresses in the rubber article. These internal stresses limit the ultimate performance of the article, whether in modulus or elongation.

Well-designed mixing protocols and carefully controlled cure cycles allow the development of a homogeneous distribution of the crosslinks in the final manufactured article. Mixing protocols can generate a well-dispersed accelerator system in the raw rubber. The NMR imaging technique can experimentally measure the homogeneity.

Solvent adsorption and swelling behavior have been used to determine the cross-link density in elastomeric systems(75–78). The basis of the method is that with a higher degree of cross-link density, the less solvent will be imbided into the system and the lower the degree of swell-

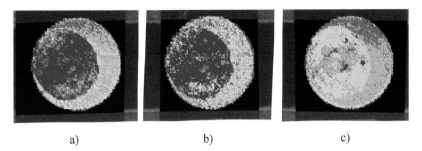

a)　　　　　　b)　　　　　　c)

Figure 10.22. Proton-NMR images of 1,4-dioxane in swollen plybutadiene rubber. The rubber is fully swollen. The inside diameter of the vial is approximately 18.5 mm. Slice thickness is approximately 2 mm. (a) T_e = 30 ms, (b) T_e = 70 ms; and (c) a computer T_2 image generated from the images in (a) and (b). (Reproduced with permission from ref 75. Copyright 1989 John Wiley & Sons.)

ing. NMRI imaging allows one to pursue this idea further by examining the homogeneity of the swelling process. The intensities of the mobile protons of the swelling agent probe the homogeneity and spatial distribution of the cross-links of the network system (75). Figure 10.22 shows proton images of 1,4-dioxane produced from a spin echo pulse sequence of a highly cross-linked sulfur-vulcanized rubber sample. The black spot in the image is an air bubble artifact. A three-level contour map is shown in Figure 10.23. This contour plot indicates that there is a benzene background; regions of an intermediate level of benzene indicate a moderate level of cross-linking; and regions of little benzene indicates a high level of cross-linking. This rubber sample obviously has considerable inhomo-

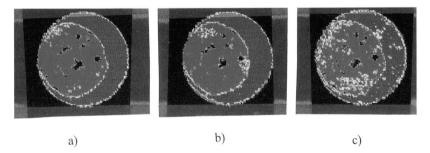

a)　　　　　　b)　　　　　　c)

Figure 10.23. Contour plots of the images presented in Figure 10.22. Images in Figure 10.22 first undergo an averaging process. Signal intensities and T_2 relaxation times are scaled from 0 to 256 for each figure, with 0 being the lowest value and 256 the highest value. (a) Contour of Figure 10.22a, where dark = 0–160, medium = 161–197, and light = 198–256, (b) contour of Figure 10.22 where dark = 0–105, medium = 106–177, and light = 178–256; and (c) contour of Figure 10.22, where dark = 0–95, medium = 96–165, and light = 166–256. (Reproduced with permission from ref 75. Copyright 1989 John Wiley & Sons.)

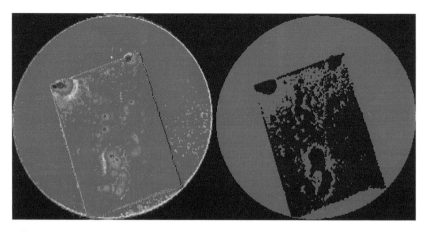

Figure 10.24. Proton image of a vulcanized butyl rubber sample cured for 25 min and swollen in cyclohexane. The right image is a two-color plot of the left image cut at 100 (on a 0–256 color scale). The sample was swollen in a 20-mm-diameter tube, and the resolution was 78 μm/pixel. (Reproduced from ref 77. Copyright 1991 American Chemical Society.)

geneity in cross-linking. Such inhomogeneities could arise from improper mixing, thermal gradients, or variations in vulcanization chemistry.

Samples of cured butyl rubber swollen in cyclohexane have been obtained that demonstrate poor mixing of the formulation (Figure 10.24). The image in Figure 10.24 shows regions of high cross-link density and entrapped air bubbles. This sample exhibits a broad range of cross-link density indicating a highly inhomogeneous sample.

NMRI of swollen elastomers can reveal the presence of thermal gradients resulting from the molding process (77). Figure 10.25 shows such an example. A T_1-weighted image of a butyl rubber sample cured for 10 minutes and swollen in cyclohexane is shown. The right-hand image is a two-color contour plot cut at 180 (on a 0–256 color scale).

The dark regions in the left-hand image are the result of highly cured elastomer present at the walls of the mold, where the temperature is highest. The two-color contour plot shows the further dispersion in the cross-link density of the sample (77).

NMR imaging has been successful in studying the dispersion of cross-link density in traditional elastomers, both along the chains as well as spatially (down to a resolution of 20 μm) (78). Because of the high mobility and narrow proton line width of swollen elastomer samples, images may be obtained directly from the rubber portion of the sample. Single and multiecho (T_2-resolved) images have been obtained of samples of *cis*-1,4-polybutadiene highly vulcanized with tetramethylthiuram disulfide (TMTD). In preparation for imaging, the samples were extracted to re-

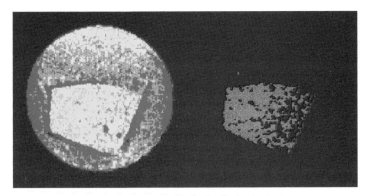

Figure 10.25. A T₁-weighted image of a butyl rubber sample cured for 10 min and swollen in cyclohexane. The right image is a two-color contour plot cut at 180 (on a 0–256 color scale). (Reproduced from ref 77. Copyright 1991 American Chemical Society.)

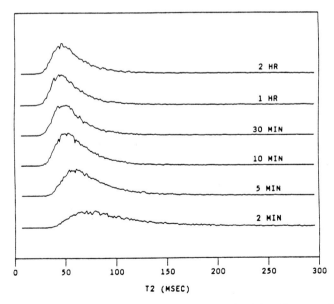

Figure 10.26. Histograms of T₂ values for TMTD-cured polybutadiene as a function of cure time. (Reproduced from ref 78. Copyright 1991 American Chemical Society.)

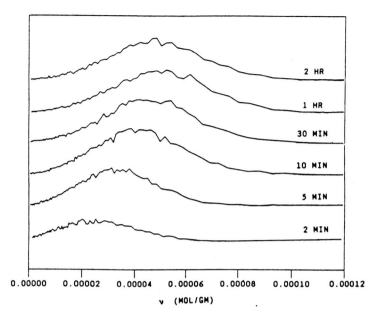

Figure 10.27. Number of voxels with local cross-link density values versus local cross-link density values for a cured polybutadiene at different cure times. (Reproduced from ref 78. Copyright 1991 American Chemical Society.)

move the non-network material and then swollen in deuterated cyclohexane. Consequently, NMR images of the swollen materials were obtained through spatial variations in the proton signal intensities of the cross-linked rubber. For T_2-weighted images, the signal intensity of each pixel characterizes the degree of segmental motion of its contents. Histograms can be obtained that yield the number of pixels with a particular T_2, as shown in Figure 10.26 (78). The relationship of the voxels to the cross-link density is shown in Figure 10.27.

It is also possible, using NMR imaging, to examine the homogeneities and degree of curing for different vulcanization formulations (79). NMRI has been carried out on vulcanized high-vinyl polybutadiene using different ratios of N-tert-butylbenzothiazolesulfenamide–sulfur. Figure 10.28 shows the proton images obtained by the spin echo technique for the swollen (cyclohexane) samples cured for 20 min at 150 °C for the conventional, semiefficient, and efficient formulations. The intensities of the mobile protons of the solvent probe the homogeneities and spatial distribution of the cross-linked network system. The nonuniformity of the signal indicates that the system under investigation has a nonhomogeneous cross-link distribution. The color scale used in the displayed images is from 0 to 255, corresponding to a spectrum from blue to red. The lighter

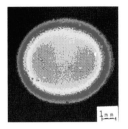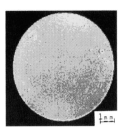

Figure 10.28. Images obtained of high-vinyl polybutadiene cured samples for different cure recipes with the same cure time. (Reproduced from ref 79. Copyright 1994 American Chemical Society.)

blue area represents the lowest signal intensity, while the red area indicates the high signal intensity. The differences in the state of cure are reflected in the color gradients in the samples.

It is also possible to obtain direct rubber proton images for this system by swelling it with a protonated solvent which yields no proton signal. The relatively high chain mobility and consequent narrow proton line widths of the swollen elastomers allow one to acquire the direct image of the rubber. Representative images are shown in Figure 10.29. Histograms of the number of pixels with a particular T_2 are shown in Figure 10.30 (74). These histograms clearly show the differences in the distribution of the cross-links for the different vulcanization recipes. The results are also consistent with observations made on the bulk rubber samples.

In quite a different approach in terms of interpretation, a study of the cross-linked structure of polystyrene has been reported (80). In this case, the image structure was interpreted in terms of the materials properties of the cross-linked polystyrene. This approach shows promise for the future.

Detection of Nonuniform Dispersion of Fillers

For improved performance of most polymers, an increase in the stiffness of the system is a common requirement. Fillers such as carbon, glass, and silica are often used for this purpose. Some interaction between the polymer and the filler particles is usually desirable. This is accomplished most easily if there is a large polymer–filler interface, and this interface is maximized if the particles are of colloidal dimensions. Spherical particles 1 μm in diameter have a specific surface area of 6 m^2/cm^3. This is the lower limit, and more often specific surface areas on the order of 300–400 m^2/cm^3 are achieved. It is generally perceived that, under tension, the weakly bonded matrix will pull away from the filler particles and filler-dewetting

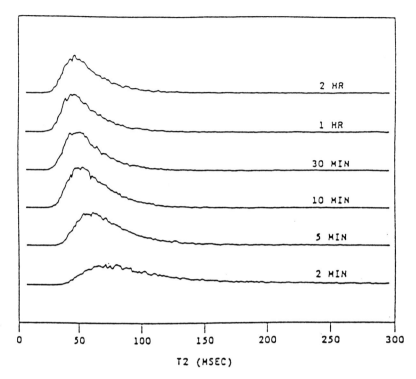

Figure 10.29. Histogram of the number of pixels with a particular T_2 value for the conventional, semiefficient, and efficient vulcanization images. (Reproduced with permission from ref 74. Copyright 1994 Huthig & Wepf Verlag.)

will occur. The stronger interfaces will remain intact to produce a higher level of stiffness of the system.

Filled systems are complex and heterogeneous; it is difficult to quantify their behavior in analytical terms. Particulate fillers affect the strength properties according to their amounts, sizes, shapes, size distribution, and interfacial bondings. In general, the modulus increases with increased filler loading. Larger particles cause greater stress concentrations in the matrix than do smaller particles. All other conditions being equal, smaller average particle sizes provide greater strengths. When the aspect ratio of the filler becomes greater, that is, when the filler is more asymmetric, the modulus will tend to increase.

The aspect ratio, a major parameter of the filler particles, is used to indicate their capability to provide reinforcement of the resin and develop high product strength. Defined as the ratio of the particle diameter to its thickness, the aspect ratio represents the degree of platelike structure to be found in the average particle. Particles having a low aspect ratio are nearly spherical. Thin, flat, sheet-shaped particles have a high aspect ratio

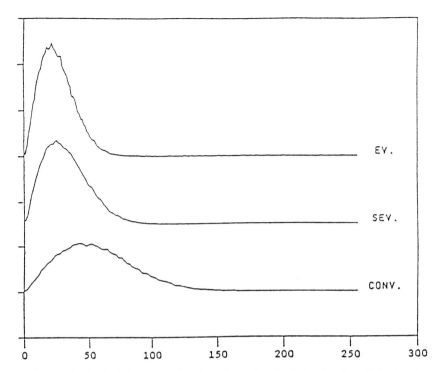

Figure 10.30. Pixel diagram as function of cure time for different recipes. (Reproduced from ref 79. Copyright 1994 American Chemical Society.)

and are always able to stack up against each other in mass numbers. The result of such a flat sheet-shaped structure is a material that has significantly higher flexural moduli, tensile yield strengths, Young's moduli, and heat deflection temperatures. High-aspect-ratio materials are more rigid and more heat resistant than products made from lower aspect ratios.

The improvement in mechanical properties by inorganic fillers is considerably reduced if there is a nonuniform dispersion of particles in the polymer matrix by formation of agglomerates. NMRI can produce visual pictures of the spatial variation of the organic phase distribution. This is accomplished by observing the proton images of the elastomers as a function of proton density and spin–spin T_2 relaxation times. These NMR parameters provide a measure of the molecular mobility, which in turn is related to the spatial variation of the polymer and the filler in the sample.

Differences in magnetic susceptibility between the polymer matrix and embedded particles broaden the line width; however, this problem can be minimized by using a variety of methods. Gradient echo imaging methods do not remove susceptibility effects, but spin echo imaging methods at least partially remove the susceptibility effects.

Samples of poly(dimethylsiloxane) (PDMS) that were reinforced by in situ precipitated silica have been examined by NMRI (*81–83*). The images were obtained with a spin echo technique by using a slice thickness of 500 μm and a digital resolution of 185 μm, and they required a time of 25.6 min. A dark rim is observed near the surface of the sample, which indicates a reduced mobility of the network chains compared to the sample core. This gray-level difference is ascribed to a high concentration of SiO_2 in this region leading to increased immobilization of the elastomer by adsorption on the surface.

NMRI has also been used for the study of carbon black distribution in tire composites (*84, 85*). A non-steel-belted, bias-ply tire was used, and the tread samples were cut from sections of the tire about 10 mm × 10 mm square and 4 mm thick. Figure 10.31 shows 10 contiguous transverse slices, each 200 μm thick, with an in-plane resolution of 100 μm. The tread is to the left and the interior portion of the tire is to the right. The tread layer at the left can be distinguished from the second region of higher intensity containing rigid fibers (which appear as dark spots in each slice). A large number of defects, including voids, aggregates of carbon black, and broken or misaligned fibers, can be observed in these images (*84*).

The carbon black dispersion can be studied by NMRI (84). The images of two finished tire tread sections—one with good and other with poor carbon black dispersion—are shown in Figure 10.32. The tread grooves are well-defined in both sections. The image in Figure 10.32B has a highly uneven appearance and consists of regions of poor dispersion of carbon black several hundred microns in diameter (*84*).

Physical Aging Process

NMRI has been used to study the physical aging of cross-linked natural rubber filled with carbon black (86). Physical aging results in a change in the molecular mobility of the polymer chains, and contrast is produced in the image which increases with aging. The nondestructive character of NMRI provides a method of monitoring various changes in the material properties of a single sample without destroying the sample during analysis, which is a drawback of the usual methods. A cylinder sample (5-mm diameter) of natural rubber filled with carbon black was used. The sample was removed after each measurement and aged for a predetermined period of time in a dry box at 130 °C. The samples were imaged using a conventional multiecho pulse sequence. The gradient strength was 250 mT/m, and the spatial resolution was 80 × 80 μm with a slice thickness of 1 mm. The images revealed air bubbles resulting from the molding process. When the sample was aged, inhomogeneities of varying sizes were

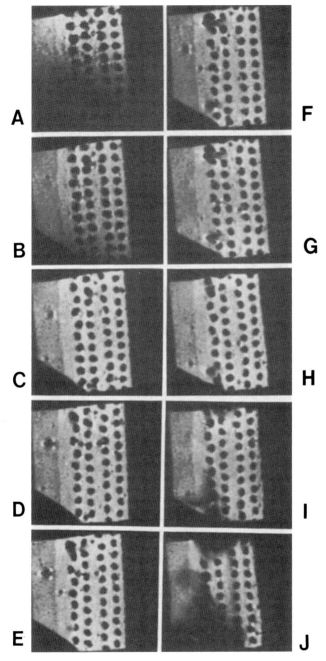

Figure 10.31. Images of 10 contiguous transverse slices of a tire, each 200 μm thick, with an in-plane resolution of 100 μm. The tread is to the left and the interior portion of the tire is to the right. (Reproduced from ref 84. Copyright 1992 American Chemical Society.)

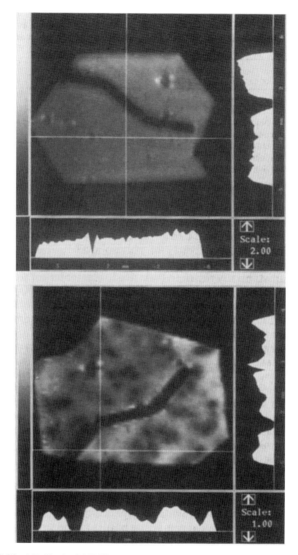

Figure 10.32. (A) *Typical NMR image of an experimental tire tread with good dispersion of carbon black, at 150 × 150 × 350 μm^3; and (B) NMR image of an experimental tire tread with poor carbon black dispersion. (Reproduced from ref 84. Copyright 1992 American Chemical Society.)*

observed as dark spots with bright shadows around them. The shadows arise from the difference in susceptibility of the inhomogeneities in comparison to the surrounding rubber. A ring in the aged surface layer was observed at the interface of the unaged material in the interior of the sample. This ring may arise from the presence of stabilizers such as stearin or paraffin, which diffuse to the reaction front.

The onset of aging in the natural rubber can be observed by NMRI after only 2 h. The thickness of the aged layer shows the asymptotic behavior expected for a radially protected film. If the aging reaction is modeled (for purposes of NMRI) as

$$U + O_2 \rightarrow A \tag{10.22}$$

where U is the soft rubber reacting with oxygen at elevated temperature, yielding a hardened, aged rubber, A. In NMR terms, the sample has only two possible internal states, that is, soft and hard. These two states have two different T_2 s: 5.8 ms for the unaged rubber and 0.3 ms for the aged portion. Since the relaxation times are different by an order of magnitude, the first echo results only from the unaged rubber U and can be used to determine the concentration. The concentration dependence of the unaged rubber U on the aging time, t_a, can be expressed as

$$U(t_a) = U_0 \exp(-kt_a) \tag{10.23}$$

Then the amplitude of the echo becomes proportional to the concentration of U. The inverse rate constant (k^{-1} is determined to be 8 h $\pm$ 30%). Additional NMRI studies have been reported for the degradation of rubber tubing and pipe (*87*).

Fiber–Matrix Composites: Interfaces and Interphases

NMRI can be used to probe interfacial and interphase structure and dynamics (88, 89). (The interphase region is the region intermediate between that of the interface and the bulk matrix.) Differences in the mobility of molecules in the interphase region can be observed in the T_2-weighted images (*90*).

The interphase region has been detected by the change in mobility of the epoxy polymerization (*91*). The NMRI sampling technique is shown in Figure 10.33. The fibers were braided, epoxy was placed in the cell, and polymerization occurred as shown in Figure 10.34. The dark ring surrounding the Kevlar fibers is the more highly cured fiber-surface-ini-

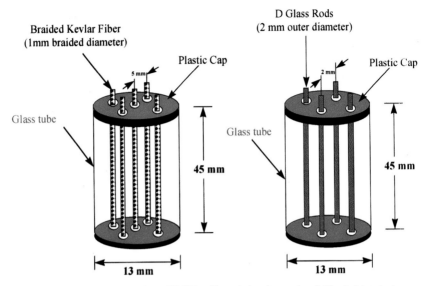

Figure 10.33. Diagram of an NMRI cell used for the study of fiber-initiated epoxy polymerization. (Reproduced with permission from ref 91. Copyright 1993 Gordon & Breach.)

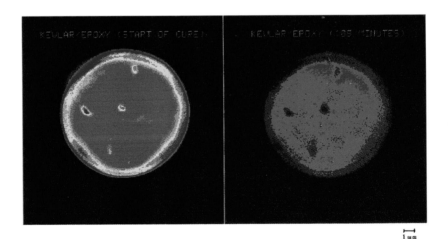

Figure 10.34. NMR image of Kevlar fibers that have been polymerized in the presence of epoxy. (Reproduced with permission from ref 91. Copyright 1993 Gordon & Breach.)

COMPOSITES PREPARED WITH HEAT-TREATED GLASS FIBERS

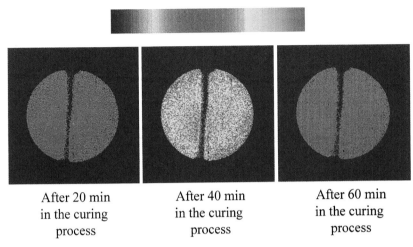

| After 20 min in the curing process | After 40 min in the curing process | After 60 min in the curing process |

Figure 10.35. NMR images of one heat-treated E-glass fiber in epoxy. Images are shown after 20 min in the curing process (left); 40 min in the curing process (middle); and 60 min in the curing process (right). (Reproduced with permission from ref 92. Copyright 1995 Gordon & Breach.)

tiated epoxy polymer, corresponding to the interphase region postulated (*91*).

Similar studies have been carried out for glass fibers (*92*). Figure 10.35 shows the gradient in the cure of the epoxy along the interface of the fiber as a function of cure time. The effect of the coupling agent can be seen by comparison with Figure 10.36, which shows the effects of the treated fiber on epoxy polymerization (*92*). The coupling agent reduces the effects of fiber surface on the curing of the epoxy.

Applications of NMRI Utilizing Spatial Motion of Molecules

Recent advances in technologies have produced a continuing and growing need for materials that have longer useful lifetimes than those now available and are more environmentally resistant. Polymeric materials are used in many engineering applications, such as packaging, where their function is protection of the contents from environmental chemical substances. Molecular migration or diffusion of chemical agents in these materials is the rate-determining mechanism limiting the useful lifetime of the con-

| COMPOSITES PREPARED WITH SILANE-TREATED GLASS FIBERS |

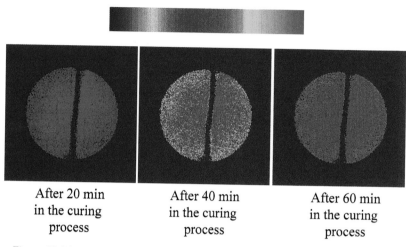

| After 20 min in the curing process | After 40 min in the curing process | After 60 min in the curing process |

Figure 10.36. NMR images of four heat-treated E-glass fibers with a silane coupling agent, gamma-aminopropyl silane(γ-APS) in epoxy. Images are shown after 20 min in the curing process (left); 40 min in the curing process (middle); and 60 min in the curing process (right). (Reproduced with permission from ref 92. Copyright 1995 Gordon & Breach.).

tents. Nearly all polymers of present commercial interest suffer from chemical decompositions arising from their exposure to harsh environmental conditions. There is a continuing need for methods of characterizing the rate at which substances diffuse into and out of polymeric materials. NMRI is a particularly powerful method for evaluating these diffusion processes (99–102).

Using the NMRI Method to Study Diffusion

NMR imaging techniques have been used for the study of sorption and diffusion as well as desorption of multiple chemical substances in polymeric materials (93,94). NMR imaging can directly provide diffusion coefficients as a characteristic quantity of a liquid component in a sample, making it possible to map molecular migration on a microscopic scale. NMR imaging also provides additional information on the microdynamic and structural properties of heterogeneous systems, such as subregion diameters, exchange times, and phase-boundary resistances.

The principal advantage of NMR imaging is the possibility of making spatially localized diffusion measurements nondestructively, noninvasively, and continuously. One can examine, by NMR imaging, the concentration and location of a permeating liquid in a solid sample. A diffusion parameter image can be calculated such that the diffusion coefficients are encoded into an intensity scale. The diffusion coefficients are quantitatively evaluated from a series of images recorded with different gradient field strengths (*103*). Analysis involves the simulation of the effects of diffusion using the dynamic magnetization equations to calculate, for each pixel, the magnetization, which ultimately yields an image whose intensities represent the spatially resolved diffusion coefficients. Finally, a true diffusion constant image can be obtained where the calculated diffusion coefficients are encoded into an intensity scale (*103*). In this scale, high intensities correspond to fast diffusion. In this manner, the spatial diffusion of a liquid into a solid material can be characterized in a quantitative fashion (*107*).

One of the obvious advantages of NMR imaging for the study of diffusion is the visual presentation of the data in the form of images. Such a presentation allows one to directly view the concentration and location of the penetrant and ignore extraneous factors influencing the diffusion. Other advantages of NMR imaging are that it allows (1) the study of samples of virtually any shape, and (2) the detection of any initial imperfections, such as voids or cracks, in the sample being studied. It is generally difficult to interpret liquid sorption measurements in solids because the samples being examined are not perfect, that is, they initially contain cracks and voids, which increase both diffusion and uptake of the liquid. Also, adsorption can induce volumetric changes which, though small, can cause microcracking or void formation. For example, Turner calculated, from changes in density (*95*) in high-molecular-weight poly(methyl methacrylate) (PMMA), that the increase in volume accounted for only half the volume change due to the uptake of water. The discrepancy was attributed to accommodation of about 50% of the water uptake in microvoids (*95*).

The NMR imaging technique also allows the system to be studied dynamically, since measurements can be made on the solid sample immersed in the penetrant. The measurements are rapid. Using the FLASH pulse sequence techniques (*96,97*), an image can be obtained in a few minutes. In this fashion, it is possible to study the dynamics of the diffusion process. Of course, the sample–penetrant system can be studied under isothermal conditions.

Finally, a primary advantage of NMR imaging is the fact that all of the NMR parameters of the sample can be measured and used to interpret the diffusion or sorption process. Images obtained utilizing different pulse sequences and interrupt times can be used to calculate the spin-lattice T_1 and spin–spin T_2 as well as the spin density. These additional parameters relate to the bonding and environment of the penetrant in the polymer system. These types of measurements have been useful in understanding the morphological changes that are observed (98).

The images reflect the spatial distribution and concentration of the substances. Fick's law can be written:

$$J = -Dx \qquad (10.24)$$

where J is the rate of diffusion, D is the rate of change of the concentration of the diffusing substance, and x is the thickness of the plane across which diffusion occurs. Eq. 10.24 becomes, in a time-dependent form, Fick's second law:

$$\frac{\delta C}{\delta t} = D \frac{\delta^2 C}{\partial \chi^2} \qquad (10.25)$$

where C is concentration and t is time.

In the ideal case of a semi-infinite sheet, and assuming that D is independent of concentration, the solution is given by

$$C = C_0 \operatorname{erfc} \frac{\chi}{2\sqrt{(Dt)}} \qquad (10.26)$$

where

$$\operatorname{erfc}(\chi) = 1 - \frac{2}{\pi^2} \int_0^\pi \exp - \eta^2 d\eta \qquad (10.27)$$

where η is viscosity.

Thus, the concentration profiles of solvents diffusing into plane sheets of polymer should be smooth curves governed by the mathematical function erfc x.

However, the diffusion coefficient for polymer–solvent systems displaying Fickian behavior is concentration dependent, usually having the form:

$$D = D_0 \exp(kC) \qquad (10.28)$$

A Boltzmann transformation can be used in this case to give:

$$\left(\frac{\partial x}{\partial t^c}\right)_c = f(D,C)^{1/2} \qquad (10.29)$$

Therefore, for a given value of C/C_0 the distance diffused is proportional to the square root of time, irrespective of the form of the concentration dependence of the diffusion coefficient. The time dependence of the distance diffused can be expressed by

$$\text{distance} = (\text{time})\log \text{distance} = n\log(\text{time}) \qquad (10.30)$$

where n theoretically lies between 0.5 for pure Fickian diffusion and 1.0 for pure case II diffusion.

Determination of the Mode of Diffusion

Diffusion in polymers can be categorized into three general types: Fickian, anomalous, and case II. Fickian diffusion describes the relationship between the mass flux of a penetrant and the concentration gradient present, and it can be characterized by an exponential decay in concentration with penetration into a material. The main feature of case II diffusion is the constant concentration front throughout the imbibed region, whereas anomalous diffusion falls between the two extremes. Fickian diffusion often occurs in polymers that are at temperatures above their glass-transition temperatures. This is because the polymer chains are in the rubbery state and possess sufficient mobility in which to allow solvent penetration. Fickian diffusion also commonly occurs when the activity of the solvent is sufficiently low and the diffusion therefore occurs only in the free volume of the polymer. Case II and anomalous diffusion are found primarily in polymers that are below their glass-transition temperatures. At these lower temperatures, the polymer chains are not sufficiently mobile to immediately accommodate the solvent. Therefore, the polymer dynamics become important for the transport of small molecules through a matrix.

NMR imaging can be used to determine the mode of diffusion (*103–106*). One example has been used the study of methanol diffusing into PMMA (*103*). Figure 10.37 shows the image from the PMMA immersed in methanol initially and after 48 h. The diffusion coefficient can be calculated by measuring the thickness of the sorbed layer as a function of time. With data processing techniques, it is possible to simplify the measurements by giving the images a three-level gray scale and then drawing a profile across the sample, as shown in Figure 10.38. The diffusion results are shown in Figure 10.39, where the thickness of the sorbed layer is plotted versus time. The linearity of this plot with time confirms that case II diffusion is occurring. The nonzero intercept at time zero is indicative of an initial Fickian diffusion process followed by case II diffusion. The constant level of methanol (as indicated by the uniformity of color in the sorbed layer) in the penetrant front is also indicative of case II diffusion.

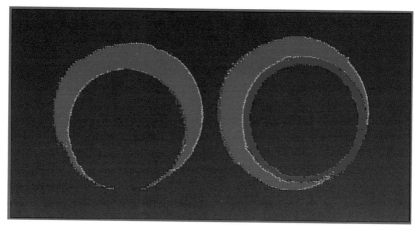

Figure 10.37. Proton-NMR images of a 30-mm PMMA sphere initially submersed in methanol (left); and after 48-h exposure to methanol at 30 °C. (Reproduced from ref 103. Copyright 1990 American Chemical Society.)

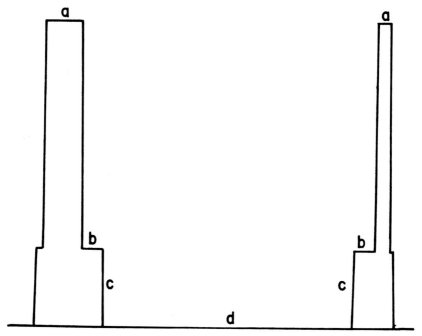

Figure 10.38. A three-level gray scale of the image in Figure 10.37 (right), obtained by drawing a profile across the sample. (a) Solvent; (b) imbibed region; (c) concentration of methanol in the imbibed region; and (d) core region of the PMMA rod. (Reproduced with permission from ref 100. Copyright 1989 John Wiley & Sons.)

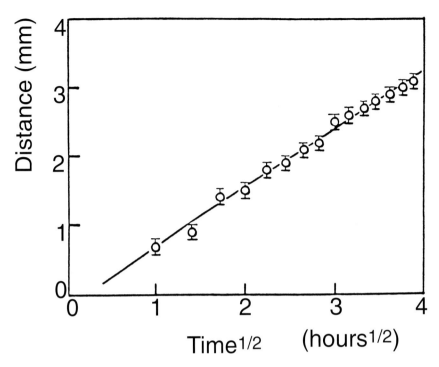

Figure 10.39. Plot of thickness of the layer of methanol in PMMA versus time. The linearity of this plot with time confirms that case II diffusion is occurring. The nonzero intercept at time zero indicates an initial Fickian diffusion process followed by case II diffusion. (Reproduced from ref 103. Copyright 1990 American Chemical Society.)

Molecular Mobility of the Penetrant

NMR relaxation parameters are useful probes of molecular mobility of the penetrant in polymers. Each correlation time represents the average value of the penetrating system with some distribution around that average. NMR imaging permits the determination of the spatial distribution of NMR relaxation times. This distribution provides information concerning the local motions of the system. When the polymer is partially swollen with solvent, the spatial distributions of relaxation times reveal the existence of interactions between the solvent and the polymer in the diffusion process (*107,108*).

Figure 10.40 shows a series of images of methanol in PMMA at different echo times from a multiecho experiment. The intensities are represented here by colors, with red being the highest intensity and blue being the lowest. The T_2 attenuation of the intensity is more apparent for

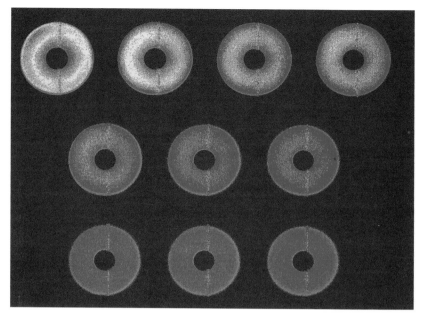

Figure 10.40. A series of images of methanol in PMMA at different echo times from the multiecho experiment. Echo times (from upper left to lower right): (A) 5.25 ms; (B) 10.50 ms; (C) 15.75 ms; (D) 21.00 ms; (E) 26.25 ms; (F) 31.50 ms; (G) 36.75 ms; and (H) 42.00 ms. (Reproduced from ref 103. Copyright 1990 American Chemical Society.)

the longer echo times. Figure 10.41 is the calculated T_2 image from the images in Figure 10.40.

In this case red represents longer T_2 s and blue the shorter T_2 s. The gray-level profile of the image of Figure 10.41 shows an almost linear decrease of T_2 s to the center core. Since the volume available to the methanol molecules is the same from the surface to the core, the relaxation times should be constant throughout the region if no other interactions exist. The change in T_2 from the surface to the core indicates that some physical effect other than the presence of free volume is affecting the mobility of the solvent. The polymer core effectively fixes the position of the chains at the interface between the swollen and the glassy regions of the polymer. The polymer chain dynamics change from the usual anisotropic motions with a variety of frequencies and amplitudes to anisotropic motions, which depend on the position along the polymer chain. Frequencies and amplitudes decrease toward the fixed point of the chain, that is, the glassy core. These changes in polymer motion influence the motion of the solvent in PMMA. The result is a change in the T_2 s with distance from the glassy core.

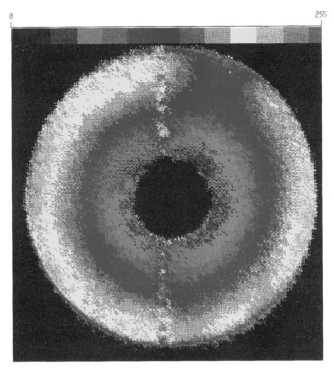

Figure 10.41. T_2 image calculated from the images in Figure 40. The color scale at the top of the image represents T_2 s ranging from 7 ms (red) to 24 ms(dark blue). Reproduced from ref 103. Copyright 1990 American Chemical Society.)

Self-Diffusion Coefficient Images

Self-diffusion coefficient images can be generated using the magnitude images. The images are fitted to the equation:

$$I = I_0 \exp(d/B) \tag{10.31}$$

where I is intensity, I_0 is intensity at the start, and d is the gradient attenuation factor, d. This generates an image based on a dummy variable: $B = 1/D$. The inverse of the image is taken to generate the self-diffusion coefficient image based on D. Figure 10.42 is a series of images of methanol in PMMA taken at different gradient strengths (*103*). Each successive image is the result of a more intense motion probing gradient. The first image has a more intense signal on the outer regions of the rod, with intensity decreasing toward the core. Figure 10.43 is the calculated self-diffusion image generated from the images in Figure 10.42 (*103*). The region within 100 μm of the glassy core exhibits a self-diffusion coefficient

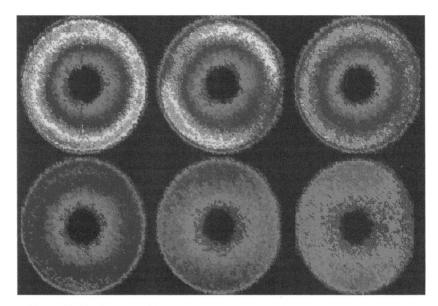

Figure 10.42. A series of images of methanol in PMMA taken at different gradient strengths (from upper left to lower right): (A) 0 G/cm; (B) 13.9G/cm; (C) 27.6 G/ cm; (D) 41.4 G/ cm; (E) 55.1 G/cm; and (F) 68.8 G/cm. Each successive image is the result of a more intense motion probing gradient. The first image has a more intense signal on the outer region of the rod with intensity decreasing toward the core. (Reproduced from ref 103. Copyright 1990 American Chemical Society.)

of $3.2 \pm 0.9 \times 10^{-7}$ cm^2 s^{-1}. The outer region of the swollen polymer exhibits a self-diffusion coefficient of $9.2 \pm 0.9 \times 10^{-7}$ cm^2 s^{-1}. Acetone swells PMMA to a greater extent than methanol, and the self-diffusion coefficients of the system are about two orders of magnitude greater than those of the methanol–PMMA system (*103*). This is apparently due to the increased volume available to the acetone molecules. The self-diffusion coefficients decrease by 35% from equilibrium in the outer regions to the region near the glassy core. The decreasing motions of the polymer chains as the core is approached reduce the solvent mobility as reflected in the self-diffusion coefficients.

NMRI Studies of Desorptions

Desorption is one diffusion process that has been given little attention, primarily because of the lack of adequate analytical techniques. Desorption measurements above the glass-transition temperature, T_g, of a swollen polymer are expected to follow Fickian characteristics. Likewise, a polymer swollen so that T_g is below the experimental temperature initially exhibits

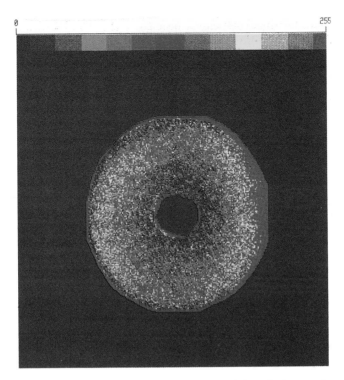

Figure 10.43. Calculated self-diffusion image generated from the images of methanol in PMMA shown in Figure 10.42. The region within 100 μm of the glassy core exhibits a self-diffusion coefficient of 3.2 ± 0.9 × 10^{-7} cm² s⁻¹. The outer region of the swollen polymer exhibits a self-diffusion coefficient of 9.2 ± 0.9 × 10^{-7} cm² s⁻¹. Reproduced from ref 103. Copyright 1990 American Chemical Society.)

Fickian desorption. The solvent is thought to desorb rapidly from the surface of the polymer and raise T_g of the surface layer. After the surface T_g is above the experimental temperature, the desorption process slows, and the process is controlled by the diffusion through the glassy surface layer. NMR imaging provides the spatial distribution of solvent in the polymer as well as the spatial distribution of the rate of desorption (*109*).

The desorption process can be related to T_d, which is the inverse of the rate of net solvent loss for a given pixel, through the equation:

$$M = M_0 \exp(-\exp T/T_d) \qquad (10.32)$$

where M is concentration at time T and M_0 is the initial concentration. A nonlinear least-squares fit of the experimental data is used to calculate a T_d image on a pixel-by-pixel basis (*109*).

Images of the desorption of methanol from swollen rods of PMMA

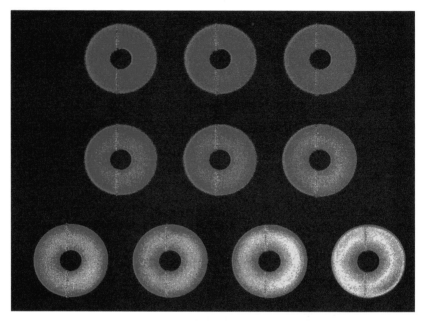

Figure 10.44. A representative set of spin echo images acquired after different exposure times to cyclohexane (from upper left to lower right): (A) 6 min;(B) 10 h; (C) 20 h; (D) 30 h; (E) 40 h; (F) 50 h; (G) 60 h; (H) 70 h; (I) 80 h; and (J) 90 h. The signal intensities decrease with time. The decrease in the maximum intensity from the first image to the last is 50% over the 90-h interval. (Reproduced from ref 103. Copyright 1990 American Chemical Society.)

have been obtained (*109*). The initial methanol volume fraction was 0.26. The rods were then placed in fully deuterated cyclohexane for the desorption experiment. The first image acquisition began 6 min after initial submersion. Images were collected in 1-h increments over a 104-h period. A portion of the series of images acquired after submersion in cyclohexane is shown in Figure 10.44.

In these images, red represents the highest intensity and blue the lowest, with the various shades of color representing the intermediate intensity levels. The signal intensity decreases with time, as seen from a comparison of the two images obtained 100 h apart. The maximum intensity of the latter image is only 50% of that of the initial image. The diameter of the rod decreases by 816 ± 68 mm over the 100-h period as measured from the images.

A T_d image calculated from 20 images taken at 5-h intervals over 100 h is shown in Figure 10.45. In this image, red represents the largest T_d and blue the smallest T_d, which correspond to the slowest and fastest intensity decreases, respectively. This image shows that the faster intensity decreases are near the surface of the rod, while the slower intensity de-

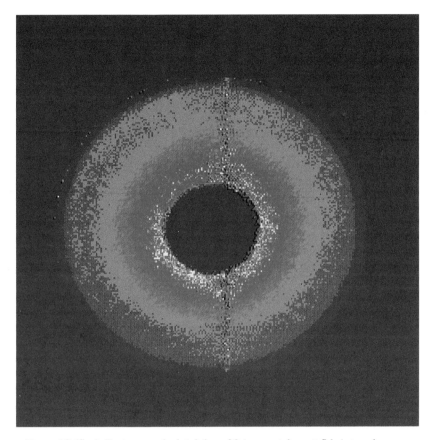

Figure 10.45. A T_d image calculated from 20 images taken at 5-h intervals over a period of 100 h. This image shows that the faster intensity decreases are near the surface of the rod and the slower intensity decreases are near the glassy PMMA core. (Reproduced from ref 103. Copyright 1990 American Chemical Society.)

creases are near the glassy PMMA core. The value of T_d at the surface is 58 h, and the value of T_d at the glassy core is 450 h as determined from this image. These values agree with the Fickian characteristics and indicate that imbibed solvent near the surface desorbs quickly, with the desorption rate decreasing toward the sample core. However, for this system there is no evidence of a glassy skin developing on the polymer surface (*109*).

NMRI Studies of Multicomponent Diffusion

Most of the analytical methods for studying diffusion (with the exception of FTIR) have little capability for differentiating between two or more

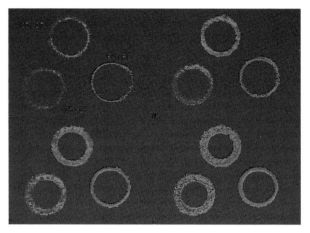

Figure 10.46. Images of PC in acetone with perdeuterated methanol mixtures of 75:25, 65:35, and 50:50 v/o AC:MeOH-d_4 after 2.25, 10.5, 22.25, and 30.33 h. (Reproduced from ref 110. Copyright 1992 American Chemical Society.)

penetrants. However, there are three different approaches to studying multicomponent diffusion using NMRI:

- Measurements at different chemical shifts
- Measurements using differences in NMR relaxation times
- Measurements on systems with isotopic labels.

Multicomponent diffusion NMRI measurements have been made on the basis of isotopic labeling method (*110*). In this method, only one component generates a proton signal, as the other component is deuterated. Figure 10.46 shows images for polycarbonate (PC) in acetone with perdeuterated methanol (AC:MeOH-d_4) mixtures of 75:25, 65:35 and 50:50 v/o AC:MeOH-d_4. The images show the movement of the acetone diffusion front within the PC rod after 2.25, 10.5, 22.25, and 30.33 h. The images include only the imbibed portion of the PC rods (shown in green) and the unexposed glassy core (shown in blue). The contributions of the free solvent mixtures surrounding the PC rods in the images were removed by computer processing. This was done to reduce the problems that occur when observing the diffusion fronts caused by the large dynamic range of intensities between the free solvent and the imbibed solvent. As can be seen in these images, the 75:25 AC:MeOH-d_4 mixture diffuses most rapidly, while diffusion of the 50:50 AC:MeOH-d_4 mixture is the slowest.

Perdeuterated acetone:hydroxydeuterated methanol (AC-d_6:MeOD) mixtures of 75:25, 65:35, and 50:50 v/o AC-d_6:MeOD were also studied. The images were obtained from the signal collected from the methyl protons of methanol after 1.55, 7.55, 14.88, and 27.63 h. These images include only the imbibed portion of the PC rods and the unexposed glassy core. The solvent mixtures surrounding the PC rods were again removed. The signal-to-noise ratio is lower for the AC-d_6:MeOD mixtures than the AC:MeOH-d_4 mixtures because there are twice as many hydrogen nuclei in the acetone of AC:MeOH-d_4 as in the methyl group of methanol in AC-d_6:MeOD. From these two sets of images, one can monitor the solvent-front movement of acetone and methanol into the PC rods (*110*).

The diffusion into PC increases as the acetone content of the mixture increases. In addition, there is a linear dependence between the solvent-front movement and the square root of time. This indicates that the diffusion of AC:MeOH-d_4 into PC is Fickian (*110*). This result was anticipated, since the diffusion of both pure methanol and pure acetone has previously been determined to be Fickian (*111*). (Note, however, that the diffusion of pure acetone was found to be Fickian after an initial period in which the diffusion was reported to be anomalous.) This may seem contrary to what is usually found for the diffusion of solvent into a glassy polymer that is below its glass-transition temperature, T_g. Although T_g of dry PC is 149 °C, acetone and methanol diffuse into PC in a Fickian manner because T_g is reduced to approximately -9 °C by the solvent ingress. In this way, the plasticized PC is able to relax and accommodate the solvent almost immediately, allowing solvent movement, and is not inhibited by the rate of polymer relaxation.

The solvent front was also monitored for the AC-d_6:MeOD mixtures. For the mixtures of 75:25, 65:35, and 50:50 AC-d_6:MeOD into PC, similar behavior was observed when the methyl resonance of methanol was monitored. There is an increase in the rate of the front movement as the acetone content increases, as well as a linear dependence of the front movement with the square root of time. This linear dependence indicates Fickian diffusion.

A question is whether the solvent fronts move at the same or different rates for the same acetone-to-methanol ratio. Comparison of the rates of the solvent-front movements of the 65:35 AC:MeOH-d_4 and the 65:35 AC-d_6:MeOD mixtures revealed that they were equal within experimental error. There is an overlapping of the front movement when plotted as a function of the square root of time. This result indicates that acetone and methanol are diffusing jointly through the PC rod in a Fickian manner and do not appear to separate. Similar behavior is also found in comparing the 50:50 AC:MeOH-d_4 and the 50:50 AC-d_6:MeOD mixtures.

It may be proposed that this joint diffusion through PC is the result of the formation of a weak "complex," possibly a weak hydrogen bonding interaction between acetone and methanol (*110*). The 65:35 AC:MeOH

mixture has approximately a 1:1 molar ratio; all the acetone may thereby
be complexed with methanol. This would account for the overlap of the
front-movement plots within error. In the 50:50 AC:MeOH mixture, the
molar ratio is approximately 35:65 AC:MeOH. If all the acetone com-
plexes with methanol, that leaves approximately equal amounts of AC:
MeOH complex and free methanol, which would diffuse together. As
shown by the 65:35 AC:MeOH mixture diffusing at a greater rate than the
50:50 AC:MeOH mixture, the "complex" may have a greater interaction
with PC than the free methanol would in the 50:50 mixture. The complex
may be able to reduce T_g further than the free methanol, allowing an
increased rate of penetration of solvent. The 75:25 AC:MeOH-d_4 and 75:25
AC-d_6:MeOD mixtures start to show a slight deviation from each other in
the solvent-front movements through the PC rods as the diffusion time
increases. This may be explained by the fact that the 75:25 composition
corresponds to a molar ratio of 62:38 AC:MeOH, so a portion of the
acetone exists that has not "complexed" with methanol, allowing "free"
acetone to proceed first into the PC rod. It would then be expected that
the acetone would diffuse at a slightly faster rate than the methanol that
is "complexed."

By plotting the front movements as a function of the square root of
time for the AC:MeOH mixtures, the translational diffusion coefficients,
D_{amp}, can then be determined as being equal to the square of the slope
of the resulting plots. The slopes of the resulting lines in the 65:35
mixtures are nearly equal. It was therefore not unexpected that the 65:35
AC:MeOH mixtures have the same translational diffusion coefficients
within experimental error. The same was found for the 50:50 AC:MeOH
mixtures, which also have corresponding diffusion coefficients. A difference in the diffusion coefficients begins to be seen with the 75:25 AC:
MeOH mixtures. This can be explained by the fact that the acetone diffuses at a greater rate than the complexed methanol.

NMRI Studies of Absorption–Desorption Cycling

We have studied the absorption–desorption cycling of water and methanol
into and out of the PMMA rods to determine its effect on diffusion characteristics (112). The effects of cycling on water diffusion were measured
for 0.6, 0.75, and 1.0 wt% water contents in PMMA. It was observed that
there is an increase in the rate of weight gain with cycling. The increase
in the rate of weight gain becomes even more evident as the initial water
content in PMMA increases. This may be caused by the increased plasticization of PMMA by the water for higher water contents followed by
higher methanol contents. This may have created increased porosity
within the PMMA as the solvent contents were increased.

A similar effect can be seen for the cyclic diffusion of methanol into water-soaked PMMA with water contents of 0.6 and 0.75 wt%. The percent weight gain versus time of diffusion is linear with time except for an initial lag time, which is anomalous. The lag time may be attributed to methanol desorbing a thin layer of water from the PMMA, which causes the layer to be in tension from the shrinkage back to the unswollen state. This decreases the mobility of the polymer in this layer and makes the polymer less accommodating to methanol diffusion (*113*).

For polymers, which are in the glassy state, the removal of diluents or small molecules may cause an increase in sorption and mass transport properties of the polymer (*114*). It has been found that the removal of liquids from a polymer below its T_g can increase the porosity within the polymer (*114*). This would result in an increase in the rate of diffusion of diluents such as water and MeOD into PMMA. It should be noted that there was no reduction in dry weight with each successive absorption–desorption cycle of water and MeOD, which would indicate initially that unreacted monomer and other small molecules were not leaching out. It would also indicate that the initial drying was sufficient to remove unreacted monomers and other small molecules from the PMMA rods, and none were further released by the absorption–desorption of methanol and water.

In a sense, this conditioning of the PMMA rods allows the uptake of increased amounts of solvent with each successive cycle (*112*). These increased uptakes may be attributed to reduced packing or lack of reconsolidation, when the solvent is removed from the swollen polymer in the glassy state (*114, 115*). It may be suggested that an increase in uptake is due to an increase in free volume (*115*), although an increase in dry volume was not observed. It is possible that not all the methanol was removed in the drying process, and there may have been some loss of other low-molecular-weight materials such as unreacted monomer from the PMMA sample. It has been well established that it is difficult to remove solvents or monomers from polymers (*116*). PMMA rods were kept in a vacuum at 60 °C until no further weight loss was apparent.

It has been shown in previous experiments (*114*) that when PMMA is exposed to solvents such as MeOH, unreacted monomer present within the material will leach out. It is expected that after the polymer is dried to remove the solvent that has diffused into PMMA, the zero weight of the sample should decrease. However, the zero weight of the samples analyzed in this experiment remained the same, indicating that not all of the methanol that diffused into the PMMA is removed during the drying periods between the absorption cycles.

In addition to using weight gain measurements to describe the effects of cycling on the diffusion characteristics of water and methanol in PMMA, NMRI was used to observe the solvent-front movements within the

PMMA rods. Proton–NMR images were obtained of the diffusion of MeOD into three different PMMA samples. These three samples had previously been subjected to 0, 1, and 2 cycles of the absorption and desorption of MeOD. The images show that the diffusion of MeOD increases with an increase in the number of cycles. A linear dependence of the solvent-front movement with time can be seen for the PMMA samples subjected to 0 and 1 cycles. This indicates that case II diffusion is occurring after an initial time period of anomalous diffusion. However, the MeOD front movement for the second cycle is not linear with time but rather linear with the square root of time. After an initial lag period of anomalous diffusion, the front movement for the second cycle behaves in a Fickian manner, indicating that the diffusion characteristics of the PMMA are altered with absorption–desorption cycling.

The slope of the curves for the samples subjected to 0 and 1 cycles yields the solvent front velocities for case II diffusion of 12.6 and 14.5 nm/s, respectively. These values are more than twice the magnitude of those found by Thomas and Windle (*117*) and Weisenberger and Koenig (*108,109*) for methanol diffusion in PMMA. This can be explained by the fact that the water content and drying treatment of the PMMA were never taken into account in the previous experiments. The removal of water and the leaching out of unreacted monomer were also not considered. The increases in diffusion of MeOD may be accounted for by the plasticization by water of PMMA, which can be demonstrated between the 0 and 1 cycles. The increased plasticization allows for the increased rate of relaxation of the polymer chains and a greater solvent-front velocity (*118*).

Diffusion in Polymer Gels

A gel is composed of a three-dimensional network of synthetic polymers swollen in a liquid. Gels can be considered as cross-linked networks of polymers with microencapsulated fluids. Gels reversibly swell and shrink by as much as several hundred times in response to infinitesimal changes in environmental conditions such as temperature, solvent composition, electric field, stress, and light (*119*). Gels exhibit discontinuous phase transitions as a function of temperature and liquid composition. Volume phase transitions of gels can be classified into three types according to their temperature behavior: thermoswelling, thermoshrinking, and convexo. The thermoswelling type expands with increasing temperature, the thermoshrinking type contracts with temperature, and the convexo type expands or contracts depending upon conditions (*120*). These changes can occur discontinuously at a specific stimulus level (a phase transition), or gradually over a range of stimulus values. The gels can function as valves which turn solute flux on and off in response to changes in environment—turning "on" when the gel swells, and "off" when the gel shrinks. Selectivity and separation of solutes can be designed on the basis of size exclusion of a solute, which is based on the direct relationship between the swelling of a gel and its permeability to solutes, that is, the average pore size of the gel.

All of these changes are reversible with no inherent limit in lifetime. In fact, cycles have been demonstrated to be repeatable many times over periods of many months (*121*). Applications have been developed for responsive gels including drug delivery systems, novel separation systems, artificial muscles, and devices such as switches and sensors.

The volume phase transition of a gel brings about dramatic changes in physical properties, and this phenomenon can be applied to the creation of new types of materials with sensing and switching ability. Molecular interactions rule the macroscopic size and shape of gels and networks. Since these interactions are functions of temperature, polymer concentration, solvent composition (if a mixture of solvents is used), pH, and salt concentration (for gels capable of ionization), the volume phase transition can be induced by controlling one or more of these parameters. The polymer gel dramatically changes its volume in response to a change in environment: temperature, solvent composition, mechanical strain, electric field, exposure to light, etc.

Before a responsive gel can respond to a chemical stimulus, that stimulus must permeate the gel itself. The rate of swelling or shrinking of a gel is influenced by a number of different processes, including stimulus rate, diffusion, and polymer relaxation. When a polymer network is initially in the dry state, solvent must penetrate into the network by diffusion. When the polymer is rubbery, this diffusion process is rate limiting. If the polymer is in the form of a thin slab, then solvent uptake will initially be correlated with the square root of time. When the polymer is in an initially glassy state, swelling kinetics become more complicated. While solvent diffusion into the polymer still initiates the swelling process, the polymer requires a finite time to relax to its final state. If the solvent is unable to plasticize the polymer, very little change in volume will occur. On the other hand, at a sufficiently high solvent concentration, the polymer will change from the glassy to the rubbery state. In this case, a rather well defined "swelling front" is observed separating a rubbery periphery from a glassy core in a swelling polymer. The rate of advancement of the front is inversely related to the characteristic glass-to-rubber relaxation time of the polymer in response to changes in solvent concentration. The ratio of this characteristic relaxation time to the characteristic diffusion time, called the Deborah number (De), determines the shape of the swelling kinetics curve. When relaxation is faster than diffusion (De $\gg$ 1), the cumulative swelling of a slab initially follows Fickian or $t^{1/2}$ kinetics, where t denotes time. When diffusion is much faster than relaxation (De $\ll$ 1), cumulative swelling also appears Fickian, but with a much slower rate. In the intermediate case, where De ~ 1, swelling takes the anomalous form t^α, where $1/2 < \alpha < 1$. The limiting case, where $\alpha = 1$, is called "case II" or zero-order swelling.

Diffusion of probe particles through an entangled polymer gel is characterized by a mesh size, ξ. The particle can move through the mesh of the gel with little hindrance if the particle is significantly smaller than ξ, but with increasing difficulty as the particle size (radius R) becomes comparable to ξ. The following equation has been proposed:

$$D/D_0 = \exp(-bc^k) \qquad (10.33)$$

where D_0 is the diffusion coefficient in a gel of zero concentration, b is a constant that depends on the radius of the probe molecules and the molecular weight of the polymer, c is the concentration of the gel/solution, and k is a constant expected to be in the range of 0.5–1.0 and whose exact value depends largely on the polymer species. Gels of rodlike polymers should have a value of k of 0.5 (122).

For any given gel, the amount of solvent uptake is dependent on the chemical nature of the gel and its network structure. Using the Flory excluded volume theory, the degree of swelling, η, is given by

$$\eta = 1 - e^{-cv}$$

where c is the cross-link density and v is the solute volume (123). The gel volume is proportional to $-2/3$ power of the degree of cross-linking. As the cross-link density decreases, the polymer chain length between crosslinks increases, yielding a looser structure, which the solvent can penetrate.

The time needed for swelling and shrinking is proportional to the square of the gel diameter (124). These gel systems have been studied by NMRI (125,126) with useful results. For cross-linked polystyrene (with divinylbenzene) cross-linked to 2.5%, Figure 10.47 shows the FLASH images (doped with chromium(III) acetylacetonate [Cr(acac)$_3$] to shorten T_1) for four different stages of exposure to dioxane. The light areas surrounding the sample correspond to the contribution of the bulk dioxane. The diffusion front shows up in different gray tones depending on the concentration of the dioxane (126). The diffusion behavior is a function of the cross-link density for the cross-linked polystyrene: at the lowest cross-linking (1%), case II diffusion occurs, but at the highest cross-linking (5%), Fickian diffusion occurs, with the 2.5% cross-linked sample showing intermediate behavior. This suggests that there is a strong interaction of the highly mobile phenyl group with the dioxane in the case of the 1% crosslinked material, which results in case II diffusion. For the higher crosslinking leading to decreased mobility, the phenyl group motion is de-

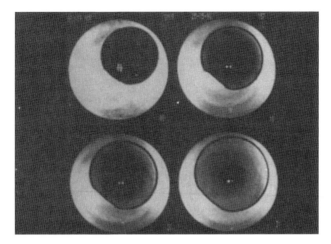

Figure 10.47. FLASH images of cross-linked polystyrene (with divinylbenzene) cross-linked to 2.5%, for four different stages of exposure to dioxane. Doping with chromium(III) acetylacetonate was used to shorten T_1 Image parameters: slice thickness, 1.5 mm; spatial resolution, 115 μm; and field of view, 28 mm. (Reproduced from ref 126. Copyright 1994 American Chemical Society.)

creased, and so is the magnitude of the interaction with dioxane; hence, Fickian diffusion.

Limitations of NMRI for Diffusion Studies

The major limitation of NMRI for diffusion studies is the long measurement times, which are required to obtain a sufficient S/N ratio per pixel. These long measurement times restrict the measurements to slowly diffusing systems. When diffusion is too fast, the spins in the diffusion gradients are misregistered, and artifacts are produced.

A number of approaches have been suggested for fast imaging. The goal of fast imaging is to allow rapid acquisition of data for either real-time imaging (i.e., for processes whose correlation times are on the same order as the time to perform one image), or to improve throughput. Long measurement times lead to motional artifacts and give no access to the study of fast chemical processes.

Proton images can be acquired at small tip angles and short echo time values, leading to high speeds of image acquisition. The rapid stimulated-echo imaging technique, termed STEAM, allows the simultaneous recording of a number of directly neighbored slices (*96*). The magnetization of the entire volume is prepared by two leading nonselective radio-frequency (rf) pulses. Then only the slice-selective readout pulse is repeated with different irradiation frequencies to select magnetizations of individual slices. Each readout pulse creates a corresponding slice-selective "STEAM" signal.

Echo-planar methodology involves data digitization during programmed gradient sweeping (x and y axes simultaneously). This is a time-efficient way to sweep k-space. In one acquisition or "shot," enough k-space data points are sampled and collected (typically in 40 milliseconds/image) to later reconstruct a two-dimensional image. This opens up the possibility of real-time MRI at rates of 10–20 images per second. Although high-resolution images have not yet been forthcoming, it is possible to use new reconstruction techniques or to cover k-space with several extra scans to improve the resolution significantly. With the utilization of these fast imaging techniques, the use of NMRI will increase for the study of faster diffusing systems.

Conventional NMR imaging methods rely on the use of static field gradients and generally involve spin echoes or gradient echoes for refocusing nuclear magnetization and avoiding artifacts due to gradient switching. Any echo has some magnetization loss due to the transverse relaxation. The use of static field gradients results in effects due to the switching process, including eddy currents created in the materials surrounding the probe and spectrometer perturbations due to finite rise and fall times of B_0 gradients (127).

The use of rf field gradients is based on nutation around the B_1 rf field and not precession with respect to the static field B_0 as in conventional NMR imaging techniques (128). The advantages of this method are that it is not subject to the effects of:

- variation of magnetic susceptibility (no echo is needed),
- rapid transverse relaxation, or
- translational diffusion.

The experiment is based on the Fourier zeugmatography image-reconstruction algorithm and is carried out in the following way: first, the B_1 gradient is applied for an incremented time interval t_1, producing a (rotating-frame) nutation of nuclear magnetization whose angle depends on the location of the relevant molecules along the axis of the gradient coil, say the X direction. The sample is then rotated by an angle of exactly $90°$. The stored longitudinal magnetization is then nutated according to a read B_1 gradient along the Y direction during a time t_2. The double-modulated signal $S(t_1, t_2)$ acquired at the end of this two-dimensional NMR experiment directly provides the spatial spin density distribution by means of a double Fourier transformation applied to both variables t_1 and t_2. This approach allows image acquisition in a measurement time of 2 h with a resolution of 100 μm. This technique appears to be particularly useful for diffusion measurements (128).

NMRI of the Ingression of Moisture Through Coatings

NMRI can be used to evaluate water resistance of adhesive joints and has been specifically used to compare water penetration at 25 °C in polyurethane dispersion coatings with water penetration in a plasticized poly(vinyl acetate) (PVAc) emulsion and a 5-min-cure epoxy adhesive (*46*). The water penetration is followed by the acquisition of cross-sectional images of the samples. Wood rods with coatings of 300 and 50 μm were immersed in water, and images were obtained at different times after immersion. The NMR images of PVAc- and polyurethane-dispersion-coated rods are shown in Figures 10.48 and 10.49. The images of the epoxy-coated rods did not show any water penetration through the coating within 24 h. The PVAc coating shows rapid water penetration into the coating (Figure 10.48a). The coating swells, and separate layers of the coating partially delaminate, which shows poor interfacial adhesion of the layers (Figure 10.48 b–d). After an immersion of 6 h, the coating shows deformation (Figure 10.48d) that continues to grow. Although the coating further swells and degrades, the wood reaches saturation at about 10 h of immersion.

The non-cross-linked polyurethane dispersion shows much better water resistance than the PVAc emulsion. The differences in water penetra-

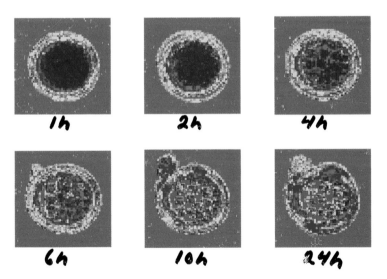

Figure 10.48. Proton images of water penetration through a 300-μm film of PVAc into wood. The NMR images were acquired after the following times: (a) 1 h; (b) 2 h; (c) 4 h; (d) 6 h; (e)10 h; and (f) 24 h. White represents the highest signal intensity. (Reproduced with permission from ref 46. Copyright 1990 Gordon & Breach.)

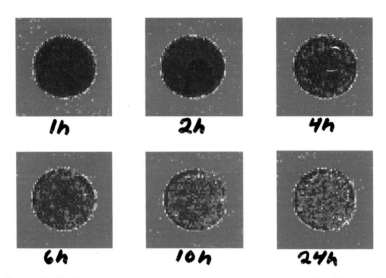

Figure 10.49. Proton images of water penetration into wood through a 300-μm film of non-cross-linked polyurethane dispersion coating (top), and PVAc coating (bottom). The NMR images were acquired after the times shown. White represents the highest signal intensity. (Reproduced with permission from ref 46. Copyright 1990 Gordon & Breach.)

tion mechanisms arise from two factors. First, the PVAc contains polyvinyl alcohol, which is water soluble and allows the rapid intrusion of water. Second, the average particle size of the colloidal polyurethane dispersion is much smaller, which leads to a tighter packing of the dried film. The addition of the polyaziridine cross-linker decreases the water absorption of the polyurethane coating by a factor of approximately 14.

NMRI Studies of Deformation of Cross-Linked Polymers

The NMRI technique can be used to study the stress–strain behavior of water-swollen polymer gels (*129*). In this case the water is imaged, as no signal is obtained from the polymer network. If stress is applied to the gel sample in compression by a piston, the image is changed, reflecting the restraint of the molecular motion of the water due to the strain induced by the stress. Figure 10.50 shows the cross-sectional profile parallel to the direction of the applied stress, for a PMMA gel with a degree of swelling of 22 without stress and under a stress of 4.8 kPa. It is apparent that the spatial distributions of proton spin density and the molecular motions of the water molecules in the gel are changed by the stress. As the stress

¹H NMR Imaging Study on a Polymer Gel

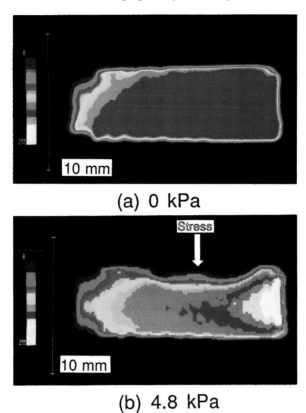

(a) 0 kPa

(b) 4.8 kPa

Figure 10.50. T₂-enhanced images of a PMMA gel with a degree of swelling of 22 without stress (a), and under a stress of 4.8 kPa (b). (Reproduced from ref 129. Copyright 1992 American Chemical Society.)

increases, the molecular motions of the water in the compressed regions decrease relative to the surrounding uncompressed regions (*129*).

NMRI Studies of Flow Behavior

NMRI provides the means for investigating flow phenomena noninvasively (*130*). In its simplest form, the NMRI technique labels the spins with a particular frequency corresponding to their initial position in the field gradient aligned along the flow axis, and determining their spatial distribution at a given time later. The distance traveled by a voxel will be proportional to the flow rate. A three-dimensional equivalent of the flow ex-

a)   b)

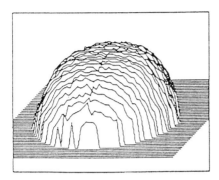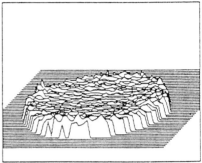

Figure 10.51. Maps of water solvent velocity (a) and diffusion (b) for the 4.5% w/v poly(ethylene oxide) solution. (Reproduced with permission from ref 30. Copyright 1991 Clarendon Press.)

periment allows one to display not only orthogonal velocity profiles but also planes of equal flow velocities. Flow velocities must be below 100 mm s^{-1} if the excited slice is to remain within the rf coil over the duration of the spin echo (*130*). Therefore, in order to observe shear rates of 100 s^{-1}, it is necessary to use submillimeter-diameter capillaries. The flow profiles obtained for different concentrations of poly(ethylene oxide) in water are shown in Figure 10.51 (*131*). Examination of Figure 10.51 reveals a transition from Newtonian to non-Newtonian behavior as the concentration increases.

NMRI of Rigid Polymeric Solids

For a given magnetic field gradient strength, the spatial resolution in NMR imaging is determined by the line width. The difficulties of solid-state imaging arise because the solid-state line width is approximately 1000 times broader than its solution counterpart owing to the presence of dipolar couplings. Dipolar couplings are averaged out in liquids, leading to narrow lines. Increasing the gradient by three or four orders of magnitude to maintain spatial resolution in solids imaging is experimentally very difficult (*20, 132–135*).

Attempting to perform solids imaging by using classical instrumentation is limited because rigid solids have short T_2 values on the order of 100 μs, and instruments based on medical applications generally fail because of the long echo times (>1 ms) they usually require.

Spatially resolving a given volume in an NMR image is equivalent to doing an NMR experiment on that volume; therefore, spatial resolution is limited by the smallest amount of sample that can be detected by NMR. The line broadening in solids decreases the sensitivity. This is a consequence of having to increase filter bandwidths to admit the broader signal. The S/N ratio decreases one to two orders of magnitude for solid materials as compared to liquids. Therefore, the inherent sensitivity is lower for solids (*132*).

Efforts to overcome the problems of NMR imaging of solids associated with line broadening involve techniques such as increased gradient strength, phase encoding methods, and line narrowing techniques such as magic angle spinning (MAS) (*134*) and multiple pulse methods (*135,136*). Classical back-projection can also be used where gradient switching during the pulse is not required. Slice selection can be accomplished by presaturation of the out-of-plane magnetization (*137*). Another technique for slice selection is the use of a spin-lock pulse to capture the magnetization of the selected slice (*138*). PVC has been imaged using this method.

Another approach is the use of "magic sandwich echoes" (MSEs) (*139*). Data are sampled stroboscopically. The sandwiches denote the phases of the rf irradiation that are applied to the sample. The sign of the magic echo is changed every 6 cycles. In this case, an effective T_2, $T_{2\text{eff}}$, is recorded for each spatial dimension in a stretched polycarbonate sample (*140*). In the drawn sample there are regions of stressed and unstressed material. The $T_{2\text{eff}}$ characterizes the molecular mobility. In the stressed region, $T_{2\text{eff}}^{-1}$ is higher than in the unaffected material, indicating a reduced mobility.

Large gradients have been used to image solids (*141,142*). It was realized that there is a strong linear gradient present near the edge of a superconducting magnet coil. STRAFI (Stray Field I imaging) resulted. Numerical estimations of the field profile show that there is a planar surface with a constant gradient and constant field near the edge of the coil. Gradients of 40 to 80 T/m were found in magnets with a 4.7- to 9.4-T central field. The surface just described can be used to perform selective excitation, where the slice thickness depends on G and the bandwidth of the rf pulses applied. This surface is called the sensitive plane. Then the sample is translated or rotated to bring the desired slice into the sensitive plane. Because of the large G values, one can excite very thin slices. The spatial resolution of STRAFI is limited only by the step size and precision of the mechanical motion, as long as the echo time in the pulse sequence can be kept short compared to T_2. Images with 60-μm resolution have been obtained for rigid polymeric solids (PMMA) (142), as well as for poly(oxymethylene), and polycarbonate(143).

Two types of line narrowing procedures have been demonstrated for imaging of solids: magic angle spinning (MAS)(144) and multiple pulse cycles (145). MAS averages the spin interactions in the same manner as the motion in liquids. MAS rapidly rotates the sample about an axis inclined at an angle of 54.7° to the magnetic field. The motion of the sample averages spatially the chemical shift anisotropy to the isotropic chemical shift with a corresponding decrease in line width. In order to perform MAS imaging, a rotating magnetic field gradient must be synchronized with the rotation of the sample. The proton image of a blend of polystyrene and polybutadiene has been obtained using MAS. Only the polybutadiene can be imaged, as MAS is unable to reduce the dipolar-line broadening (134). For most solids, it is not possible to spin fast enough to average the dipolar interactions.

Multiple pulse cycles manipulate the spin states in order to average the dipolar interactions to zero. A limitation of the multiple pulse cycles arises from the interference of the magnetic field gradients with the line narrowing of multiple pulse cycles. Using the multiple pulse cycles, the image of Ultem [poly(ether imide)] has been obtained (143).

Three-Dimensional Reconstruction of Images

It has been demonstrated that one can generate a three-dimensional image of polycarbonate using multiple-pulse homonuclear decoupling in combination with three-dimensional back-projection (146). The spatial resolution was 150 × 150 × 150 μm. The multiple pulse sequence narrows the polycarbonate proton line from 35 to 0.85 kHz. A series of frequency-encoded, one-dimensional projections of the object in space parallel to the magnetic field gradient are obtained. The projections were acquired by electronically rotating the magnetic field gradient in classical spherical coordinate space about the sample. The images were subsequently calculated from the series of one-dimensional phase projections using three-dimensional Radon transform inversion.

NMRI of ^{13}C Chemical Shifts

The advantages of ^{13}C NMR is the extremely high sensitivity of the chemical shifts to small changes in chemical structure. Encoding this chemical information into an NMR image would be a great advantage over other imaging techniques. It would appear at first sight to be easier to obtain solid-state ^{13}C chemical-shift images than the corresponding ^1H chemical-shift images because of the larger chemical-shift dispersion and the lack of homonuclear dipolar interactions for ^{13}C (*147*). But the sensitivity of ^{13}C is much lower than for protons. The natural abundance of ^{13}C is only 1.1%. In addition, to achieve high spatial resolution, the magnetic field gradient must be four times larger than that for protons because the γ of ^{13}C is about four times smaller than that of ^1H. There is also the technical problem that the ^{13}C signal intensity depends on ^1H frequency offsets in ^1H decoupling, requiring special attention. However, reports of solid-state ^{13}C have appeared (*148–151*).

Solid-state ^{13}C chemical-shift images have been achieved with MAS using a rotating magnetic field gradient synchronized with the rotor and a composite heteronuclear broadband decoupling sequence (*147*). A spatial resolution of 370 μm was achieved in a measurement time of 24 h. A phantom consisting of Delrin, hexamethylbenzene, and adamantane was studied. It was possible, by chemical selective imaging, to image the hexamethylbenzene alone, the adamantane separately, and, finally, Delrin. A transverse-plane, two-dimensional, spatial-spectral ^{13}C image of the three-component phantom was obtained.

The ^{13}C chemical-shift images of composites of organic matrix and carbon fiber have been reported(151). The image obtained required 26.7 h and had a resolution of approximately 0.5 mm. Thus, the potential is there, but much work remains to produce such images routinely.

Current Capabilities and Future Prospects

The applications of NMRI to materials and chemistry have not grown as rapidly as the applications of NMRI in the medical field. There are a variety of reasons for this slower growth, but primarily it is a matter of spatial resolution. In materials science, resolution approaching that of the optical microscope is widely expected, and anything less is unacceptable. In the medical field, the resolution is much lower, but the resolution requirements are also lower.

The highest potential advantage of NMRI is the in situ examination of internal defects in materials systems such as polymer adhesives and composites. There is a need for a nondestructive, noninvasive test for the quality of such systems. NMRI offers a newer method, but it is limited by instrument cost and resolution.

References

1. Bovey, F. A. *Chain Structure and Conformation of Macromolecules;* Academic: New York, 1982.
2. *NMR and Macromolecules;* Randall, J. C., Ed.; ACS Symposium Series 247; American Chemical Society: Washington, D.C., 1984.
3. Tonelli, A. E. *NMR Spectroscopy and Polymer Microstructure;* VCH Publishers: New York, 1989.
4. Koenig, J. L. *Spectroscopy of Polymers;* American Chemical Society: Washington, D.C., 1992.
5. Pham, G. T.; Petaud, R.; Waton, H. *Proton and Carbon NMR Spectra of Polymers;* CRC Press: London, 1991.
6. Wehrli, F. W. *Prog. NMR Spectroscopy* **1995**, *28*, 87.
7. Mansfield, P.; Morris, P. G. *NMR Imaging in Biomedicine;* Academic: New York, 1982.
8. Parikh, A. M. *Magnetic Resonance Imaging Techniques;* Elsevier: New York, 1992.
9. Komoroski, R. A. *Anal. Chem.* **1993**, *65*, 1068.
10. Bailes, D. R.; Bryant, D. J. *Contemp. Phys.* **1984**, *25*, 441.
11. Gadian, D. *Nuclear Magnetic Resonance and its Applications to Living Systems;* Clarendon Press: Oxford, England, 1982.
12. Mansfield, P.; Hahn, E. L. *NMR Imaging;* Royal Chemistry Society: London, 1990.
13. King, K. F.; Moron, P. R. *Med. Phys.* **1984**, *11*, 1.
14. Herman, G. T. *Image Reconstruction from Projections: The Fundamentals of Computerized Tomography;* Academic: New York, 1980.
15. Miller, J. D. *Trends Anal. Chem.* **1991**, *59.*
16. Gerstein, B. C.; Dybowski, C. R. *Transient Techniques in NMR of Solids: An Introduction to Theory and Practice;* Academic: New York, 1985.
17. *High Resolution NMR of Synthetic Polymers in Bulk;* Komoroski, R. A., Ed.; VCH Publishers: Deerfield Beach, FL, 1986.
18. Saito, H.; Ando, I. *Annu. Rep. NMR Spectrosc.* **1989**, *21*, 209.
19. *Solid State NMR of Polymers;* Mathias, L., Ed.; Plenum: New York, 1991.
20. Cory, D. G. *Annu. Rep. NMR Spectrosc.* **1992**, *24*, 87.
21. Dixon, W. T. *Radiology* **1984**, *153*, 189.
22. Hall, L. D.; Sukumar, S.; Talagala, S. L. *J. Magn. Reson.* **1984**, *56*, 275.
23. Hall, L. D.; Rajanayagam, V. *J. Magn. Reson.* **1987**, *74*, 139.
24. Volk, A.; Tiffon, B.; Mispelter, J.; Lhoste, J. M. *J. Magn. Reson.* **1987**, *71*, 168.
25. Liu, J.; Nieminen, A. O. K.; Koenig, J. L. *Appl. Spectrosc.* **1989**, *43*, 1260.
26. Cory, D. G.; Reichwein, A. M.; Veeman, W. S. *J. Magn. Reson.* **1988**, *80*, 259.
27. Nieminen, A. O. K.; Koenig, J. L. *Appl. Spectrosc.* **1989**, *43*, 153.
28. Garrido, L.; Ackerman, J. L.; Mark, J. E. *Mater. Res. Soc. Symp. Proc.* **1990**, *71*, 65.
29. *Magnetic Resonance Microscopy;* Bluemich, B.; Kuhn, W., Ed.; VCH: Basel, Switzerland, 1992.
30. Callaghan, P. *Principles of Nuclear Magnetic Resonance Microscopy;* Claredon Press: Oxford, England, 1991.
31. Blumich, B.; Bluminer, P.; Gunther, E.; Schauss, H.; Spiess, H. W. *Makromol. Chem. Macromol. Symp.* **1991**, *44*, 37.
32. Jackson, P.; Clayden, N. J.; Barnes, J. A.; Carpenter, T. A.; Hall, L. D.; Jezzard, P. *Int. SAMPE Symp. Exhib.* **1991**, *36*, 246.
33. Jackson, P.; Clayden, N. J.; Barnes, J. A.; Carpenter, T. A.; Hall, L. D.; Jezzard, P. *Mater. Sci. Monogr.* **1991**, *72*, 277.
34. Blumler, P.; Blumich, B. L. *NMR;* **1994**, *30*, 209.

35. Kuhn, W. *Angew. Chem. Int. Ed. Engl.* **1990,** *29,* 1.
36. *Magnetic Resonance Microscopy: Methods and Applications in Materials Science, Agriculture, Biomedicine;* Blumich, B.; Kuhn, W., Eds.; VCH: Weinheim, Germany, 1992.
37. *Multidimensional Solid-State NMR and Polymers;* Schmidt-Rohr, K.; Spiess, H., Eds.; Academic: London, 1994.
38. Hoh, K. P.; Perry, B.; Rotter, G.; Ishida, H.; Koenig, J. L. *J. Adhes.* **1989,** *27,* 245.
39. Perry, B. C.; Koenig, J. L. *J. Polym. Sci. Part A: Polym. Chem.* **1989,** *27,* 3429–3438.
40. Fleer, G. J.; Cohen, S. M. A.; Scheutjens, J. M. H.; Cosgrove, T.; Vincent, B. *Polymers at Interfaces;* Chapman and Hall: New York, 1993.
41. Wool, R. P. *Structure and Strength of Polymer Interfaces;* Hanser: New York, 1994.
42. Nieminen, A. O. K.; Koenig, J. L. *J. Adhes. Sci. Technol.* **1988,** *2,* 407.
43. Nieminen, A. O. K.; Koenig, J. L. *Int. J. Adhes. Adhes.* **1991,** *11,* 5.
44. Nieminen, A. O. K.; Koenig, J. L. *Adhes. Age* **1989,** *32,* 17.
45. Nieminen, A. O. K.; Liu, J.; Koenig, J. L. *J. Adhes. Sci. Technol.* **1989,** *3,* 455.
46. Nieminen, A. O. K.; Koenig, J. L. *J. Adhes.* **1989,** *32,* 105.
47. Nieminen, A. O. K.; Koenig, J. L. *J. Adhes.* **1989,** *43,* 1.
48. Nieminen, A. O. K.; Koenig, J. L. *J. Adhes.* **1989,** *30,* 47.
49. Fondeur, F.; Koenig, J. L. *J. Adhes.* **1993,** *43,* 289–308.
50. Nieminen, A. O. K.; Koenig, J. L. *Appl. Spectrosc.* **1989,** *43,* 1358.
51. Liu, J.; Nieminen, A. O. K.; Koenig, J. L. *Appl. Spectrosc.* **1989,** *43,* 1260.
52. Nieminen, A. O. K.; Koenig, J. L. *J. Adhes. Sci. Technol.* **1988,** *2,* 407.
53. Nieminen, A. O. K.; Liu, J.; Koenig, J. L. *J. Adhes. Sci. Technol.* **1989,** *3,* 455.
54. Nieminen, A. O. K.; Koenig, J. L. *Adhes. Age* **1989,** *32,* 17.
55. Nieminen, A. O. K.; Evans, M.; Koenig, J. L. *New Polym. Mater.* **1990,** *2,* 197.
56. Jackson, P.; Clayden, N. J.; Walton, T. A.; Hall, L. D.; Jezzard, P. *Polym. Int.* **1990,** *24,* 139–143.
57. Balcom, B.; Carpenter, T.; Hall, L. *Macromolecules* **1992,** *25,* 6818.
58. Gunther, U.; Albert, K.; Grossa, M. *J. Magn. Reson.* **1992,** *198,* 593.
59. Jackson, P. I. *J. Mater. Sci.* **1992,** *27,* 1302.
60. Krejsa, M. R.; Koenig, J. L. *Elastomer Technology Handbook;* Cheremisinoff, N. P., Ed.; CRC: Boca Raton, FL, 1993; Chap. 11, pp 475–491.
61. Krejsa, M. R.; Koenig, J. L. *Rubber Chem. Technol.* **1993,** *66,* 376–410.
62. Koenig, J. L.; Patterson, D. J. *Materials Characterization;* Cahn, R. W., Lifshin, E., Eds.; Pergamon: New York, N.Y.,1992; pp 99–101.
63. Koenig, J. L.; Patterson, D. J. *Elastomers* **1986,** *118,* 21.
64. Koenig, J. L. *Prog. Polym. Spectrosc. Proc. Eur. Symp. Polym. Spectrosc. 7th* **1986,** *9,* 28.
65. Koenig J. L.; Andreis, M. *Solid State NMR of Polymers;* Mathias, L. J., Ed.; Plenum: New York, 1991; p 201.
66. Andreis, M.; Koenig, J. L. *Adv. Polym. Sci.* **1989,** *89,* 71.
67. Krejsa, M.; Koenig, J. L. *Rubber Chem. Technol.* **1992,** *65,* 427.
68. Smith, S. R.; Koenig, J. L. *Rubber Chem. Technol.* **1992,** *65,* 176–200.
69. Krejsa, M. J.; Koenig, J. L. *Rubber Chem. Technol.* **1991,** *64,* 40.
70. Krejsa, M. R.; Koenig, J. L. *Rubber Chem. Technol.* **1992,** *65,* 956–964.
71. Andreis, M.; Koenig, J. L. *Polym. Prepr.(Am. Chem. Soc. Div. Polym. Chem.)* **1988,** *29,* 29.
72. Rana, M. A.; Koenig, J. L. *J. Appl. Polym. Sci. Appl. Polym. Symp.* **1994,** *53,* 87–101.
73. Andreis, M.; Liu, J.; Koenig, J. L. *Rubber Chem. Technol.* **1989,** *62,* 82.
74. Koenig, J. L. *Makromol. Chem. Symp.,Macromol. Chemie, Mainz, Germany* **1994,** *86,* 283–297.

75. Clough, R. S.; Koenig, J. L. *J. Polym. Sci. Lett.* **1989**, *27*, 451.
76. Chang, C.; Komoroski, R. A. *Macromolecules* **1989**, *22*, 600.
77. Krejsa, M. R.; Koenig, J. L. *Rubber Chem. Technol.* **1991**, *64*, 635.
78. Smith, S. R.; Koenig, J. L. *Macromolecules* **1991**, *24*, 3496.
79. Rana, M. A.; Koenig, J. L. *Macromolecules* **1994**, *27*, 3727–3734.
80. Kuhn, W.; Barth, P.; Hafner, S.; Simon, G.; Schneider, H. *Macromolecules* **1994**, *27*, 5773.
81. Garrido, L.; Mark, J. E.; Sun, C. C.; Ackerman, J. L.; Chang, C. *Macromolecules* **1991**, *24*, 4067.
82. Garrido, L.; Ackerman, J. L.; Vevea, J. M.; Mark, J. E. *Polymer* **1992**, *33*, 1826.
83. Garrido, L.; Mark, J. E. *Polym. Reprints, (Am. Chem. Soc. Div. Polym. Chem.):* **1989**, *30*, 217.
84. Sakar, S. N.; Komoroski, R. A. *Macromolecules* **1992**, *25*, 1420.
85. Komoroski, R. A.; Sakar, S. N. *Mater. Res. Soc. Symp. Proc.* **1991**, *217*, 3.
86. Bluemler, P.; Bluemich, P. *Macromolecules* **1991**, *24*, 2183.
87. Catwalk, J. A.; Hunter, G. J. *Mater. Sci. Lett.* **1992**, *11*, 222.
88. Koenig, J. L. In *Characterization of Composite Materials;* Ishida, H.; Fitzpatrick, L. E., Eds.; Butterworth-Heinemann: Boston, 1994; Ch 3, pp 44–62.
89. Koenig, J. L. In *Controlled Interphases in Composite Materials;* Ishida, H., Ed.; Elsevier: New York, 1990.
90. Hoh, K.; Ishida, H.; Koenig, J. L. *Polym. Compos.* **1990**, *11*, 192.
91. Mavrich, A.; Fondeur, F.; Ishida, H.; Wagner, H. D.; Koenig, J. L. *J. Adhes.* **1993**, *41*, 649.
92. Arvanitopoulos, C.; Koenig, J. L. *J. Adhes.* **1995**, *53*, 15.
93. Weisenberger, L. A.; Koenig, J. L. *Polym. Prepr. (Am. Chem. Soc. Div. Polym. Chem.)* **1988**, *29*, 98.
94. Weisenberger, L. A.; Koenig, J. L . *Solid State NMR of Polymers;* Mathias, L. J., Ed.; Plenum: New York, l991; p 377.
95. Turner, D. T. *Polymer* **1987**, *28*, 293.
96. Haase, A.; Frahm, J.; Mattaei, D.; Hanicke, W.; Merboldt, K. D. *J. Magn. Reson.* **1986**, *67*, 258.
97. Merboldt, K. D.; Hanicke, W.; Gyngell, M. L.; Frahm, J.; Bruhn, H. *J. Magn. Reson.* **1989**, *82*, 115.
98. Callaghan, P. In T. *NMR Spectroscopy of Polymers;* Ibbett, R. N., Ed.; Academic: Glasgow, Scotland,1993.
99. Rothwell, W. P.; Holecek, D. R.; Kershaw, J. A. *J. Polym. Sci. Polym. Lett. Ed.* **1984**, *22*, 241.
100. Weisenberger, L. A.; Koenig, J. L. *J. Polym. Sci.* **1989**, *27*, 55.
101. Blackband, S.; Mansfield, P. *Solid State Phys.* **1986**, *19*, 49.
102. Iig, M.; Pfielderer, B.; Albert, K.; Bayer, E. *Mater. Res. Soc. Symp. Proc.* **1991**, *217*, 27.
103. Weisenberger, L. A.; Koenig, J. L. *Macromolecules* **1990**, *23*, 2445.
104. Webb, A. G.; Hall, L. D. *Polym. Commun.* **1990**, *31*, 422.
105. Webb, A. G.; Hall, L. D. *Polym. Commun.* **1990**, *31*, 425.
106. Webb, A. G.; Hall, L. D. *Polymer* **1991**, *32*, 2926.
107. Weisenberger, L. A.; Koenig, J. L. *Polym. Mater. Sci. Eng.* **1989**, *61*, 16.
108. Weisenberger, L. A.; Koenig, J. L. *Appl. Spectrosc.* **1989**, *43*, 1117.
109. Weisenberger, L. A.; Koenig, J. L. *Macromolecules* **1990**, *23*, 2454.
110. Grinsted, R. A.; Koenig, J. L. *Macromolecules* **1992**, *25*, 1229.
111. Ware, R. A.; Tirtowidjojo, S.; Cohen, C. *J. Appl. Polym. Sci.* **1981**, *26*, 2975.
112. Grinsted, R. A.; Clark, L.; Koenig, J. L. *Macromolecules* **1992**, *25*, 1235.
113. Burchill, J.; Stacewicz, R. H. *J. Mater. Sci. Lett.* **1982**, *1*, 448.
114. LaBarre, E. E.; Turner, D. T. *J. Polym. Sci. Polym. Phys. Ed.* **1982**, *20*, 557.
115. Berens, A. R.; Hopfenberg, H. B. *J. Polym. Sci. Polym. Phys. Ed.* **1979**, *17*, 1757.

116. Mandelkern, L.; Long, F. A. *J. Polym. Sci.* **1951**, *6*, 457.
117. Thomas, N. L.; Windle, A. H. *Polymer* **1978**, *19*, 255.
118. Tirrell, M. *Rubber Chem. Technol.* **1984**, *57*, 523-556.
119. Tanaka, T. *Phys. Rev. Lett.* **1978**, *40*, 820.
120. Saito, S.; Konno, M.; Inomata, H. In *Responsive Gels: Volume Transitions;* Dusek, K., Ed.; Springer, Berlin, 1993; Vol. 109, p 208.
121. Lyu, L. H.; Gehrke, S. H. Presented at the AIChE Annual Meeting, San Francisco, CA, 1989; Paper 98J.
122. Cukier, R. *Macromolecules* **1984**, *17*, 252.
123. Gehrke, S. H.; Andrews, G. P.; Cussler, E. L. *Chem. Eng. Sci.* **1986**, *41*, 2153.
124. Suzki, A. In *Responsive Gels: Volume Transitions; Advan. Polym. Sci.*:Dusek, K., Ed.; 1993; Springer, Berlin, Vol. 110, p 201.
125. Smith, E. G.; Rockliffe, J. W.; Riley, P. I. *J. Coll. Interface Sci.* **1989**, *131*, 2.
126. Pfieiderer, M. I. B.; Albert, K.; Rapp, W.; Bayer, E. *Macromolecules* **1994**, *27*, 2778.
127. Fyfe, C. A.; Randall, L. H.; Burlinson, N. E. *Chem. Mater.* **1992**, *4*, 267.
128. Maffei, P.; Kiene, L.; Canet, D. *Macromolecules* **1992**, *25*, 7114.
129. Yasunga, H.; Kurosu, H.; Ando, I. *Macromolecules* **1992**, *25*, 6506.
130. Callaghan, P. T.; Xia, Y. *J. Magn. Reson.* **1991**, *91*, 326.
131. Xia, Y.; Callaghan, P. T. *Macromolecules* **1991**, *24*, 4777.
132. Miller, J. B. *Trends Anal. Chem.* **1991**, *10*, 59.
133. Jezzard, P.; Attard, J. J.; Carpenter, T. A.; Hall, L. D. *Prog. NMR Spectrosc.* **1990**, *23*, 1.
134. Veeman, W. S.; Bijl, G. *Magn. Reson. Imaging* **1992**, *10*, 755.
135. Cory, D. G.; Miller, J. B.; Garroway, A. N. *Mol. Phys.* **1990**, *70*, 331.
136. Miller, J. B.; Cory, D. G.; Garroway, A. N. *Magn. Reson. Imaging* **1992**, *10*, 789.
137. Dodrell, D. Bulsing, J.; Galloway, Brooks W.; Field, J.; Irving, M.; Baddeley, H.; *J. Magn. Reson.* **1986**, *70*, 319.
138. Rommel, E.; Kimmich, R. *J. Magn. Reson.* **1989**, *83*, 299.
139. Weigand, F.; Blumich, B.; Spiess, H. *Solid State Nucl. Magn. Reson.* **1994**, *3*, 59.
140. Weigand, F.; Spiess, H. *Macromolecules* **1995**, *28*, 6361.
141. Samolienk, A. A.; Zick, K. *Bruker Report*, Billerica, MA, **1990**, *1*, 40.
142. Samoilenko, A. A.; Yu, D.; Artemov; Sibeldina, L. A. *JETP Lett. (Engl. Transl.)* **1988**, *47*, 348.
143. Muller, D.; *Bruker Report*, Billerica, MA, **1990**, *5*, 28.
144. Cory, D. G.; Van Os, J. W. M.; Veeman, W. S. *J. Magn. Reson.* **1988**, *76*, 543.
145. Chingas, G. C.; Miller, J. B.; Garroway, A. N. *J. Magn. Reson.* **1986**, *66*, 530.
146. Dieckman, S. L.; Pizo, P.; Gopalsami, N.; Heeschen, J. P.; Botto, R. E. *J. Am. Chem. Soc.* **1992**, *114*, 2717.
147. Sun, Y.; Lock, H.; Shinozaki, T.; Machiel, G. E. *J. Magn. Reson. Ser. A* **1995**, *83*, 165.
148. Szeverenyl, N. M.; Maciel, G. E. *J. Magn. Reson.* **1984**, *60*, 460.
149. Cory, D. G.; Veeman, W. S. *J. Phys. E. Sci. Instrum.* **1989**, *22*, 180.
150. Scheler, U.; Blumich, B.; Spiess, H. W. *Solid State Nucl. Magn. Reson.* **1993**, *2*, 105.
151. Scheler, U.; Titman, J. J.; Blumich, B.; Spiess, H. W. *J. Nucl. Magn. Reson. A* **1994**, *109*, 251.
152. Fry, C. G.; Lind, A. C.; Davis, M. F.; Duff, D. W.; Maciel, G. E. *J. Magn. Reson.* **1989**, *83*, 656.

Glossary

Abbe criterion A definition of resolution in the Fourier sense. The resolution of optical microscopy is determined by the highest object spatial frequency that can be accommodated. This limiting Fourier component can be considered as a sort of sinusoidal grating in the object. When this grating is sufficiently fine that the first-order maxima occur at angles greater than the angle, θ_{max}, subtended by the lens, the limit is reached. This corresponds to a spatial period $\lambda/\sin \theta_{max}$. So the resolution is given by

$$r \approx \lambda/\sin \theta_{max} \simeq 2 \pi/\kappa_{max}$$

aberrations Imperfections or processes that prevent an optical lens from producing an exact geometric (and chromatic) pattern between an object and its image.

accuracy The difference between the average of a series of measurements and the "true" value given by a standard or a series of standards. Accuracy involves the consistency between the results obtained and the actual concentration of the analyte in a particular sample.

achromats Lenses that are chromatically corrected for blue and red wavelengths and show no spherical aberration in the yellow.

acousto-optic tunable filter (AOTF) A computer-controlled notch filter that can provide random wavelength access, wide spectral coverage, and moderate spectral resolution. AOTFs are prepared from optically transparent birefringent crystals such as TeO_2 to which an array of piezoelectric transducers are bonded.

adaptive optics Optics that correct the wave-front distortions in real time.

Airy disk The two-dimensional diffraction pattern observed when the three-dimensional optical diffraction pattern of a lens is sectioned in the focal plane. The resolving power of an objective lens can be determined by examining the size of the Airy disk formed by that lens.

aliasing When a sine wave of a higher frequency than the Nyquist frequency exists, it will be folded back into the spectrum at a frequency exactly as much lower than the Nyquist frequency as the sine wave is higher than the Nyquist frequency. This phenomenon is referred to as *aliasing*.

aliasing angle of incidence   The angle between the path of the light as it approaches a surface and the perpendicular to the surface at the point of incidence.

angle of refraction   The angle between the ray, after it has passed the interface, and the perpendicular to the interface.

apochromats   Lenses that exhibit spherical-aberration control throughout the visible and are chromatically corrected at three wavelengths: in the blue, yellow, and red.

array detector   Area light-detection device. Light causes the device to generate charge that is read out quickly and interpreted to construct images.

array $1/f$ noise   Spatial noise due to something external that makes all the pixels unstable, and the effect of the instability is not the same for every pixel. The effect of array $1/f$ noise on images is to introduce a pattern in the image where the pattern itself changes little in time but the magnitude of the pattern is variable.

artifact contrast   Deviation from proportionality between signal intensity and concentration.

aspect ratio   The ratio of the diameter of a particle to its thickness.

axial resolution (r_z)   A measure of the ability of the confocal optical system to discriminate signals from adjacent layers of the sample in the direction of the optical axis of the objective.

autoscaling   Subtracting the mean and dividing by the standard deviation of each feature converts each feature in the data set to a mean of 0 and a standard deviation of 1.

ballistic photons   Photons that result from the coherent interference of light scattered in the forward direction. They propagate nearly straight through the medium. The ballistic component is always present, and only its intensity is reduced by scattering (away from the forward direction).

bias   The systematic uncertainty or the nonrandom component of measurement error.

binary image   An image composed of only two pixel-brightness values.

binning   A process whereby the output from adjacent pixels is noiselessly co-added on the charge-coupled device (CCD) chip prior to read. Binning involves moving the charge from a number of neighboring detector elements (e.g., a single column or row of the CCD chip) into a single "bin" and then transferring the charge from the bin to the output mode. To do this, each pixel of the CCD performs two functions: it collects the photoelectrons during the integration period and transfers the charge out of the chip during the readout period.

birefringence The difference between the refractive index of the extraordinary wave, n_e, and that of the ordinary wave, n_o. The linear birefringence, Δn(given by $= |n_e - n_o|$), the linear retardance, δ, and the sample thickness, d, are related by

$$\Delta n(\lambda) = (\lambda/2\,\pi d)\delta(\lambda).$$

birefringent material A material that is doubly refracting, that is, a material whose index of refraction depends on the direction of the light traveling through it.

bright-field illumination Light irradiation of a specimen from behind, leading to a diffracted or refracted image of the objective on a bright background.

brightness For an image, the magnitude of the gray levels, which is determined by the intensity of illumination and the change in light introduced by the sample.

calibration Calibration establishes an empirical model relating the instrument response to the concentration. An empirical calibration model is used with instrument measurement of an unknown sample to estimate the analyte concentration for which the calibration was developed.

carpet plot A derivative of a surface plot that makes use of solid area fills in conjunction with three-dimensional curves.

case II diffusion A type of diffusion that exhibits a constant concentration front throughout the imbibed region rather than the exponential decay in concentration found for *Fickian diffusion.*

charge-coupled device (CCD) detector Optical-array detector based on silicon metal–oxide semiconductor technology. A CCD detector is based on collecting and storing photon-induced charge on a continuous silicon substrate divided into individual elements (pixels) in a two-dimensional array by a series of electrodes that are used to manipulate the charge. Exposure of this two-dimensional imaging area to light leads to charge separation at the n–p junction, and an image of charge accumulates localized by potential wells established by electrodes on the detector surface. This charge image can be propagated, row by row, across the CCD by a series of potentials applied to the electrodes.

chemical shift The resonance NMR frequency of a chemically shielded nucleus measured relative to that of a suitable reference compound. The chemical shift values d are typically on the order of 10^{-6} and are therefore commonly specified in parts per million (ppm). These units are independent of magnetic fields, thereby allowing direct comparison of results from different instruments. In absolute frequency terms, the separation between nuclei with different chemical shifts increases with increasing magnetic field, yielding better dispersion of resonances at high field.

coma   An optical aberration in which the off-axis beams do not form a single focused spot but rather comet-shaped patterns.

confocal microscopy   Excitation light is focused onto a small spot within the sample, and light originating at the in-focus point is imaged onto a detector, where a small pinhole is used to reject light from other planes within the sample. Confocal microscopes have a very high level of discrimination against light from outside the image plane, and they have shown themselves to be capable of providing high-quality images from significant depths below the surface of highly light scattering materials. By coupling the confocal microscope with a spectrophotometric detection system, it is possible to construct wavelength-selective images of any thin slice through a semitransparent sample. By collecting and analyzing a series of such slices, it should be possible to reconstruct the pattern of absorption at any particular wavelength in three dimensions.

conjugate points   Points having a one-to-one geometric correspondence in the object space and the image space.

contact noise   See *modulation noise.*

contour plots   See *topographic plots.*

chromatic aberrations   Blurring due to different wavelengths. Chromatic aberrations are common because each wavelength transmitted through a lens-based optical system has its own index of refraction.

condenser lens   The lens that handles the light before it interacts with the specimen.

conjugate variables   Variables in frequency and time with inverse dimensions.

contrast   A measure of the gradation in luminance that provides gray scale (or color) information. Contrast is expressed as the ratio (difference in luminance)/(average luminance) in adjoining areas of the image. Under optimum conditions, the human eye can just detect the presence of 2% contrast.

dark current   Current produced by the detector when it is not exposed to any external radiation source. Dark current originates within the detector and is a property of the detector material and its temperature. When the magnitude of the source signal falls below that of the dark current, a signal can no longer be detected.

dark current noise   Noise that is not affected by the magnitude of the incoming radiation and is characteristic of the detector and its manufacturing history.

dark-field microscopy   Dark-field microscopy may be carried out using either transmitted or reflected light. In contrast to bright-field illumination, directly transmitted or reflected light is prevented from entering the objective. In transmission, this is accomplished by placing a circular stop above the condenser to block the central beam. In reflection, special objective apertures are used to provide very oblique illumination of the sample. When viewed in dark field, objects that reflect or scatter light are visible against a dark background, and marked gains in contrast are often obtainable.

defectoscopy   Determination of the structure, composition, and spatial distribution of defects or inhomogeneities.

delta function   The delta function, $\delta(x, y)$, is an idealized two-dimensional input function, since it has infinitesimal width in both dimensions and has an integrated volume of unity, that is,

$$\int_{-\infty}^{\infty} \int \delta(x, y) = 1$$

depolarization ratio   The intensity ratio of the two polarized components of scattered light that are parallel and perpendicular to the direction of propagation of the (polarized) incident light. The polarization of the incident beam is perpendicular to the plane of propagation and observation. For this geometry, the depolarization ratio is defined as the intensity ratio, ρ, by

$$\rho = VH/VV$$

where for the right-angle scattering experiment, V is perpendicular to the scattering plane and H is in the scattering plane.

depth of field   The distance between the closest and farthest objects in focus within an image as viewed by a lens at a particular focus and with given settings. The depth of field varies with the focal length of the lens, and its f-stop setting of numerical aperture, and the wavelength of light.

derivatization   Modification of an analyte that renders it amenable to analysis by a particular analytical procedure or that improves the analysis for the analyte as a result of enhanced selectivity or detectability.

detection limit   The detection limit, c_L, is experimentally defined as the analyte concentration that yields a net analyte signal, x_A, equal to k times the standard deviation (s_B) of the background, x_B:

$$c_L = k s_B / [x_A / c_0]$$

where c_0 is the concentration yielding a net analyte signal x_A. The right side of the equation is the quotient of the net signal at the detection limit $k s_B$ and the sensitivity x_A / c_0. The value of $k = 3$ is recommended.

detectivity   A measure of the minimum detectable incident light power for a set of given conditions.

diffuse photons   Photons that undergo multiple scattering and follow a random walk through the medium. Diffuse photons lose all the signal information they carried on entering the medium and form noise at the image plane.

digital filter   A discrete array of possibly complex numbers whose effect is to alter the spatial frequency content of an image. The filtering operation can be in either image space or Fourier space. Digital filters are direct or spectral if they are applied, respectively, in image space or Fourier space.

dilation   A process whereby objects are enlarged by additional pixels around their perimeters.

discrete-element images   Images for which some prior knowledge about the spatial structure of objects in the image is available.

electronic noise   Noise generated by the electronics, amplified and sampled together with the signal for digital transmission, but independent of the magnitude of the arriving radiation.

ergodicity   An assumption that the spatial statistics taken in the field of view are unbiased estimates.

erosion   Reduction in the size of an object by removal of layers of pixels from the boundary of the object.

etendue   Optical conductance of an optical device.

$1/f$ noise   See *modulation noise.*

f-number   The focal length divided by the aperture, which is a direct measure of the light-gathering power of an optical system.

feature measurements   Measurements that determine each individual object in a field of view and produce a unique set of data. Feature or object measurements commonly provided are for area, perimeter, length, and width.

Fickian diffusion   Diffusion that is used to describe the relationship of the mass flux of a penetrant to the concentration gradient present. It can be characterized by an exponential decay in concentration with penetration into a material.

field measurements   Measurements that provide information for an entire field of view and do not provide information for individual objects within the field. Field measurements provide the total number, average size, etc., of objects in the field of view.

field of view   The area of the object that will be viewed by the optical system.

fixed pattern noise   See *spatial noise.*

flicker noise   See *modulation noise.*

fluorescence   Luminescence that occurs when the excitation line is partially absorbed and reemitted. Fluorescence is usually produced with short-wavelength incident light, and it depends on the excitation wavelength as well as the material itself.

fluorescence microscope   An instrument that is like a conventional microscope, except that its light source, usually a mercury or xenon arc lamp, produces ultraviolet, blue, and green light. This light is passed through a monochromator or interference filter to select the excitation wavelengths that induce fluorescence in the sample being examined. By spectrally filtering the emission from the sample, the localization and concentration of the probe target can be mapped.

foot-candle (fc) A unit of illuminance expressed in lumens per square foot. It is the amount of illumination from 1 international candle (the candela) falling on a 1-ft^2 surface at a distance of 1 ft. In SI units, 1 fc = 10.764 lux (lx).

footlambert (fl) A unit of luminance or brightness equal to $1/\pi$ candela per square foot. In SI units, which are preferred, 1 fl = 3.426 nit, where 1 nit is 1 candela per square meter.

fractal objects Objects exhibiting noninteger dimensionality. Whereas a straight line is one-dimensional, a randomly coiled polymer chain in solution has a dimension somewhere between 1 and 2. A square or triangle is a two-dimensional, but the surface of a solid catalyst usually has a dimension between 2 and 3.

Fraunhofer diffraction Scattering that occurs when particles much larger than the wavelength of incident light are struck by the light. Scattering occurs at small angles, and the scattering intensity patterns consist of concentric annular rings of light and dark intensity.

glitches Anything in a signal having the character of discontinuities.

gradient system A system that produces time-varying magnetic fields of controlled spatial nonuniformity.

gray-image operators Image-processing operators that use a grayscale image as input and, through mathematical calculations, output another gray-scale image.

gyromagnetic ratio, γ A physical property of the nuclei. Different chemical elements have different gyromagnetic ratios. The gyromagnetic ratio of protons is 4.26×10^7 Hz T^{-1} (or 2.68×10^8 rad s^{-1} T^{-1}).

heteroscedastic noise Noise containing different amounts of variance due to error at different concentration levels.

homoscedastic noise Noise containing a constant amount of random–normal variance due to error at all levels of concentration in the data.

illumination Manner in which light impinges on a sample.

bright-field illumination Irradiation of a specimen from behind, leading to a diffracted or refracted image of the objective on a bright background.

critical illumination Illumination in which a small extended light source is directly imaged by a condenser lens onto the object plane of the microscope.

dark-field illumination Illumination in which a highlighted image of the specimen is generated on a dark background. A hollow cone of light with an angle greater than the acceptance angle of the objective is used.

In this way, only light diffracted or scattered from a specimen at the apex of the light cone can enter the objective.

diascopic illumination   Illumination with a beam split into two components.

epi-ilumination   Illumination in which an annular beam splitter is used to form a hollow tube of light that passes through a donut-shaped condenser encircling the objective. In this lighting scheme the objective is bypassed altogether.

Kohler illumination   Illumination in which a field lens is used to form a real image of the light source at the primary focus of the condenser lens.

oblique illumination   Illumination that is off axis from the plane of focus.

vertical illumination   Illumination in which the light passes through or around the objective to reach the specimen. This requires a beam splitter or prism to direct light down the optical axis of the objective.

image analysis   The study of shapes and structures in two-dimensional images. Image analysis is the counting, measuring, and classifying of features or objects in an image.

image convolution   Convolution of an image by use of a masking function, with coefficients chosen to perform the desired convolution.

image correlation   A means of comparison of a reference and a sample image generated using identical imaging systems.

image enhancement   Use of techniques to process a given image so that the result is more interpretable than the original image for utilization in specific applications. Enhancement techniques are basically procedures designed to manipulate images in order to take advantage of the human visual system.

image processing   Image processing concerns the mathematical manipulation of digital imaging data.

image rectification   Image rectification concerns the use of spatial transformations that can remove positional distortions and permit images to be properly registered with respect to each other.

image restorations   Methods that attempt to reconstruct or recover a degraded image by using some prior knowledge of the degradation phenomenon. Thus, restoration methods are based on modeling the distortion process and applying the inverse process in order to recover the original image.

image reconstructions   Mathematical methods involved in the reconstruction of a three-dimensional image from its projections. The three-dimensional image is reconstructed using a matrix inversion of the projection data.

image segmentation   Recognition and classification of features for subsequent analysis. Image segmentation is generally based on gray-level and spectroscopic differences (characteristic frequencies) between classes of objects.

image understanding   Methods that generate a more global understanding of what an image means and how the features detected in an image are related to each other.

imaging spectroscopy   Combined analysis of both spatial and spectral information so that each pixel in a two-dimensional scene includes a third dimension of spectral information. This means that the absorption spectrum of any pixel or group of pixels within an image can be measured. Any two-dimensional target containing spectroscopically distinguishable morphological units is potentially a target for imaging spectroscopy.

infrared microscopy   A combination of a microscope and an infrared spectrometer, which allows the spectroscopic analysis and mapping of small samples.

interference microscopy   Microscopy in which monochromatic light from a single source is divided by the use of a beam splitter. One beam is reflected from or transmitted through the sample and then recombined with the reference beam. Contrast results from interference of the two beams. A pattern of bright and dark lines (fringes) results from an optical path difference between a reference beam and a sample beam.

isoplanatic system   A system in which the output image does not depend on the location of the input object.

Jacquinot advantage   See *throughput advantage.*

Johnson noise   The random signal generated by the thermal motion of electrons in the detector and its circuitry. This noise, which is proportional to the absolute temperature, exists even when no signal voltage is applied to the device. As the temperature is decreased, the thermal motion also decreases; hence, Johnson noise can be reduced.

K-matrix method   A multiple-component quantitative method of spectroscopy, which in matrix language is written

$$\mathbf{A} = \mathbf{K} \, \mathbf{C}$$

where $\mathbf{A}$ is an $n \times m$ matrix constructed from the measured absorbance spectra of m mixtures at n frequencies, and $\mathbf{C}$ is the $l \times m$ concentration matrix responding to the concentrations of each of the l components in the m known references.

kurtosis   A measure of the sharpness of a peak. For a value less than 3, the peak is sharper than a Gaussian peak with the same width at half-maximum, and if greater than 3, it is flatter than a Gaussian peak. If the kurtosis is 3, the peak is a Gaussian peak.

Larmor frequency   The Larmor frequency of the spin system, ω_0, is given by

$$w_0 = -\gamma \, H_0,$$

where γ is the gyromagnetic ratio of the nucleus and H_0 is the applied magnetic field. This is the resonant frequency that characterizes each nucleus in a static field.

lens A transparent material that converts a paraxial, parallel beam of light into a concentrated, fine spot of light at its focus.

line illumination mode Mode in which, the illuminated Raman spot is a line that traverses the sample.

luminance Photometric brightness, or the brightness of light calibrated for a sensor whose spectral response is similar to the light-adapted human eye; measured in nits, where 1 nit is 1 candela per square meter.

magic angle spinning (MAS) Rapidly rotation of a sample about an axis inclined at an angle of $54.7°$ to the magnetic field. The motion of the sample averages spatially the chemical shift anisotropy to the isotropic chemical shift with a corresponding decrease in the line width.

materials science The field that deals with the relationships between the processing, structure, and properties of materials.

mathematical morphology A mathematical method used to segment an image on the basis of a range of parameters, such as according to texture.

mean-centered method Method used to remove nonzero intercepts in models. Mean-centered values, x_i^*, are given by

$$x_i^* = x_i - x_{ave}$$

where

$$x_{ave} = (1/n) \, \Sigma \, x_i$$

microanalysis Chemical analysis of microvolumes using microscale equipment; the characterization and analysis of microscopic features using instrumental methods based on physical principles.

Mie scattering Scattering of electromagnetic radiation from a spherical particle, generally when the particle radius is of the same order of magnitude as the wavelength of the radiation.

misregistration Wrong spatial placement of the concentration value inside the analyzed volume.

modulation noise Noise due to traps that capture and release charge carriers randomly, causing random fluctuations in the current itself. The noise exists only when current is flowing through the device. The traps are created by contaminants or crystal defects at semiconductor junctions. Lower temperatures contribute to shallower traps and less trapping.

modulation transfer function Fourier transform of the impulse function, if we have an initial input function g_i, (x, y) with Fourier transform $_iG_x(f_y, f_x)$ where f_x and f_y are the spatial frequencies, and we multiply G_i by another function, $H(f_x, f_y)$. The function H defines a linear spatial filter and is the modulation transfer function.

multiplex advantage The advantage resulting from simultaneous detection of all the spectral frequencies rather than detection of a single frequency and step-scanning through all of them.

multiplexing Simultaneous detection of all spectral frequencies.

near-field optical microscopy Microscopy in which a subwavelength-sized light source is used very close to a specimen and an optical image of the specimen is built up pixel-by-pixel. The light source is an optical aperture (~25 nm in diameter) fabricated at the tapered apex of an aluminized optical fiber. Using force feedback, the tip of the probe maintains a constant separation from the sample (~5 nm). Thus, as the light emanates from the probe tip, it only illuminates a volume of the sample approximately equal to the aperture size. Any collected optical contrast, either transmitted through or reflected from the sample, originates from this small volume. Hence, resolution is limited by the size of the aperture and not by the wavelength of light.

NMR imaging (NMRI) A method whereby the stimulating NMR signal is spatially encoded so an image can be reconstructed showing the spatial distribution of nuclei in a heterogeneous sample.

noise equivalent power (NEP) Power, in watts, of sinusoidally modulated light incident on a detector that produces a signal-to-noise (S/N) ratio of 1 for a bandwidth of 1 Hz.

normalization Reduction of spectral data to a constant sum over all absorbances, which minimizes variations in peak height or areas because of differences in sample thickness or amount. Normalization removes this bias in feature magnitudes that may arise as a result of experimental variations.

normal modes Modes of vibration where the respective atomic motions of the atoms are in "harmony," that is, they all reach their maximum and minimum displacements at the same time. These normal modes can be expressed in terms of bond stretches and angle deformation (termed internal coordinates) and can be calculated by using a procedure called normal coordinate analysis.

numerical aperture (NA) The angle, $\rangle$, over which the objective lens of the microscope collects light. For an objective lens,

$$NA_{obj} = n \sin \theta,$$

where n is the refractive index of the medium between the specimen and the objective lens, and θ is the half-cone angle of light captured by the objective lens.

Nyquist frequency The sampling theorem for Fourier spectroscopy states that the highest-frequency sine wave in a given spectrum, from whatever source, must be sampled at a rate of at least two points per cycle in order to be represented accurately. This sampling frequency is known as the Nyquist frequency. When a sine wave of a higher frequency than the Nyquist frequency exists, it will be folded back into the spectrum at a frequency exactly as much lower than the Nyquist frequency as the sine wave is higher than the Nyquist frequency.

object deletion   A process which allows objects to be selectively removed from the binary image on the basis of object measurements such specific spectral measurements.

objective lens   The first lens that light enters after interacting with the sample.

ocular lens   The microscope lens that is viewed through using the eye.

optical conductance   The flux of radiation emitted by one line of the sample and transported through the spectrometer to the detector. The radiant flux, ϕ (power), transported by any optical system from the source to the detector is described by

$$\phi = LG\tau$$

where L is the *radiance* (power/solid angle area) of the light source, G is the optical conductance (solid angle area), and τ is the *transmission factor* of the system. The effective optical conductance of an optical instrument is determined by that component which cannot be enlarged for theoretical or technical reasons. In the case of microscopes, it is usually the objective lens.

optical microscope   A microscope in which spatial variations at a single wavelength of the refractive index, and absorption and reflectivity in a specimen, produce the modulation in light intensity necessary to form the magnified image.

orientation function   Function representing the average orientation of all chains in a sample.

outliers   Specimens that exhibit some form of discordance with the bulk of the samples in the calibration set.

paraxial beam   A beam of light that has an axially symmetric distribution around axes of propagation.

P-matrix method   This method is a frequency-limited calibration method, that is, the number of frequencies included in the analysis does not exceed the number of calibration samples used in the calibration. This inverse method minimizes the sum of the squared concentration errors rather than the spectral absorbances. In matrix language, this method can be written:

$$c = \mathbf{Pb} + e$$

where $\mathbf{c}$ is a vector of the concentrations of all the samples, $\mathbf{P}$ is the instrument response function matrix (a collection of spectra), $\mathbf{b}$ is the vector of the model parameters, and **e** is the matrix of the concentration residuals.

path accuracy The ability to move the focal spot along a specified path.

performance parameter Parameter used to compare the performance of different systems and methods. The performance parameter, P_p, is defined as the ratio of the S/N, to the square root of the total measurement time, T_t:

$$P_p = (S/N)/\sqrt{T_t}$$

phase contrast microscopy Microscopy that depends on the fact that differences in the refractive indices of two transparent phases produce small phase shifts in the light exiting from each component. By the use of an annular ring condenser and a phase-retarding absorption ring, these phase differences are converted to intensity variations in the observed image. The intensity variations are qualitatively proportional to optical path variations in the object.

phase encoding A form of encoding in which the spatial information in the phase is encoded into one dimension of a two-dimensional data set.

phosphorescence Luminescence that continues for a significant period of time (in the millisecond to second time regime) after excitation. Many compounds that do not fluoresce do exhibit phosphorescence.

photomultiplier tube A tube that consists of an evacuated glass envelope containing a surface coated with an active metal. Incident radiation causes emission of electrons from this surface by the photoelectric effect; these electrons are collected by a positively charged plate. The plate current is proportional to the intensity of the radiation. The photomultiplier effect arises if the electrons emitted from the cathode of a phototube are accelerated by a large potential and then allowed to strike another active surface; it is then possible to get multiple emission of electrons from the second surface for each of the original electrons. These secondary electrons may, in turn, be accelerated, and upon striking the next surface, can give rise to a larger number of electrons. By combining 10 or 12 such amplification stages, it is possible to get stable amplification of the original signal by a factor of a million or more.

photon detectors Detectors, such as photodiodes and photoconductors, which convert photons directly to mobile charge carriers; these are then measured as a current or voltage.

photon noise Noise that originates from the random arrival of photons at the detector surface. Because of this randomness, the number of photons collected within a given time interval varies around some statistical average, and the variance depends on the light source. Coherent laser light approximates a Poisson distribution, while incoherent black-body radiation follows a broader geometrical distribution called Bose–Einstein. Thermal light sources are inherently noisier than lasers.

pixel 1/f noise   The nonstationary spatial noise associated with each pixel. When present, this noise causes each pixel to drift with respect to the other pixels on the array in a spatially uncorrelated fashion.

pixel resolution   Resolution determined by the dimensions of the field of view divided by the number of pixels measured. Only the total resolution is influenced by the width of the pixels.

pleochroism   Variation in absorption color with the direction of light vibration in an anisotropic colored particle. An isotropic particle or a colorless anisotropic particle will show no color variation.

polarization   Polarization describes the orientation of the **E** vector in spectroscopy relative to a given laboratory-defined direction.

  s-polarization   (also known as TE polarization) is perpendicular to the place of incidence

  p-polarization   (also known as TM polarization) is parallel to the plane of incidence. The plane of incidence is the plane including the direction of incident light and the surface normal direction.

polarized light microscopy   Microscopy in which the illuminating source is plane-polarized before it impinges on the sample and a second polarizer is inserted in the reflected or transmitted beam. In most cases, the two polarizers are crossed at 90° to each other and only materials that cause a partial depolarization of the light are visible.

polymorphism   The existence of a substance in different forms. The substance has the same chemical composition in each form, but the solid-state structure is different in each form. Only a single form is thermodynamically stable, but different polymorphs can be kinetically stable and exist at the same temperature.

position accuracy   Ability to statically locate the focal spot at a desired point.

precision   Variation of a value over a series of repeated measurements.

principal component analysis   The absorbance matrix, **A**, is broken down into the product of the scores matrix, $\mathbf{T}(m, a)$, by the loadings matrix, $\mathbf{P}(a, k)$:

$$\mathbf{A} = \mathbf{TP} + \mathbf{F}$$

where $\mathbf{F}(m, k)$ is the residuals matrix and a is the number of principal components. Each factor in the model can be thought of as an abstract phenomenon that is a source of variation in the spectra of the calibration samples. The scores for a single factor can be thought of as the "intensities'' of the corresponding abstract phenomenon for each sample, and the loadings for a single factor can be thought of as the "spectral signature" of the corresponding abstract phenomenon.

probe resolution Resolution defined by the case of two objects imaging as two separate peaks; moving the objects closer together will result in a single image feature whose shape cannot often be predicted.

profiling Reduction of an image to a plot of a single frequency.

projection imaging Obtaining an image by projecting the radiation from the specimen through a lens onto a uniformly spaced, discrete array of detectors.

projection Forming an image by summing all of the pixel values, according to some predefined rule, in a chosen direction.

projection–reconstruction imaging Imaging obtained by rotating the sample or source to obtain snapshots of the sample at many different angles covering an arc of at least $180°$. From such a set of data, a computation can reconstruct a cross-sectional image of the sample.

quantization noise Noise introduced by digitizing the detector signal. This type of noise is the basic limitation of digital systems in determining the true value of the signal, just as random noise is the basic limitation of analog systems. The quantization error is inversely proportional to the square of the number of quantization levels. For $K = 8$, the root mean square magnitude of the quantization noise is 4.5×10^{-3}. This quantization noise can be reduced by using a larger number of quantization levels, that is, a larger word size.

quantum efficiency The ratio of the photons incident on the detector to the electrons emitted.

radiance Optical power per unit area per solid angle.

radiation noise Noise that arises from the fluctuation in the amplitude of light itself, which results from the random nature of photoemission. This radiation noise is the ultimate limitation in light detection, but fortunately, it is usually sufficiently low to be undetectable except for special types of measurements.

radiant sensitivity Amount of electric current produced by a given amount of radiant flux, and described in units of amperes per watt (A/W) at a specific wavelength.

radio-frequency (rf) transmitter A system which delivers radio-frequency magnetic fields to the sample in NMR.

Raman effect A phenomenon that occurs when a sample is irradiated by monochromatic light, causing a small fraction of the scattered radiation to exhibit shifted frequencies that correspond to the sample's vibrational transitions.

Rayleigh criterion Criterion for optical resolution which corresponds to the condition $D = r$, where D is the peak-to-peak distance of the intensity distribution curves (or the center-to-center distance of the

Airy disk images), and r is the disk radius, for well-corrected objective lenses with a uniform circular aperture. For the Rayleigh criterion, two adjacent points are just resolved when the centers of their Airy disks are separated by r, given by

$$r = 1.22 \, \lambda_0 / (2 \, NA_{obj})$$

where λ_o is the wavelength of light in air and NA_{obj} is the numerical aperture of the objective lens:

$$NA_{obj} = n \sin \theta$$

where n is the refractive index of the medium between the specimen and the objective lens, and θ is the half-cone angle of light captured by the objective lens.

Rayleigh range   Propagation distance over which the irradiance decreases by a factor of 2. The Rayleigh range, Z_r, is given by

$$Z_r = \pi \omega^2 n / \lambda$$

where ω is the focused beam radial spot size, n is the index of refraction, and λ is the wavelength of the light.

Rayleigh scattering   Elastic collisions of photons and sample result in an intense, unshifted component of the scattered light.

reflectance, diffuse   The radiation which has been transmitted and /or refracted through one or more sample particles and is finally reflected onto the detector.

reflectance, specular   The impinging light reflects from the sample surface and does not penetrate the sample. If the surface is smooth, the reflection and the incidence angles are equal and the reflected beam retains the polarization characteristics of the impinging beam.

reflection–absorption and translection   Occurs when a thin absorbing layer of a material is on the surface of a more reflecting substrate. The ideal case is a thin absorbing layer on a polished metallic substrate. The incident radiation passing through the absorbing layer is reflected at the metal's surface and then passes through the absorbing layer a second time before emerging as reflected radiation. Reflection-absorption spectra is often referred to as double pass transmission spectra.

reflected light microscopy   Microscopy in which illumination is usually provided through the objective lens itself but only secularly reflected light is allowed to reenter the objective. Contrast arises from variations in surface reflectivity and may be used to examine variations in surface structure.

refractive index   The refractive index of a substance is a measure of the speed of light in that substance compared to the speed of light in a vacuum.

refraction The process whereby light passing obliquely from one transparent medium into another is bent from its initial path.

registering The translation–rotation alignment process by which two images of like geometries and of the same set of objects are positioned coincident with respect to each other so that corresponding elements of the same area appear in the same place on the registered images. Registration allows the same sample area to be measured under different imaging conditions and compared.

repeatability Closeness of agreement of multiple measurements under the same conditions of measurement, observer, measuring instrument, and environment. It is equal to the standard deviation.

representativeness Consistency between the results and the analyzed sample as well as between the results and the analytical problem.

reproducibility Closeness of agreement of multiple measurements under the same conditions of measurement, measuring instrument, and environment, but with repeat readings taken by multiple operators.

resolution A measure of how fine a detail can be detected in terms of distance in space.

retardation Phase differences for each wavelength of the light entering a particle when the two components are recombined by the analyzer. Retardation is the product of the thickness and birefringence. The phase difference will be different for each wavelength and will result in various degrees of constructive interference for each color; hence, the particle will show colors related directly to its retardation.

roughness parameter The root mean square deviation from the surface plane. The roughness parameter depends on the scale of measurement.

scanning imaging Obtaining an image by scanning the detector (or source) or sample so that different spatial positions in the sample are rastered in sequence.

scanning probe microscopy A family of techniques that provide images of surface topography and, in some cases, surface properties, on the atomic scale.

selective excitation A magnetic field gradient is applied along the z direction and the object is irradiated with a $90°$ pulse having a narrow spectral width, corresponding to a narrow range of magnetic field values. Only those nuclei which fall within this narrow range of field values will be excited in a slice of the object perpendicular to z. Thus, only a slice of the object is excited, and the remainder of the object does not respond.

Shannon sampling theorem The correspondence between abrupt changes in the intensity of an image and high spatial frequencies in its transform. If an image's transform contains no high frequencies, then the image contains no abrupt gray-level transitions, so the image need not be quantized very finely. It indicates that one can reproduce the image by centering a sinc function every $W/2$ units along the coordinate x axis, where W is the bandwidth. The greater the bandwidth, W, the smaller the sampling interval $W/2$ required.

Shot (Shottky) noise   Noise that arises whenever current flows through a detector or circuit. It originates from the random arrival of charge carriers at any point within the device. Shot noise generated in the detectors is affected in magnitude by the presence of the arriving radiation.

skewness   A measure of the symmetry of a peak's distribution. If the skewness is positive, the peak's distribution is shifted to the right of its mean, and if negative, to the left its mean.

Snell's law   For a given wavelength of light, the ratio of the sine of the angle of incidence, I, to the sine of the angle of refraction, r, is constant and is the ratio of the refractive indices of the two media:

$$(\sin I)_1 / (\sin r)_2 = n_2 / n_1$$

spatial autocorrelation   A description of the behavior of the image at one point relative to its behavior at another point.

spatial coherence   Spatial coherence characterizes the distance below which the interference of the harmonic signal is constructive.

spatial frequency   The frequency in space in which a recurring feature appears.

spatial invariance   The principle requiring that the impulse response be the same for all input points. Then the impulse response merely shifts its position for different input points but does not change its functional behavior.

spatial noise   Noise resulting from some residual nonuniformity in the images. A random additive offset for each pixel is an example of a predominantly spatial noise source. Spatial noise is also known as *fixed pattern noise*. Spatial noise appears as a pattern that remains regardless of the image; hence the name fixed pattern noise.

spatial resolution   Resolution in space which determines the detectable feature size.

spatial stationarity   Spatial stationarity occurs when the parameters of the underlying function of which the image is a part do not vary with spatial position.

spectral response   Parameter that describes the useful spectral range of a device or detector.

spectroscopy   The measurement of the frequency dependence of absorption or emission of energy by a system. It is the analysis of the emitted or absorbed energy by matter across the electromagnetic spectrum. Spectroscopy can be used to identify a sample from its wavelength data and to determine concentrations, which are proportional to energy intensity.

specular reflectance Fresnel reflection from the exterior surface of a material, without the beam penetrating the sample. Specular reflection obeys the law of reflection, that is, the angle of reflection equals the angle of incidence.

spin–lattice relaxation time The spin–spin relaxation time, T_1, governs the evolution of the longitudinal magnetization, M_z, toward the equilibrium value, M_0. The physical process involved in this relaxation is the dissipation of energy from the collection of nuclei, the "spin system" to the atomic and molecular environment of the nuclei, the "lattice".

spin–spin relaxation time The spin–spin relaxation time, T_2, governs the evolution of the transverse magnetization toward its equilibrium value, which is zero. The physical processes for T_2 include the same processes as for T_1 as well as the magnetic coupling between the neighboring nuclei. This process of relaxation can be thought of as the randomization of the individual spins so that the sum of the fields of the nuclei in the collection goes to zero.

standard deviation An estimate of the precision of a measurement attributable to random uncertainty. The standard deviation, σ, is calculated by taking the square root of the variance estimate, σ^2, given by

$$\sigma^2 = \Sigma(x_i - X)^2/(n - 1)$$

$$I = 1$$

where x_i is an individual data point in the set, X is the data set mean, and n is the number of elements in the data set.

standards Standards are selected that possess a measurable property that correlates quantitatively with the property of interest, such as spectral absorption with concentration. The standard's measurable property must scale in a predictable fashion over the entire calibration range.

   reference standards are the highest level of laboratory standards and are calibrated with respect to a **primary standard**.

   certified standards are reference materials for which the parameters of interest are certified to be within specified limits of accuracy and precision by a laboratory like the National Institute of Standards and Technology (NIST).

   working standards are calibrated using reference materials available to spectroscopists.

   stereology The study of projections and sections of multidimensional objects into spaces of lower dimension.

Stokes and anti-Stokes lines Raman lines shifted to energies lower than the source (Stokes lines) are produced by ground-state molecules, whereas the slightly weaker lines at higher frequency (anti-Stokes lines) are due to molecules in excited vibrational states.

Stokes shift   In fluorescence spectroscopy, the difference between the maxima of the absorption and emission spectra.

stray light   Light that is diffracted or scattered and reaches the detector but does not originate from the conjugate object point. The presence of stray light leads to a nonlinearity in the measured absorbance with concentration in spectroscopy.

surface plasmon   A strong local electric field at the particle surface.

synchrotron radiation   High-intensity radiation generated from accelerated electrons in the frequency range from UV to X-rays. A coherent light beam 1000 times brighter than conventional laboratory light sources is obtained.

***t* test**   A test based on the average or mean of the measurements over the fields of view, the standard deviations of the measurements, and the number of observations. In this method, the calculated *t* value is compared with published values that indicate the level of significance of the difference between the measurement means. For these tests, the 95% level of significance may be selected. The *t* table states that for 10 observations, the calculated *t* values must fall within ± 2.101 to establish with a 95% confidence that the two measurement means are not significantly different.

thermal detectors   Detectors including thermocouples, bolometers, and pyroelectric instruments, that produce an output signal proportional to the increase in temperature of the sensing element induced by exposure to light.

thermocouple   A heat-sensitive detector produced by a junction of dissimilar metals, which develops an electromotive force that changes with temperature. Thermocouples have a constant response independent of wavelength, except at longer wavelengths (beyond 30 μm) and are usually operated in a vacuum to increase sensitivity and decrease noise. The sensitivity of thermocouples is roughly proportional to the inverse of the receiver area and heat capacity.

thresholding   Separating an image into a binary image, which is composed of only two pixel-brightness values. Ideally, the objects to be measured will have a different range of gray-scale (brightness) values than the remainder of the image, so the pixels within the image can be separated into two gray-scale groups, one representing the features to be measured and another representing the remainder of the image.

throughput advantage   Increase in signal-to-noise arising from increased amount of energy passing through the optical element.

tolerance   The minimum acceptable combination of systematic and random errors resulting from the measurement system.

tomography   The process that selects a given slice or volume of a sample to become activated and generates a signal while the remainder of the sample is not affected.

topographic plots   Plots that display lines of constant value as a function of the coordinate axis. Topographic plots are particularly useful for displaying intensity values.

trans-crystallization   Crystallization induced near a surface or interface.

transmitted light microscopy   Microscopy in which a collimated light beam passes directly through the sample and into the objective lens of the microscope. The magnified image is viewed through a microscope eyepiece. Morphological features that scatter light or give rise to optical density variations are visible. The sample must be thin enough that an appreciable fraction of the light is transmitted.

transverse resolution   Spatial resolution within the x-y image plane of a wide-field microscope.

virtual state   The virtual state corresponds to a quantum level relating to electron cloud distortion created by the electric field of incident light. A virtual state does not correspond to a real eigenstate (vibrational or electronic energy level) of the molecule, but rather is a sum over all eigenstates of the molecule.

vulcanization   A chemical cross-linking process that generates a rubber material with useful properties.

wavenumbers,   Wavenumbers, $\upsilon(\text{cm}^{-1})$, are the number of waves per centimeter. The relationship between υ and the wavelength, λ, is given by

$$\upsilon(\text{cm}^{-1}) \; = \; [10^4]/\lambda(\mu m)$$

which can also be written as

$$\upsilon(\text{cm}^{-1}) \; = \; 3 \text{ x } 100^{10} \text{ Hz}$$

The wavenumber scale is directly proportional to the energy and vibrational frequency of the absorbing unit.

zeptoscopists (zepto = 10^{-21})   Microscopists working with zepto-resolution capabilities.

zero filling   When the resolution of a spectrum is degraded by taking more data points, one can simply add a block of zeros to the acquired data and transform the data with this block of zeros included. This process, called *zero filling*, is mathematically justified as long as the block of zeros does not exceed the length of the original sampled data block. In practice, data are acquired in blocks of powers-of-two numbers of data points because these are easiest to transform. Zero filling is usually up to the next power-of-two number of points.

zeugmatography   An NMR imaging method which is based on the successive application of orthogonal pulsed field gradients prior to and during sampling of the free induction decay (fid).

Index

A